清华开发者书库

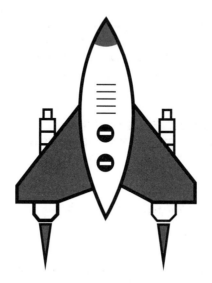

Dart Language in Action
Program Development Based on Flutter Framework

Dart语言实战
基于Flutter框架的程序开发

亢少军◎编著
Kang Shaojun

清华大学出版社
北京

内 容 简 介

本书系统阐述了跨平台Dart编程语言基础知识、面向对象编程,以及网络编程和异步编程等高级知识。

全书共分为4篇:第1篇为Dart基础(第1～9章),第2篇为面向对象编程(第10～14章),第3篇为Dart进阶(第15～21章),第4篇为商城项目实战(第22～36章)。书中主要内容包括:Dart语法基础、Dart编码规范、数据类型、运算符、流程控制语句、函数、面向对象基础、继承与多态、抽象类与接口、枚举类、集合框架、集合与泛型、异常处理、元数据、Dart库、单线程与多线程、网络编程和异步编程等。

书中包含大量应用示例,读者不仅可以由此学会理论知识还可以灵活应用。书中示例基于Flutter环境开发,读者在学习到Dart语言知识的同时还可学会Flutter框架技术。书中通过接近商业的一个商城App案例详细阐述了如何使用Flutter开发App,内容完整,步骤清晰,提供了工程化的解决方案。

本书可作为Dart和Flutter初学者的入门书籍,也可作为从事跨平台移动开发的技术人员及培训机构的参考书籍。

本书封面贴有清华大学出版社防伪标签,无标签者不得销售。
版权所有,侵权必究。侵权举报电话: 010-62782989 13701121933

图书在版编目(CIP)数据

Dart语言实战:基于Flutter框架的程序开发/亢少军编著. —北京:清华大学出版社,2020.4
(清华开发者书库)
ISBN 978-7-302-55233-8

Ⅰ. ①D… Ⅱ. ①亢… Ⅲ. ①程序语言-程序设计 Ⅳ. ①TP312

中国版本图书馆CIP数据核字(2020)第054660号

责任编辑:赵佳霓
封面设计:李召霞
责任校对:徐俊伟
责任印制:杨 艳

出版发行:清华大学出版社
网　　址:http://www.tup.com.cn, http://www.wqbook.com
地　　址:北京清华大学学研大厦A座　　邮　编:100084
社 总 机:010-62770175　　邮　购:010-62786544
投稿与读者服务:010-62776969, c-service@tup.tsinghua.edu.cn
质量反馈:010-62772015, zhiliang@tup.tsinghua.edu.cn
课件下载:http://www.tup.com.cn,010-83470236
印 装 者:三河市龙大印装有限公司
经　　销:全国新华书店
开　　本:186mm×240mm　　印　张:34.75　　字　数:795千字
版　　次:2020年5月第1版　　印　次:2020年5月第1次印刷
印　　数:1～1500
定　　价:98.00元

产品编号:086685-01

前言
PREFACE

近些年来，利用跨平台技术来开发App无论是在移动端还是在桌面端都备受欢迎。开源的跨平台框架也是百花齐放，Flutter是最新的跨平台开发技术，可以横跨Android、iOS、macOS、Windows、Linux等多个系统。Flutter还可以打包成Web程序运行在浏览器上。Flutter采用了更为彻底的跨平台方案，即自己实现了一套UI框架，然后直接在GPU上渲染UI页面。

笔者最早接触的跨平台技术是Adobe Air技术，写一套ActionScript代码便可以运行在Windows、Android及iOS三大平台上。目前，笔者与朋友开发视频会议产品及开源项目，需要最大化地减少前端的开发及维护工作量，我们先后研究过Cordova、React Native及Flutter等技术。我们觉得Flutter方案更加先进，效率更高，后来就尝试用Flutter开发了全球第一个开源的WebRTC插件（可在GitHub上搜索Flutter WebRTC）。

Flutter的开发语言是Dart，所以本书重点介绍Dart语言的相关知识。写本书的目的是想传播Flutter知识（因为Flutter确实优秀），想为Flutter社区做点贡献的同时也为我们的产品打下坚实的技术基础。在编写本书的过程中，笔者查阅了大量的资料，知识体系扩大了不少，收获良多。

本书主要内容

第1章为Dart语言简介，介绍Dart语言的发展及支持的平台。

第2章介绍Dart语言的两个开发环境的搭建过程，包括Windows及macOS的开发环境搭建。

第3章简单介绍如何使用IDE在Flutter环境下运行第一个Dart程序。

第4章介绍Dart语言的语法基础，包括关键字、变量和常量等。

第5章介绍Dart语言的编码规范，包括样式规范、文档规范，以及各种使用规范。

第6章介绍Dart语言的常用数据类型，包括数字、字符串、布尔、List、Map以及Set类型的定义及使用方法。

第7章介绍Dart语言的运算符，包括算术、关系、逻辑、类型测试以及级联操作符等。

第8章介绍常用的流程控制语句，包括条件分支、循环语句，以及断言assert等。

第9章介绍函数的定义、参数传递方法、可选参数的使用，以及匿名函数的使用方法等。

第10章介绍面向对象的基本概念、类的声明、成员变量与成员方法，以及枚举类型等相关知识。

第 11 章介绍对象的创建与使用，以及 Dart 里各个构造方法的定义及使用。

第 12 章介绍继承与多态，通过示例详细讲解方法重写的知识点。

第 13 章介绍抽象类与接口的概念，以及如何声明抽象类与接口，如何实现抽象类和接口。

第 14 章介绍 Dart 语言里 Mixin 混入的概念及特性、Mixin 的使用、重命名方法处理，以及 Mixin 对象类型。

第 15 章介绍 Dart 异常的概念，抛出异常及捕获异常的使用方法，如何自定义异常并使用，最后通过 Http 异常处理的实例综合运用异常。

第 16 章介绍集合的概念，详细讲解了 Dart 语言中 List、Set 以及 Map 等常用集合的概念及使用方法。

第 17 章介绍泛型的概念及作用，通过示例详解泛型在集合、类、抽象类以及方法里的使用方法。

第 18 章介绍单线程与多线程的概念、事件循环机制、Future 概念以及异步处理，同时介绍 Stream 概念及 Bloc 设计模式，另外还介绍 Isolate 的高级用法。

第 19 章通过多个示例详细介绍 Http 网络请求、Dio 网络请求，以及 WebSocket 的用法。

第 20 章介绍元数据定义、常用元数据以及自定义元数据，另外还通过 Json 生成实体类的方法详细介绍元数据的应用场景。

第 21 章介绍常用开发库及第三方库的使用，如库的导入、导出，以及命名与拆分等。

第 22 章对商城项目进行一个总体的功能介绍，包括所使用的前端技术、后端技术、后台管理技术以及所使用的数据库，同时详细讲解后端及数据库的安装步骤。

第 23 章介绍商城项目创建、项目框架搭建、目录结构分析，以及项目的数据流程分析等内容。

第 24 章介绍商城项目的颜色、图标、字符串以及数据接口等配置项。

第 25 章介绍商城项目中用到的工具，如路由工具、Http 请求工具、本地存储工具，以及字符串处理工具等。

第 26 章介绍商城项目中封装的组件，如缓存图片组件、没有数据提示组件、加载数据组件、图文组件，以及分割线组件等。

第 27 章介绍商城项目的入口程序处理、路由配置、状态管理配置，以及加载页面实现。

第 28 章介绍商城项目的首页模块的数据模型、数据服务、轮播图、首页分类、首页产品，以及首页组件组装的实现过程。

第 29 章介绍商城项目的分类模块的数据模型、数据服务、一级分类组件、二级分类组件，以及分类页面组装的实现过程。

第 30 章介绍商城项目登录及注册的数据模型、数据服务，以及页面实现过程，并分析 token 的获取与使用。

第 31 章介绍商城项目中商品详情复杂页面布局、商品相关的数据模型、数据服务，以及商品详情页面的实现过程。

第32章介绍商城项目中的购物车模块的实现过程,同时讲解了购物车与其他模块的关系及购物商品数量组件的使用。

第33章介绍商城项目中订单列表及订单详情的实现过程,同时详细介绍了订单状态及订单详情复杂页面的布局。

第34章介绍商城项目中地址列表及地址编辑模块的实现过程,同时介绍了地址与填写订单的关系及它们之间是如何传递数据的。

第35章介绍商城项目中商品收藏模块的数据模型、数据服务,以及页面的实现。

第36章介绍商城项目中"我的"页面的实现过程,同时介绍"我的"页面与其他页面的关系。

阅读建议

本书是一本基础入门加实战的书籍,既有基础知识,又有丰富示例,包括详细的操作步骤,实操性强。由于Dart语言内容较多,所以本书对Dart语言的基本概念讲解很详细,包括基本概念及代码示例。每个知识点都配有小例子,力求精简,还提供完整代码,读者复制完整代码就可以立即看到效果。这样会给读者信心,在轻松掌握基础知识的同时能够快速进入实战。

本书共分四篇,建议读者先把第1篇Dart的基础理论通读一遍,并搭建好开发环境,在第3章编写出第一个Dart程序。

第2篇是Dart语言面向对象的一些知识,掌握这一部分内容可以写出结构清晰的程序,同时还能掌握Dart语言的Mixin混入等特性。

第3篇属于Dart进阶内容,包括异常处理、集合以及泛型的使用。这里的异步编程属于Dart的核心知识,可通过示例详细了解Bloc设计模式及程序是如何解耦的。

第4篇属于项目实战部分,读者在掌握了前面的基础知识后,可以通过一个接近商业应用的商城案例项目来全面掌握Flutter的开发过程。这里建议读者在开发过程中,遇到不熟悉的组件或者第三方库,先运行一下小示例后再进行使用。

关于随书代码

本书所列代码力求完整,但由于篇幅所限,代码没有全部放在书里。完整代码可扫描下方二维码下载。

本书源代码下载

致谢

首先感谢清华大学出版社赵佳霓编辑给编者提出了许多宝贵的建议,以及推动了本书的出版。

  还要感谢我的家人，特别感谢我的母亲及妻子，在编者写作过程中承担了全部的家务并照顾孩子，使我可以全身心地投入写作工作之中。

  由于时间仓促，书中难免存在不妥之处，请读者见谅，并提宝贵意见。

亢少军

2019年12月24日

目录
CONTENTS

第 1 篇　Dart 基础

第 1 章　Dart 语言简介 ········· 3
  1.1　移动端开发 ········· 3
  1.2　Web 开发 ········· 4
  1.3　服务端开发 ········· 4

第 2 章　开发环境搭建 ········· 5
  2.1　Windows 环境搭建 ········· 5
  2.2　macOS 环境搭建 ········· 11

第 3 章　第一个 Dart 程序 ········· 17

第 4 章　Dart 语法基础 ········· 21
  4.1　关键字 ········· 21
  4.2　变量 ········· 22
  4.3　常量 ········· 22
    4.3.1　final 定义常量 ········· 22
    4.3.2　const 定义常量 ········· 22
    4.3.3　final 和 const 的区别 ········· 23

第 5 章　编码规范 ········· 24
  5.1　样式规范 ········· 24
  5.2　文档规范 ········· 27
  5.3　使用规范 ········· 29
    5.3.1　依赖 ········· 29
    5.3.2　赋值 ········· 29

5.3.3 字符串	30
5.3.4 集合	31
5.3.5 参数	32
5.3.6 变量	33
5.3.7 成员	34
5.3.8 构造方法	35
5.3.9 异常处理	37

第 6 章 数据类型 … 38

- 6.1 Number 类型 … 38
- 6.2 String 类型 … 39
- 6.3 Boolean 类型 … 41
- 6.4 List 类型 … 42
  - 6.4.1 定义 List … 42
  - 6.4.2 常量 List … 43
  - 6.4.3 扩展运算符 … 43
- 6.5 Set 类型 … 43
- 6.6 Map 类型 … 44

第 7 章 运算符 … 46

- 7.1 算术运算符 … 47
- 7.2 关系运算符 … 48
- 7.3 类型测试操作符 … 49
- 7.4 赋值操作符 … 49
- 7.5 逻辑运算符 … 50
- 7.6 位运算符 … 50
- 7.7 条件表达式 … 51
- 7.8 级联操作 … 51

第 8 章 流程控制语句 … 53

- 8.1 if 和 else … 53
- 8.2 for 循环 … 53
- 8.3 while 和 do-while … 54
- 8.4 break 和 continue … 55
- 8.5 switch 和 case … 56
- 8.6 断言 assert … 56

第9章 函数 ... 57

- 9.1 函数的概念 ... 57
- 9.2 可选参数 ... 57
  - 9.2.1 命名参数 ... 58
  - 9.2.2 位置参数 ... 61
- 9.3 参数默认值 ... 62
- 9.4 main 函数 ... 63
- 9.5 函数作为参数传递 ... 63
- 9.6 匿名函数 ... 64
- 9.7 词法作用域 ... 65

第 2 篇 面向对象编程

第 10 章 面向对象基础 ... 69

- 10.1 面向对象概述 ... 69
- 10.2 面象对象基本特征 ... 69
- 10.3 类声明及构成 ... 70
  - 10.3.1 类声明 ... 71
  - 10.3.2 成员变量 ... 71
  - 10.3.3 成员方法 ... 72
- 10.4 静态变量和静态方法 ... 73
  - 10.4.1 静态变量 ... 73
  - 10.4.2 静态方法 ... 75
- 10.5 枚举类型 ... 78

第 11 章 对象 ... 79

- 11.1 创建对象 ... 79
- 11.2 对象成员 ... 81
- 11.3 获取对象类型 ... 81
- 11.4 构造方法 ... 82
  - 11.4.1 声明构造方法 ... 82
  - 11.4.2 使用构造方法 ... 83
  - 11.4.3 命名构造方法 ... 87
  - 11.4.4 调用父类的非默认构造方法 ... 87
  - 11.4.5 初始化列表 ... 89

11.4.6 重定向构造方法 ·· 90
11.4.7 常量构造方法 ·· 91
11.4.8 工厂构造方法 ·· 92
11.5 Getters 和 Setters ·· 93

第 12 章 继承与多态

12.1 Dart 中的继承 ·· 95
12.2 方法重写 ·· 96
    12.2.1 基本使用 ·· 96
    12.2.2 重绘 Widget 方法 ································ 98
    12.2.3 重写高级示例 ······································ 99
12.3 操作符重写 ··· 102
12.4 重写 noSuchMethod 方法 ··························· 104
12.5 多态 ·· 105

第 13 章 抽象类与接口

13.1 抽象类 ·· 108
    13.1.1 抽象类定义格式 ································ 108
    13.1.2 数据库操作抽象类实例 ······················· 109
    13.1.3 几何图形抽象类 ································ 110
13.2 接口 ·· 112

第 14 章 Mixin 混入

14.1 Mixin 概念 ·· 115
14.2 Mixin 使用 ·· 115
14.3 重名方法处理 ··· 121
14.4 Mixin 对象类型 ·· 124

第 3 篇 Dart 进阶

第 15 章 异常处理

15.1 异常概念 ·· 129
15.2 抛出异常 ·· 129
15.3 捕获异常 ·· 131
    15.3.1 try-catch 语句 ··································· 131
    15.3.2 try-on-catch 语句 ······························ 132

15.4	重新抛出异常 ···	134
15.5	finally 语句 ··	135
15.6	自定义异常 ··	136
15.7	Http 请求异常 ···	137

第 16 章 集合 ·· 141

16.1	集合简介 ···	141
16.2	List 集合 ···	141
	16.2.1 常用属性 ··	142
	16.2.2 常用方法 ··	143
	16.2.3 遍历集合 ··	144
16.3	Set 集合 ··	146
	16.3.1 常用属性 ··	147
	16.3.2 常用方法 ··	148
	16.3.3 遍历集合 ··	149
16.4	Map 集合 ···	150
	16.4.1 常用属性 ··	150
	16.4.2 常用方法 ··	151
	16.4.3 遍历集合 ··	152

第 17 章 泛型 ·· 154

17.1	语法 ···	154
17.2	泛型的作用 ··	154
	17.2.1 类型安全 ··	154
	17.2.2 减少重复代码 ··	155
17.3	集合中使用泛型 ··	156
17.4	构造方法中使用泛型 ··	161
17.5	判断泛型对象的类型 ··	161
17.6	限制泛型类型 ···	161
17.7	泛型方法的用法 ··	162
17.8	泛型类的用法 ···	163
17.9	泛型抽象类的用法 ···	165

第 18 章 异步编程 ·· 167

18.1	异步概念 ···	167
	18.1.1 单线程 ···	167

- 18.1.2 多线程 168
- 18.1.3 事件循环 168
- 18.2 Future 169
  - 18.2.1 Dart 事件循环 169
  - 18.2.2 调度任务 171
  - 18.2.3 延时任务 172
  - 18.2.4 Future 详解 173
  - 18.2.5 异步处理实例 178
- 18.3 Stream 188
  - 18.3.1 Stream 概念 188
  - 18.3.2 Stream 分类 189
  - 18.3.3 Stream 创建方式 189
  - 18.3.4 Stream 操作方法 193
  - 18.3.5 StreamController 使用 201
  - 18.3.6 StreamBuilder 204
  - 18.3.7 响应式编程 207
  - 18.3.8 Bloc 设计模式 208
  - 18.3.9 Bloc 解耦 208
  - 18.3.10 BlocProvider 实现 212
- 18.4 Isolate 216
  - 18.4.1 创建 Isolate 216
  - 18.4.2 使用场景 221

第 19 章 网络编程 222

- 19.1 Http 网络请求 222
- 19.2 HttpClient 网络请求 224
- 19.3 Dio 网络请求 226
- 19.4 Dio 文件上传 234
- 19.5 WebSocket 238

第 20 章 元数据 249

- 20.1 元数据定义 249
- 20.2 常用元数据 249
  - 20.2.1 @deprecated 250
  - 20.2.2 @override 251
  - 20.2.3 @required 254

20.3 自定义元数据 ·········· 256
20.4 元数据应用 ·········· 257

第21章 Dart库 ·········· 264

21.1 本地库使用 ·········· 264
21.2 系统内置库使用 ·········· 266
21.3 第三方库使用 ·········· 267
    21.3.1 Key-Value 存储介绍 ·········· 268
    21.3.2 shared_preferences 使用 ·········· 268
    21.3.3 shared-preferences 实现原理 ·········· 272
21.4 库重名与冲突解决 ·········· 275
21.5 显示或隐藏成员 ·········· 277
21.6 库的命名与拆分 ·········· 278
21.7 导出库 ·········· 282

第4篇 商城项目实战

第22章 项目简介 ·········· 287

22.1 功能介绍 ·········· 287
22.2 总体架构 ·········· 288
    22.2.1 前端 Flutter ·········· 289
    22.2.2 后端接口 SpringBoot ·········· 290
    22.2.3 后台管理 Vue ·········· 290
    22.2.4 数据库 MySQL ·········· 291
22.3 后端及数据库准备 ·········· 292
    22.3.1 MySQL 安装 ·········· 292
    22.3.2 JDK 安装 ·········· 294
    22.3.3 Maven 安装 ·········· 294
    22.3.4 IntelliJ IDEA 启动项目 ·········· 295
    22.3.5 Node 安装 ·········· 296

第23章 项目框架搭建 ·········· 297

23.1 新建项目 ·········· 297
23.2 目录结构 ·········· 299

第24章 项目配置 ... 303

- 24.1 颜色配置 ... 303
- 24.2 图标配置 ... 304
- 24.3 字符串配置 ... 305
- 24.4 接口地址配置 ... 307
- 24.5 导出配置 ... 308

第25章 工具集 ... 309

- 25.1 路由导航工具 ... 309
  - 25.1.1 路由参数处理 ... 309
  - 25.1.2 导航工具 ... 310
- 25.2 Http 请求工具 ... 313
- 25.3 本地存储工具 ... 315
- 25.4 字符串处理工具 ... 316
- 25.5 Toast 提示工具 ... 317

第26章 组件封装 ... 318

- 26.1 缓存图片组件 ... 318
- 26.2 数量加减组件 ... 320
- 26.3 分割线组件 ... 325
- 26.4 图文组件 ... 325
- 26.5 文本组件 ... 327
- 26.6 加载数据组件 ... 329
- 26.7 没有数据提示组件 ... 331
- 26.8 网页加载组件 ... 332

第27章 入口与路由配置 ... 334

- 27.1 入口程序 ... 334
- 27.2 路由配置 ... 336
- 27.3 加载页面 ... 341

第28章 首页 ... 344

- 28.1 索引页面 ... 344
- 28.2 首页数据模型 ... 347
- 28.3 首页布局拆分 ... 352

28.4 轮播图实现 ………………………………………………………… 353
28.5 首页分类实现 ………………………………………………………… 355
28.6 首页产品实现 ………………………………………………………… 357
28.7 首页数据服务 ………………………………………………………… 360
28.8 组装首页 …………………………………………………………… 361

第29章 分类 ………………………………………………………………… 366

29.1 分类数据模型 ………………………………………………………… 366
29.2 分类数据服务 ………………………………………………………… 371
29.3 一级分类组件实现 …………………………………………………… 373
29.4 二级分类组件实现 …………………………………………………… 377
29.5 组装分类页面 ………………………………………………………… 381

第30章 登录注册 …………………………………………………………… 383

30.1 用户数据模型 ………………………………………………………… 383
30.2 用户数据服务 ………………………………………………………… 384
30.3 登录页面实现 ………………………………………………………… 386
30.4 注册页面 ……………………………………………………………… 394

第31章 商品 ………………………………………………………………… 400

31.1 商品列表数据模型 …………………………………………………… 400
31.2 商品分类标题数据模型 ……………………………………………… 402
31.3 商品数据服务 ………………………………………………………… 404
31.4 商品分类页面实现 …………………………………………………… 406
31.5 商品详情需求分析 …………………………………………………… 414
31.6 商品详情数据模型 …………………………………………………… 415
31.7 商品详情轮播图 ……………………………………………………… 420
31.8 商品详情页面实现 …………………………………………………… 421

第32章 购物车 ……………………………………………………………… 438

32.1 购物车列表数据模型 ………………………………………………… 438
32.2 购物车数据服务 ……………………………………………………… 441
32.3 购物车页面实现 ……………………………………………………… 445

第33章 订单 ………………………………………………………………… 456

33.1 填写订单数据模型 …………………………………………………… 456

33.2 订单数据服务 ………………………………………………………… 460
33.3 填写订单页面实现 …………………………………………………… 462
33.4 我的订单数据模型 …………………………………………………… 474
33.5 我的订单页面实现 …………………………………………………… 477
33.6 订单详情数据模型 …………………………………………………… 483
33.7 订单详情页面实现 …………………………………………………… 487

第 34 章 地址 ………………………………………………………………… 498

34.1 地址数据模型 ………………………………………………………… 498
34.2 地址数据服务 ………………………………………………………… 500
34.3 我的收货地址页面实现 ……………………………………………… 502
34.4 编辑地址页面实现 …………………………………………………… 507

第 35 章 收藏 ………………………………………………………………… 519

35.1 收藏数据模型 ………………………………………………………… 519
35.2 收藏数据服务 ………………………………………………………… 521
35.3 我的收藏页面实现 …………………………………………………… 522

第 36 章 个人中心 …………………………………………………………… 528

36.1 关于我们页面实现 …………………………………………………… 528
36.2 我的页面实现 ………………………………………………………… 530

第1篇　Dart基础

第 1 章 Dart 语言简介

Dart 诞生于 2011 年 10 月 10 日,谷歌 Dart 语言项目的领导人 Lars Bak 在丹麦举行的 Goto 会议上宣布,Dart 是一种"结构化的 Web 编程"语言,Dart 编程语言在所有现代的浏览器和环境中提供高性能。

Dart 是谷歌开发的计算机编程语言,后来被 ECMA (ECMA-408) 认定为标准。它被用于 Web、服务器、移动应用和物联网等领域的开发。它是宽松开源许可证(修改的 BSD 证书)下的开源软件。

Dart 有以下三个方向的用途,每一个方向,都有相应的 SDK。Dart 语言可以创建移动应用、Web 应用,以及 Command-line 应用等,如图 1-1 所示。

Flutter
Write a mobile App that runs on both iOS and Android.

Web
Write an App that runs in any modern web browser.

Server
Write a command-line App or server-side app.

图 1-1 Dart 支持的平台

为了提高学习效率,作者提供讲解视频,网址 http://www.kangshaojun.com。本书源代码也可在作者的 GitHub 中下载,网址 https://github.com/kangshaojun。

1.1 移动端开发

Dart 在移动端上的应用离不开 Flutter 技术。Flutter 是谷歌的移动 UI 框架,可以快速在 iOS 和 Android 上构建高质量的原生用户界面。Flutter 可以与现有的代码一起工作。在全世界,Flutter 正在被越来越多的开发者和组织使用,并且 Flutter 是完全免费、开源的。简单来说,Flutter 是一款移动应用程序 SDK,包含框架、控件和一些工具,可以用一套代码同时构建 Android 和 iOS 应用,并且其性能可以达到与原生应用一样的水准。

Flutter采用Dart的原因很多，单纯从技术层面分析如下：
- Dart是AOT（Ahead Of Time）编译的，可编译成快速、可预测的本地代码，Flutter几乎可以使用Dart编写；
- Dart也可以JIT（Just In Time）编译，开发周期快；
- Dart可以更轻松地创建以60fps运行的流畅动画和转场；
- Dart使Flutter不需要单独的声明式布局语言；
- Dart容易学习，具有静态和动态语言用户都熟悉的特性。

Dart最初设计是为了取代JavaScript成为Web开发的首选语言，最后的结果可想而知，因此到Dart 2发布时，已专注于改善构建客户端应用程序的体验，可以看出Dart定位的转变。用过Java、Kotlin的人，可以很快地上手Dart。

1.2　Web开发

Dart是经过关键性Web应用程序验证的平台。它拥有为Web量身打造的库，如dart：html，以及完整的基于Dart的Web框架。使用Dart进行Web开发的团队会对速度的提高感到非常激动。选择Dart是因为其高性能、可预测性和易学性、完善的类型系统，以及完美地支持Web和移动应用。

1.3　服务端开发

Dart的服务端开发与其他的语言类似，有完整的库，可以帮助开发者快速开发服务端代码。

第 2 章 开发环境搭建

接下来我们使用Flutter的开发环境来测试Dart程序。开发环境搭建还是非常烦琐的,任何一个步骤失败都会导致不能完成最终环境搭建。Flutter支持三种环境:Windows、macOS和Linux。这里我们主要讲解Windows及macOS的环境搭建。

2.1 Windows 环境搭建

1. 使用镜像

首先解决网络问题。环境搭建过程中需要下载很多资源文件,当某个资源更新不成功时,就可能会导致后续报各种错误。在国内访问Flutter有时可能会受到限制,Flutter官方为中国开发者搭建了临时镜像,大家可以将如下环境变量加入到用户环境变量中:

```
export PUB_HOSTED_URL = https://pub.flutter-io.cn
export FLUTTER_STORAGE_BASE_URL = https://storage.flutter-io.cn
```

注意　此镜像为临时镜像,并不能保证一直可用,读者可以参考 Using Flutter in China: https://github.com/flutter/flutter/wiki/Using-Flutter-in-China 以获得有关镜像服务器的最新动态。

2. 安装 Git

Flutter依赖的命令行工具为Git for Windows(Git命令行工具)。Windows版本的下载地址为: https://git-scm.com/download/win。

3. 下载安装 Flutter SDK

去Flutter官网下载其最新可用的安装包。

注意　Flutter的渠道版本会不断更新,请以Flutter官网为准。Flutter官网下载地址: https://flutter.io/docs/development/tools/sdk/archive#windows。Flutter GitHub下载地址: https://github.com/flutter/flutter/releases。

将安装包 zip 文件解压到你想安装 Flutter SDK 的路径(如 D:\Flutter)。在 Flutter 安装目录下找到 flutter_console.bat,双击运行并启动 Flutter 命令行,接下来,你就可以在 Flutter 命令行运行 Flutter 命令了。

注意 不要将 Flutter 安装到需要一些高权限的路径,如 C:\Program Files\。

4. 添加环境变量

不管使用什么工具,如果想在系统的任意地方能够运行这个工具的命令,则需要添加工具的路径到系统环境变量 Path 里。这里路径指向 Flutter 的 bin 目录,如图 2-1 所示。同时,检查是否有名为"PUB_HOSTED_URL"和"FLUTTER_STORAGE_BASE_URL"的条目,如果没有,也需要添加它们。完成后重启 Windows 才能使更改生效。

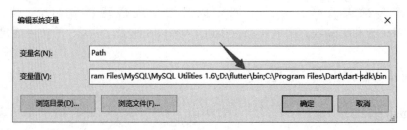

图 2-1 添加 Flutter 环境变量

5. 运行 Flutter 命令并安装各种依赖

使用 Windows 命令窗口运行以下命令,查看是否还需要安装任何其他依赖项来完成安装:

```
flutter doctor
```

该命令检查你的环境并在终端窗口中显示报告。Dart SDK 已经捆绑在 Flutter 里了,没有必要单独安装 Dart。仔细检查命令行输出以获取可能需要安装的其他软件或进一步需要执行的任务。如下显示的代码,说明 Android SDK 缺少命令行工具,需要下载并且提供了下载地址,通常这种情况只需连接网络,打开 VPN,然后重新运行"flutter doctor"命令即可。

```
[-] Android toolchain - develop for Android devices
    Android SDK at D:\Android\sdk
  ?Android SDK is missing command line tools; download from https://goo.gl/XxQghQ
    Try re-installing or updating your Android SDK,
     visit https://flutter.io/setup/#android-setup for detailed instructions.
```

注意 一旦你安装了任何缺失的依赖,需再次运行"flutter doctor"命令来验证你是否已经正确地设置,同时需要检查移动设备是否连接正常。

6．编辑器设置

如果使用 Flutter 命令行工具，可以使用任何编辑器来开发 Flutter 应用程序。输入"flutter help"可查看可用的工具，但是笔者建议最好安装一款功能强大的 IDE 来进行开发，毕竟这样开发、调试、运行及打包的效率会更高。由于 Windows 环境只能开发 Flutter 的 Android 应用，所以接下来会重点介绍 Android Studio 这款 IDE。

1）安装 Android Studio

要为 Android 开发 Flutter 应用，可以使用 macOS 或 Windows 操作系统。Flutter 需要安装和配置 Android Studio，步骤如下。

步骤 1：下载并安装 Android Studio，下载地址为 https://developer.android.com/studio/index.html。

步骤 2：启动 Android Studio，然后执行"Android Studio 安装向导"。这将安装最新的 Android SDK、Android SDK 平台工具和 Android SDK 构建工具，这是用 Flutter 为 Android 开发应用时所必需的工具。

2）设置 Android 设备

要准备在 Android 设备上运行并测试 Flutter 应用，需要安装有 Android 4.1（API level 16）或更高版本的 Android 设备。

步骤 1：在设备上启用"开发人员选项"和"USB 调试"，这些选项通常在设备的"设置"界面里。

步骤 2：使用 USB 线将手机与计算机连接。如果设备出现授权提示，请授权计算机访问设备。

步骤 3：在终端中，运行"flutter devices"命令以验证 Flutter 识别所连接的 Android 设备。

步骤 4：用 flutter run 启动应用程序。

> **提示** 默认情况下，Flutter 使用的 Android SDK 版本是基于你的 ADB 工具版本的。如果想让 Flutter 使用不同版本的 Android SDK，则必须将 ANDROID_HOME 环境变量设置为该 SDK 的安装目录。

3）设置 Android 模拟器

要准备在 Android 模拟器上运行并测试 Flutter 应用，请按照以下步骤操作。

步骤 1：启动 Android Studio→Tools→Android→AVD Manager 并选择 Create Virtual Device，打开虚拟设备面板，如图 2-2 所示。

步骤 2：选择一个设备并单击"Next"按钮，如图 2-3 所示。

步骤 3：选择一个镜像并单击"download"按钮，然后单击"Next"按钮，如图 2-4 所示。

步骤 4：验证配置信息。填写虚拟设备名称，选择"Hardware - GLES 2.0"以启用硬件加速，单击"Finish"按钮，如图 2-5 所示。

步骤 5：在工具栏选择刚刚添加的模拟器，如图 2-6 所示。

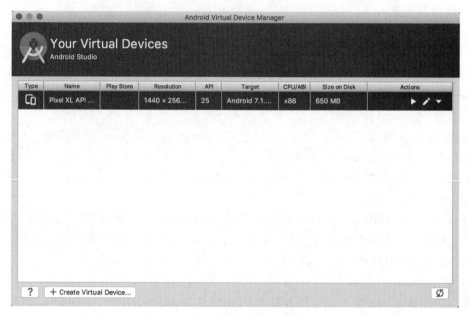

图 2-2　打开虚拟设备面板

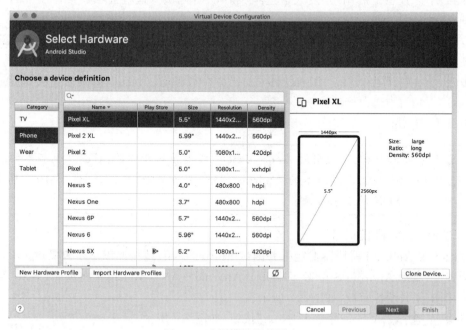

图 2-3　选择模拟硬件设备

也可以在命令行窗口运行"flutter run"启动模拟器。当能正常显示模拟器时（如图 2-7），则表示模拟器安装正常。

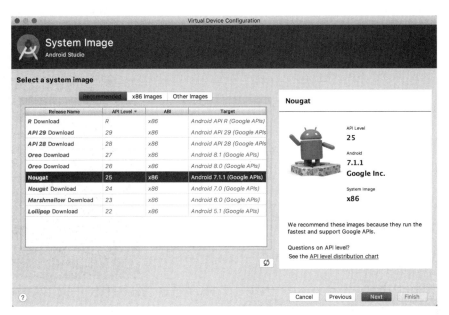

图 2-4　选择系统镜像

图 2-5　验证配置信息

> **提示**　建议选择当前主流手机型号作为模拟器,并开启硬件加速,使用 x86 或 x86_64 镜像。详细文档请参考 https://developer.android.com/studio/run/emulator-acceleration.html。

图 2-6　在工具栏选择模拟器　　　　图 2-7　Android 模拟器运行效果图

4）安装 Flutter 和 Dart 插件

IDE 需要安装两个插件：

❑ Flutter 插件：支持 Flutter 开发的工作流（运行、调试、热重载等）；

❑ Dart 插件：提供代码分析（输入代码时进行验证、代码补全等）。

打开 Android Studio 的系统设置面板，找到 Plugins，分别搜索 Flutter 和 Dart，单击安装即可，如图 2-8 所示。

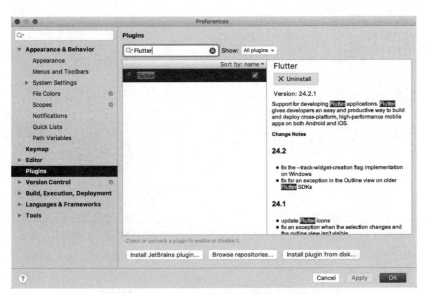

图 2-8　Android Studio 插件安装

2.2 macOS 环境搭建

首先解决网络问题,参见 2.1 节"Windows 环境搭建"。

1. 命令行工具

Flutter 依赖的命令行工具有 bash、mkdir、rm、git、curl、unzip 和 which。

2. 下载安装 Flutter SDK

请按以下步骤下载并安装 Flutter SDK。

步骤 1:去 Flutter 官网下载其最新可用的安装包。

> **注意** Flutter 的渠道版本会不断更新,请以 Flutter 官网为准。另外,在中国大陆地区,要想获取安装包列表或下载安装包有可能会遇到困难,读者也可以去 Flutter GitHub 项目下去下载安装 Release 包。
>
> Flutter 官网下载地址:https://flutter.io/docs/development/tools/sdk/archive#macOS
> Flutter GitHub 下载地址:https://github.com/flutter/flutter/releases

步骤 2:解压安装包到想安装的目录,如:

```
cd /Users/ksj/Desktop/flutter/
unzip /Users/ksj/Desktop/flutter/v0.11.9.zip.zip
```

步骤 3:添加 Flutter 相关工具到 PATH 中:

```
export PATH = 'pwd'/flutter/bin: $ PATH
```

3. 运行 Flutter 命令并安装各种依赖

运行以下命令查看是否需要安装其他依赖项:

```
flutter doctor
```

该命令检查你的环境并在终端窗口中显示报告。Dart SDK 已经捆绑在 Flutter 里了,没有必要单独安装 Dart。仔细检查命令行输出以获取可能需要安装的其他软件或进一步需要执行的任务(以粗体显示)。如下代码提示表示 Android SDK 缺少命令行工具,需要下载并且提供了下载地址,通常这种情况只需要把网络连好,VPN 开好,然后重新运行"flutter doctor"命令。

```
[-] Android toolchain - develop for Android devices
    Android SDK at /Users/obiwan/Library/Android/sdk
    ?Android SDK is missing command line tools; download from https://goo.gl/XxQghQ
    Try re-installing or updating your Android SDK,
visit https://flutter.io/setup/#android-setup for detailed instructions.
```

> **注意** 当安装了所缺失的依赖后,需再次运行"flutter doctor"命令来验证是否已经正确地设置,同时需要检查移动设备是否连接正常。

4.添加环境变量

使用 vim 命令打开 ~/.bash_profile 文件,添加如下内容:

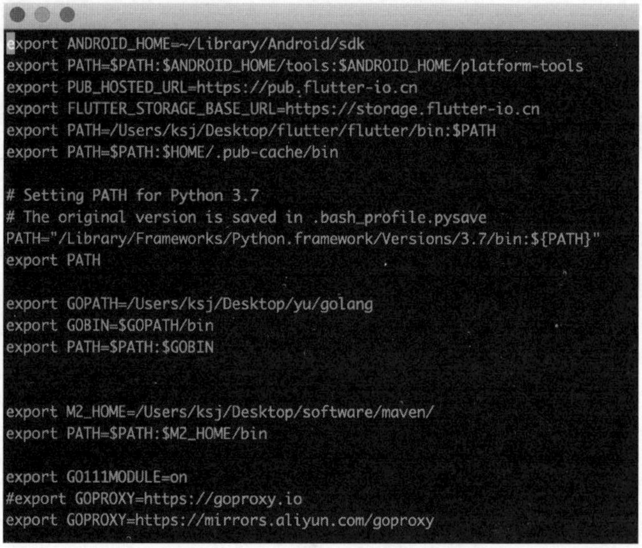

```
export ANDROID_HOME=~/Library/Android/sdk                          //android sdk 目录
export PATH=$PATH:$ANDROID_HOME/tools:$ANDROID_HOME/platform-tools
export PUB_HOSTED_URL=https://pub.flutter-io.cn                    //国内用户需要设置
export FLUTTER_STORAGE_BASE_URL=https://storage.flutter-io.cn      //国内用户需要设置
export PATH=/Users/ksj/Desktop/flutter/flutter/bin:$PATH           //直接指定 flutter 的 bin 地址
```

> **注意** 请将 PATH=/Users/ksj/Desktop/flutter/flutter/bin 更改为你的路径。

完整的环境变量设置如图 2-9 所示。

图 2-9 macOS 环境变量设置

设置好环境变量以后,请务必运行 source $HOME/.bash_profile 刷新当前终端窗口,以使刚刚配置的内容生效。

5.编辑器设置

如果使用 Flutter 命令行工具,可以使用任何编辑器来开发 Flutter 应用程序。输入"flutter help"可查看可用的工具,但是笔者建议最好安装一款功能强大的 IDE 来进行开发,毕竟这样开发、调试、运行和打包的效率会更高。由于 macOS 环境既能开发 Android 应用也能开发 iOS 应用,所以 Android 设置请参考 1.2.1 节"Windows 环境搭建"中的"安装

Android Studio",接下来仅介绍 Xcode 的使用方法。

1) 安装 Xcode

安装最新 Xcode。通过链接下载：https://developer.apple.com/xcode/，或通过苹果应用商店下载：https://itunes.apple.com/us/app/xcode/id497799835。

2) 设置 iOS 模拟器

要在 iOS 模拟器上运行并测试你的 Flutter 应用，需要打开一个模拟器。在 macOS 的终端输入以下命令：

```
open -a Simulator
```

可以找到并打开默认模拟器。如果想切换模拟器，可以打开 Hardware 并在其下的 Device 菜单选择某一个模拟器，如图 2-10 所示。

打开后的模拟器如图 2-11 所示。

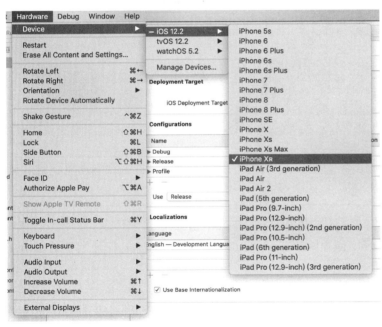

图 2-10　选择 iOS 模拟器　　　　图 2-11　iOS 模拟器效果图

接下来，在终端运行"flutter run"命令或者打开 Xcode，如图 2-12 所示，选择好模拟器。单击"Runner"按钮即可启动你的应用。

3) 安装到 iOS 设备

要在苹果真机上测试 Flutter 应用，需要一个苹果开发者账户，并且还需要在 Xcode 中进行设置。

(1) 安装 Homebrew 工具。Homebrew 是一款 macOS 平台下的软件包管理工具，拥有

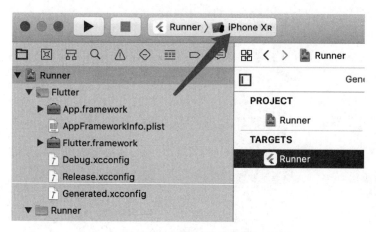

图 2-12 Xcode 启动应用

安装、卸载、更新、查看和搜索等很多实用的功能。下载地址为 https://brew.sh。

（2）打开终端并运行一些命令，安装用于将 Flutter 应用安装到 iOS 设备的工具，命令如下所示：

```
brew update
brew install --HEAD libimobiledevice
brew install ideviceinstaller ios-deploy cocoapods
pod setup
```

提示　如果这些命令中有任何一个失败并出现错误，请运行"brew doctor"并按照说明解决问题。

接下来需要 Xcode 签名。Xcode 签名设置有以下几个步骤。

步骤 1：在你的 Flutter 项目目录中双击 ios/Runner.xcworkspace 打开默认的 Xcode 工程。

步骤 2：在 Xcode 中，选择导航面板左侧中的 Runner 项目。

步骤 3：在 Runner TARGETS 设置页面中，确保在 General→Signing→Team（常规→签名→团队）下选择了你的开发团队，如图 2-13 所示。当你选择一个团队时，Xcode 会创建并下载开发证书，为你的设备注册你的账户，并创建和下载配置文件。

步骤 4：要开始你的第一个 iOS 开发项目，可能需要使用你的 Apple ID 登录 Xcode。任何 Apple ID 都支持开发和测试。需要注册 Apple 开发者计划才能将你的应用分发到 App Store，具体方法请查看 https://developer.apple.com/support/compare-memberships/这篇文章。Apple ID 登录界面如图 2-14 所示。

步骤 5：当你第一次添加真机设备进行 iOS 开发时，需要同时信任你的计算机和该设备上的开发证书。单击"Trust"按钮即可，如图 2-15 所示。

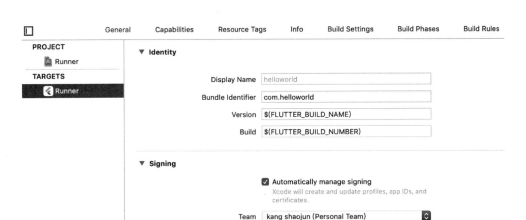

图 2-13　设置开发团队

图 2-14　使用 Apple ID

图 2-15　信任此计算机图示

步骤 6：如果 Xcode 中的自动签名失败，请查看项目的 Bundle Identifier 值是否唯一。这个 ID 即为应用的唯一 ID，建议使用域名反过来写，如图 2-16 所示。

步骤 7：使用"flutter run"命令运行应用程序。

图 2-16 验证 Bundle Identifier 值

第 3 章 第一个 Dart 程序

万事开头难,我们用输出"Hello World"来看一个最简单的 Flutter 工程,具体步骤如下。

步骤 1:新建一个 Flutter 工程,选择 Flutter Application,如图 3-1 所示。

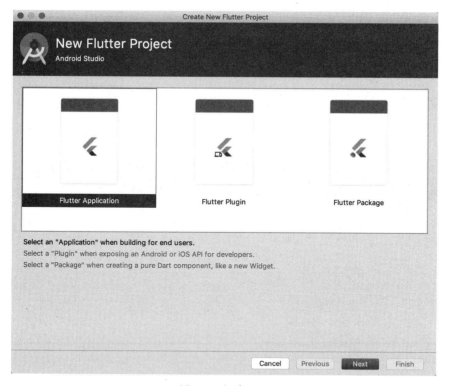

图 3-1 新建工程

步骤 2:单击"Next"按钮,打开应用配置界面,其中在 Project name 中填写 helloworld,Flutter SDK path 使用默认值,IDE 会根据 SDK 安装路径自动填写,Project location 填写为工程放置的目录,在 Description 中填写项目描述,任意字符即可,如图 3-2 所示。

步骤 3:单击"Next"按钮,打开包设置界面,在 Company domain 中填写域名,注意域名

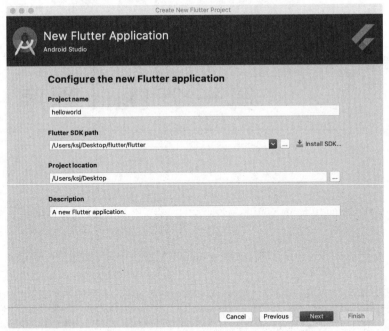

图 3-2　配置 Flutter 工程

要反过来写,这样可以保证全球唯一,Platform channel language 下面的两个选项不需要勾选,如图 3-3 所示。

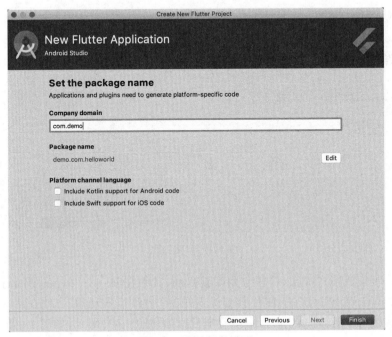

图 3-3　设置包名界面

步骤 4：单击"Finish"按钮开始创建第一个工程，等待几分钟，会创建如图 3-4 所示工程。

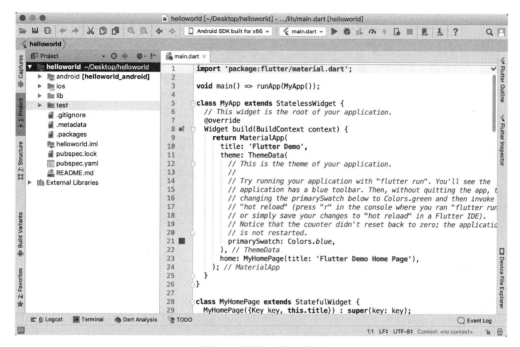

图 3-4　示例工程主界面

步骤 5：工程建好后，先运行一下，看一看根据官方推荐方案所创建的示例的运行效果，单击"Open iOS Simulator"命令打开 iOS 模拟器，具体操作如图 3-5 所示。

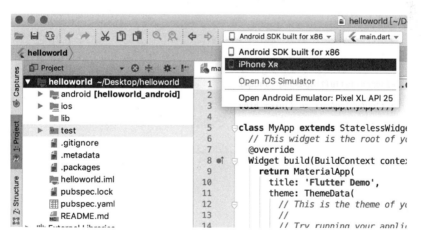

图 3-5　打开模拟器菜单示意图

步骤 6：等待几秒后会打开模拟器，如图 3-6 所示。

图 3-6　模拟器启动完成图

步骤 7：打开工程目录下的 lib/main.dart 文件，删除所有代码并替换成如下代码：

```
//main.dart 文件
//程序执行入口函数
main() {
  //定义并初始化变量
  String msg = 'Hello World';
  //调用方法
  sayHello(msg);
}
//方法
sayHello(String msg) {
  //在控制台打印内容
  print('$msg');
}
```

步骤 8：单击"debug"（调试）按钮，启动 Hello World 程序控制台输出"Hello World"，这样，第一个 Dart 程序就运行出来了。输出内容如下：

```
Performing hot restart...
Syncing files to device iPhone X_R...
Restarted application in 52ms.
flutter: Hello World
```

第 4 章 Dart 语法基础

4.1 关键字

Dart 的关键字如表 4-1 所示。

表 4-1 关键字表

abstract^2	dynamic^2	implements^2	show^1
as^2	else	import^2	static^2
assert	enum	in	super
async^1	export^2	interface^2	switch
await^3	extends	is	sync^1
break	external^2	library^2	this
case	factory^2	mixin^2	throw
catch	false	new	true
class	final	null	try
const	finally	on^1	typedef^2
continue	for	operator^2	var
covariant^2	Function^2	part^2	void
default	get^2	rethrow	while
deferred^2	hide^1	return	with
do	if	set^2	yield^3

尽量避免使用表格中的关键字作为标识符,但是如有必要可以采用关键字与其他字母组合的方式来作为标识符,如"new_user",表示"new"关键字,下画线与"user"单词组件的标识符。

带有上标 1 的单词是上下文关键字,仅在特定位置有意义。它们在任何地方都是有效的标识符。

带有上标 2 的单词是内置标识符。为了简化将 JavaScript 代码移植到 Dart,这些关键字在大多数地方是有效的标识符,但它们不能用作类或类型名称,也不能用作导入前缀。

带有上标3的单词是于 Dart 1.0 发布后添加的异步支持相关的有限保留字。不能在任何被标记 async、async * 或 sync * 标记的函数体中使用 await 或 yield 作为标识符。

4.2 变量

Dart 语言中使用"var"关键字定义变量,不是必须要指定数据类型,代码如下:

```
var name = 'kevin';
```

这里 name 变量的类型被推断为 String 类型,我们可以显式地指定其类型,代码如下:

```
String name = 'kevin';
```

如果对象不限于一个单一类型,指定它为 Object 或 dynamic 类型。

```
dynamic name = 'kevin';
```

未初始化的变量的初始值为 null,即使是数字类型的变量,最初值也是 null,因为数字在 Dart 中都是对象。下面的示例代码判断 name 是否为 null:

```
int name;
if(name = = null);
```

4.3 常量

常量就是在运行期间不会被改变的数据,例如有个存储单元这一秒存的数是 1,永远不可能会被改成 2。定义常量有两种方式,一种是用 final,另一种是用 const。

4.3.1 final 定义常量

final 数据类型 常量名 = 值;

```
//常量
void main() {
  //final 定义常量
  final String name = 'kevin';
}
```

数据类型也可以省略,赋值后就不能改了,尝试修改会有警告,强行运行就会报错。

4.3.2 const 定义常量

const 数据类型 常量名 = 值;

```
//常量
void main() {
  //const 定义常量
  const String name = 'kevin';
}
```

const 的数据类型也是可以省略的,同样 const 常量赋值后就不能改了,强行运行也是会报错的。

4.3.3　final 和 const 的区别

看起来 final 和 const 是差不多的,其实是有区别的,final 可以不用先赋值,但 const 声明时必须赋值,不然会报错,而 final 声明时没赋值不会报错。

第 5 章 编 码 规 范

编码习惯都是因人而异的,并没有所谓的最佳方案。如果你是一个人开发,当然不需要在意这些问题,但是如果你的代码需要展现给别人,或者需要与别人协同开发,编码规范就非常有必要了。

规范主要分为以下几个部分:
- 样式规范;
- 文档规范;
- 使用规范。

每个部分都有许多的例子说明,本章将会从项目代码中选取最基本、最典型和发生率较高的一些情况,作为规范说明。

5.1 样式规范

程序代码中到处都是各种标识符,因此取一个一致并且符合规范的名字非常重要。本节将从以下几点来说明样式规范。

(1) 类、枚举、类型定义,以及泛型,都需要使用大写开头的驼峰命名法。如下面代码所示,类名 Person、HttpService,以及类型定义 EventChannel 均使用了驼峰命名法。

```
//类名规范
class Person {
    //...
}

//类名规范
class HttpService {
    //...
}

//类型定义规范
typedef EventChannel<T> = bool Function(T value);
```

(2）在使用元数据的时候，也要使用驼峰命名法。如下面代码中@JsonSerializable 及 @JsonKey 遵循了此规范。

```dart
//类元数据
@JsonSerializable()
class AddressEntity extends Object {

  //变量元数据
  @JsonKey(name: 'id')
  int id;

  //变量元数据
  @JsonKey(name: 'name')
  String name;
}
```

（3）命名库、包、目录、dart 文件都应该是小写加上下画线。正确的代码写法如下：

```dart
//命名库
library peg_parser.source_scanner;

//导入 dart 文件
import 'file_system.dart';
import 'slider_menu.dart';
```

错误的命名如下所示，其中库名 pegparser.SourceScanner 中的 S 应该为小写，file-system.dart 文件名中间用下画线表示成 file_system.dart，SliderMenu.dart 文件名的大写应该改为 slider_menu.dart。

```dart
//命名库
library pegparser.SourceScanner;

//导入 dart 文件
import 'file-system.dart';
import 'SliderMenu.dart';
```

（4）将引用的库使用 as 转换的名字也应该是小写加下画线。正确的写法如下：

```dart
//导入库文件 as 命名规范
import 'dart:math' as math;
import 'package:angular_components/angular_components'
    as angular_components;
import 'package:js/js.dart' as js;
```

下面示例中是错误的写法，Math 应为小写，angularComponents 应改为 angular_components。JS 应改为 js。

```dart
//导入库文件
import 'dart:math' as Math;
import 'package:angular_components/angular_components'
    as angularComponents;
import 'package:js/js.dart' as JS;
```

(5)变量名、方法、参数名都应采用小写开头的驼峰命名法。正确的代码写法如下：

```dart
//变量名 item
var item;

//变量名 httpRequest
HttpRequest httpRequest;

//方法名 align 参数名 clearItems
void align(bool clearItems) {
  //...
}

//常量名 pi
const pi = 3.14;
//常量名 defaultTimeout
const defaultTimeout = 1000;
//常量名 urlScheme
final urlScheme = RegExp('^([a-z]+):');

class Dice {
  //静态常量名 numberGenerator
  static final numberGenerator = Random();
}
```

下面代码中常量名使用大写没有遵循规范。

```dart
//常量名 PI
const PI = 3.14;
//常量名 DefaultTimeout
const DefaultTimeout = 1000;
//常量名 URL_SCHEME
final URL_SCHEME = RegExp('^([a-z]+):');

class Dice {
  //静态常量名 NUMBER_GENERATOR
  static final NUMBER_GENERATOR = Random();
}
```

(6)花括号的用法也有一定的规范。只有一个 if 语句且没有 else 的时候，并且在一行内能够很好地展示就可以不用花括号，如下面代码就不需要花括号。

```
if (arg == null) return defaultValue;
```

但是,如果一行内展示比较勉强的话,就需要用花括号了,正确的代码写法如下:

```
if (value != obj.value) {
  return value < obj.value;
}
```

去掉花括号的写法就没有遵循规范,代码如下:

```
if (value != obj.value)
  return value < obj.value;
```

5.2 文档规范

在 Dart 的注释中,推荐使用三个斜杠"///"而不是两个斜杠"//"。至于为什么要这样做,官方表示是由于历史原因及他们觉得这样在某些情况下看起来更方便阅读。下面将从以下几点来说明文档规范。

(1)下面代码中是一种推荐的注释写法。

```
///当图片上传成功后,记录当前上传的图片在服务器中的位置
String imgServerPath;
```

下面的写法不推荐使用。

```
//当图片上传成功后,记录当前上传的图片在服务器中的位置
String imgServerPath;
```

(2)文档注释应该以一句简明的话开头,代码如下:

```
///使用传入的路径[path]删除此文件
void delete(String path) {
  ...
}
```

下面的写法显得有些啰嗦,建议简化。

```
///需要确定使用的用户授权删除权限
///需要确定磁盘上有此文件
///[path]参数如果不传将会抛出 IO 异常[IOError]
void delete(String path) {
  ...
}
```

（3）如果有多行注释，将注释的第一句与其他内容分隔开来。

```
///使用此路径[path]删除文件
///
///如果文件找不到会抛出一个IO异常[IOError]。如果没有权限并且文件
///存在会抛出一个[PermissionError]异常
void delete(String path) {
    ...
}
```

下面的几行注释代码连在一起写是不推荐使用的。

```
///使用此路径[path]删除文件。如果文件找不到会
///抛出一个IO异常[IOError]。如果没有权限并且
///文件存在会抛出一个[PermissionError]异常
void delete(String path) {
    ...
}
```

（4）在方法的上面通常会写成如下形式，这种方式在Dart里是不推荐使用的。需要使用方括号去声明参数、返回值，以及抛出的异常。

```
///使用 name 和 detail 添加一个地址
///
///@param name 用户名称
///@param detail 用户详情
///@returns 返回一个 Address
///@throws ArgumentError 如果没有传递 detail 会抛出此异常
Address addAddress(String name, String detail){
    ...
}
```

下面的示例代码中参数 name、detail 和异常 ArgumentError 都用方括号括起来了，推荐使用此方式。

```
///添加一个地址
///
///通过[name]及[detail]参数可以创建一个 Address 对象
///如果没有传递[detail]参数会抛出此异常[ArgumentError]
Address addAddress(String [name], String [detail]){
    ...
}
```

5.3 使用规范

在使用过程中也需要遵循一定的规范，例如库的依赖、变量赋值、字符串，以及集合的使用等。

5.3.1 依赖

推荐使用相对路径导入依赖。如果项目结构如下：

```
my_package
└── lib
    ├── src
    │   └── utils.dart
    └── api.dart
```

想要在 api.dart 中导入 utils.dart 可以按如下方式导入。

```
import 'src/utils.dart';
```

下面的方式不推荐使用，因为一旦 my_package 改名之后，对应的所有需要导入的地方都需要修改，而相对路径就没有这个问题。

```
import 'package:my_package/src/utils.dart';
```

5.3.2 赋值

使用"??"将 null 值做一个转换。在 dart 中"??"操作符表示当一个值为空时会给它赋值"??"后面的数据。

下面的写法是错误的。

```
if (optionalThing?.isEnabled) {
  print("Have enabled thing.");
}
```

当 optionalThing 为空的时候，上面就会有空指针异常了。

这里说明一下，"?."操作符相当于做了一次判空操作，只有当 optionalThing 不为空的时候才会调用 isEnabled 参数，当 optionalThing 为空的话则默认返回 null，用在 if 判断句中自然就不行了。

下面是正确做法，代码如下：

```
//如果为空的时候想返回false
optionalThing?.isEnabled ?? false;

//如果为空的时候想返回ture
optionalThing?.isEnabled ?? true;
```

下面的写法是错误的。

```
optionalThing?.isEnabled == true;

optionalThing?.isEnabled == false;
```

5.3.3 字符串

在 Dart 中，不推荐使用加号"＋"去连接两个字符串，而使用回车键直接分隔字符串。代码如下：

```
String str = '这个方法是用来请求服务端数据的,'
             '返回的数据格式为Json';
```

下面代码使用"＋"连接字符串不推荐使用。

```
 String str = '这个方法是用来请求服务端数据的,' +
             '返回的数据格式为Json';
```

在多数情况下推荐使用$variable或${}来连接字符串与变量值。其中${}里可以放入一个表达式。下面的示例代码就使用了这种规范。

```
String name = '张三';
int age = 20;
int score = 89;
void sayHello(){
    
    //$name 获取变量值
    String userInfo = '你好, $name 你的年龄是: ${age}.';
    print(userInfo);
    
    //${}可以加入变量及表达式
    String scoreInfo = '成绩是否及格: ${ score >= 60 ? true : false }.';
    print(scoreInfo);
    
}
```

5.3.4 集合

Dart 中创建空的可扩展 List 有两种方法：[]和 List()。创建空的 HashMap 有三种方法：{}、Map()和 LinkedHashMap()。

（1）如果要创建不可扩展的列表或其他一些自定义集合类型，那么务必使用构造函数。尽可能使用简单的字面量创建集合。代码如下：

```
var points = [];
var addresses = {};
```

下面的创建方式不推荐使用。

```
var points = List();
var addresses = Map();
```

（2）当想要指定类型的时候，推荐如下写法。

```
var points = <Point>[];
var addresses = <String, Address>{};
```

不推荐如下方式。

```
var points = List<Point>();
var addresses = Map<String, Address>();
```

（3）不要使用.length 的方法去表示一个集合是否为空。使用 isEmpty 或 isNotEmpyt。代码如下：

```
if (list.isEmpty){
    ...
}

if (list.isNotEmpty){
    ...
}
```

（4）避免使用带有方法字面量的 Iterable.forEach()。forEach()方法在 JavaScript 中被广泛使用，因为内置的 for-in 循环不能达到通常想要的效果。在 Dart 中如果要迭代序列，那么惯用的方法是使用循环。代码如下：

```
for (var person in people) {
    ...
}
```

下面的 forEach 方式不推荐使用。

```
people.forEach((person) {
  ...
});
```

（5）不要使用 List.from()，除非打算更改结果的类型。有两种方法去获取 iterable，分别是 List.from() 和 Iterable.toList()。代码如下：

```
void main(){
  //创建一个 List<int>
  var iterable = [1, 2, 3];

  //输出"List<int>"
  print(iterable.toList().toString());
}
```

下面的示例代码通过 List.from() 的方式输出 iterable，但不要这样使用。

```
void main(){
  //创建一个 List<int>
  var iterable = [1, 2, 3];

  //输出"List<int>"
  print(List.from(iterable).toString());
}
```

（6）使用 whereType() 过滤一个集合。下面使用 where 过滤一个集合是错误的用法。

```
//集合
var objects = [1, "a", 2, "b", 3];
//where 过滤
var ints = objects.where((e) => e is int);
```

正确的用法如下：

```
//集合
var objects = [1, "a", 2, "b", 3];
//whereType 过滤
var ints = objects.whereType<int>();
```

5.3.5 参数

方法参数值的设置规范有如下几种情况。

（1）使用等号"="给参数设置默认值。代码如下：

```
void insert(Object item, {int at = 0}) {
    //…
}
```

下面代码使用冒号":"是错误的用法。

```
void insert(Object item, {int at: 0}) {
    //…
}
```

(2) 不要将参数的默认值设置为 null,下面的写法是正确的。

```
void error([String message]) {
  stderr.write(message ?? '\n');
}
```

下面的代码将 message 参数设置为 null 是错误的。

```
void error([String message = null]) {
  stderr.write(message ?? '\n');
}
```

5.3.6 变量

变量的使用便于存储可以计算的值。首先看下面一个求圆的面积的例子。

```
//定义圆类
class Circle {

  //PI 值
  num pi = 3.14;

  //圆的半径
  num _radius;

  //获取圆的半径
  num get radius => _radius;
  //设置圆的半径
  set radius(num value) {
    _radius = value;
    //计算圆的面积
    _recalCulate();
  }

  //圆的面积
```

```
  num _area;
  //获取圆的面积
  num get area => _area;

  Circle(this._radius) {
    _recalCulate();
  }

  //计算圆的面积
  void _recalCulate() {
    _area = pi * _radius * _radius;
  }
}
```

上面的代码，可以简化成下面的写法。

```
//定义圆类
class Circle {
  //PI 值
  num pi = 3.14;
  //圆的半径
  num radius;

  //构造方法传入半径值
  Circle(this.radius);

  //计算并获取圆的面积
  num get area => pi * radius * radius;
}
```

5.3.7 成员

成员变量不要写没必要的 getter 和 setter。如下面代码所示，定义了一个容器类，它有两个属性：宽度和高度，由于这两个属于设置和获取值没有对值造成任何影响，所以代码量会增加一些。

```
//容器类
class Container {

  //定义宽度变量
  var _width;
  //获取宽度值
  get width => _width;
  //设置宽度值
  set width(value) {
    _width = value;
```

```
    }

    //定义高度变量
    var _height;
    //获取高度值
    get height => _height;
    //设置高度值
    set height(value) {
      _height = value;
    }

}
```

直接将上面代码简化成如下即可。

```
//容器类
class Container {

    //定义宽度变量
    var width;

    //定义高度变量
    var height;

}
```

5.3.8 构造方法

构造方法的使用规范有以下几点。

（1）尽可能使用简单的初始化形式。下面的示例代码定义了 Point 类的构造方法。

```
class Point {
  num x, y;
  //构造方法
  Point(num x, num y) {
    this.x = x;
    this.y = y;
  }
}
```

可以简化成如下形式。

```
class Point {
  num x, y;
  //构造方法
  Point(this.x, this.y);
}
```

（2）不要使用 new 来创建对象。最新的 Dart 可以全部去掉代码中的 new 关键字了。如下面示例所示，Flutter 里构建一个 Widget。

```
//构建 Widget
Widget build(BuildContext context) {
  return Row(
    children: [
      RaisedButton(
        child: Text('Increment'),
      ),
      Text('Click!'),
    ],
  );
}
```

下面的示例中使用了 new 关键字，编译器不会报错，但不推荐使用。

```
//构建 Widget
Widget build(BuildContext context) {
  return new Row(
    children: [
      new RaisedButton(
        child: new Text('Increment'),
      ),
      new Text('Click!'),
    ],
  );
}
```

（3）不要使用多余的 const 修饰对象。如下面示例代码中颜色值为常量，不需要修饰。

```
const colors = [
  Color("red", [255, 0, 0]),
  Color("green", [0, 255, 0]),
  Color("blue", [0, 0, 255]),
];
```

下面的示例代码中常量值部分不需要加任何 const 关键字。

```
const colors = const [
  const Color("red", const [255, 0, 0]),
  const Color("green", const [0, 255, 0]),
  const Color("blue", const [0, 0, 255]),
];
```

5.3.9 异常处理

异常处理中使用 rethrow 重新抛出异常。下面的代码使用 throw 抛出异常是不规范的。

```
try {
  //逻辑代码
} catch (e) {
  //抛出异常
  if (!canHandle(e)) throw e;
  handle(e);
}
```

正确的作法是使用 rethrow 重新抛出异常，代码如下：

```
try {
  //逻辑代码
} catch (e) {
  //重新抛出异常
  if (!canHandle(e)) rethrow e;
  handle(e);
}
```

第 6 章 数 据 类 型

Dart 语言常用的基本数据类型如下所示：
- Number；
- String；
- Boolean；
- List；
- Map。

6.1 Number 类型

Number 类型包括如下两类：
- int 整型；

取值范围：−2^53 到 2^53−1
- double 浮点型。

64 位长度的浮点型数据，即双精度浮点型。

int 和 double 类型都是 num 类型的子类。int 类型不能包含小数点。num 类型包括的操作包括："+"、"−"、"*"、"/"，以及位移操作">>"。num 类型有如下常用方法 abs、ceil 和 floor。

整数是没有小数点的数字。下面是一些定义整数的例子：

```
int r = 255;                    //颜色 RGB 的 R 值
int color = 0xFFFFFFFF;         //颜色值
```

如果一个数字包含小数点，那么它是一个浮点数。下面是一个定义浮点数的例子：

```
varpi = 3.1415926;
```

数值和字符串可以互相转换，代码如下：

```
void main() {
  int r = 255; //颜色 RGB 的 R 值
  int color = 0xFFFFFFFF; //颜色值

  //字符串转换成整型 String ==> int
  var valueInt = int.parse('10');
  assert(valueInt == 10);

  //字符串转换成符点型 String ==> double
  var valueDouble = double.parse('10.10');
  assert(valueDouble == 10.10);

  //整型转换成字符串 int ==> String
  String valueString = 10.toString();
  assert(valueString == '10');

  //符点型转换成字符串 double ==> String 保留两位小数
  String pi = 3.1415926.toStringAsFixed(2);
  assert(pi == '3.14');
}
```

> **提示** 示例中使用了 assert 关键字，当 assert 里的条件为 false 时系统会报错，进而终止程序的执行。

6.2 String 类型

Dart 的字符串是以 UTF-16 编码单元组成的序列。使用方法与 JavaScript 相似，具体用法如下：

1. 字符串表示方法

String 类型也就是字符串类型，在开发中大量使用。定义的例子如下：

```
var s1 = 'hello world'; //单引号
var s2 = "hello world"; //双引号
```

2. 字符串拼接

String 类型可以使用 + 操作，非常方便，具体用法示例如下：

```
var s1 = 'hi ';
var s2 = 'flutter';
var s3 = s1 + s2;
print(s3);
```

上面代码打印输出"hi flutter"字符串。

3．大文本块表示方法

可以使用三个单引号或双引号来定义多行的 String 类型变量，在 Flutter 中我们专门用来表示大文本块。示例代码如下：

```
var s1 = '''
请注意这是一个用三个单引号包裹起来的字符串,
可以用来添加多行数据.
''';

var s2 = """同样这也是一个用多行数据,
只不过是用双引号包裹起来的.
""";
```

4．转义字符处理

当一段文本需要换行时，我们可以在字符串中加入转义字符"\n"来进行处理，当需要跳到下一个 TAB 位置时需要加入"\t"来进行处理，代码如下：

```
var s = "\n这是第一行文本\n这是第二行文本\n这是第三行文本\t一个 TAB 位置";
print(s);
```

上面代码打印输出如下所示，三行文本都进行了换行处理，其中第三行有一个制表符（TAB 位置）。

```
这是第一行文本
这是第二行文本
这是第三行文本    一个 TAB 位置
```

如果想把这些带有转义字符的字符串变成普通字符串，需要添加"r"前缀来进行处理，代码如下：

```
var rs = r"\n这是第一行文本\n这是第二行文本\n这是第三行文本\t一个 TAB 位置";
print(rs);
```

上面代码输出如下内容：

```
\n这是第一行文本\n这是第二行文本\n这是第三行文本\t一个 TAB 位置
```

5．字符串插值处理

可以使用 ${expression} 将一个表达式插入到字符串中。如果这个表达式是一个变量，可以省略{}。为了得到一个对象的字符串表示，Dart 会调用对象的 toString() 方法。示例代码如下：

```
String name = '张三';
int age = 30;
//插入变量可以不用{}
String s1 = '张三的年龄是 $age';
print(s1);

int score = 90;
//插入表达式必须加入{}
String s2 = '成绩${score >= 60 ? '及格' : '不及格'}';
print(s2);
```

上面代码分别演示了插入变量及插入表达式的用法,输出如下内容:

```
flutter:张三的年龄是 30
flutter:成绩及格
```

只要所有的插值表达式是编译期常量,计算结果为 null 或者数值、字符串、Boolean 类型的值,那么这个字符串就是编译期常量。当插件为变量则不可以组成字符串常量。示例代码如下:

```
//可作为常量字符串的组成部分
const aConstNum = 0;
const aConstBool = true;
const aConstString = '常量字符串';

//不可作为常量字符串的组成部分
var aNum = 0;
var aBool = true;
var aString = '一个字符串变量';
const aConstList = [1, 2, 3];

//插值为常量可以组成字符串常量
const validConstString = '$aConstNum $aConstBool $aConstString';
//插值为变量不可以组成字符串常量
//const invalidConstString = '$aNum $aBool $aString $aConstList';
```

6.3　Boolean 类型

Dart 是强 Boolean 类型检查,只有 Boolean 类型的值是 true 才被认为是 true,但有的语言认定 0 是 false,大于 0 是 true。在 Dart 语言里则不是,值必须为 true 或者 false。下面的示例代码编译不能正常通过,原因是 sex 变量是一个字符串,不能使用条件判断语句,sex 变量必须使用 Boolean 类型才可以:

```
var sex = '男';
if (sex) {
  print('你的性别是!' + sex);
}
```

常用的检测方法如下：

```
void main() {
  //检查是否是空字符串
  var str = '';
  assert(str.isEmpty);

  //检查是否为 0
  var value = 0;
  assert(value <= 0);

  //检查是否为 null
  var isNull;
  assert(isNull == null);

  //检查是否是 NaN
  var isNaN = 0 / 0;
  assert(isNaN.isNaN);
}
```

6.4 List 类型

在 Dart 语言中，具有一系列相同类型的数据被称为 List 对象。Dart 里的 List 对象类似 JavaScript 语言的数组 Array 对象。

6.4.1 定义 List

定义 List 的例子如下：

```
var list = [1, 2, 3];
```

List 对象的第一个元素的索引是 0，最后个元素的索引是 list.lenght −1，代码如下：

```
var list = [1,2,3,4,5,6];
print(list.length);
print(list[list.length - 1]);
```

上面的代码输出长度为 6，最后一个元素值也为 6。

6.4.2 常量 List

创建常量列表,在列表字面量前加上 const 关键字即可。试图修改常量列表里的元素会引发一个错误,代码如下:

```
var constList = const [1, 2, 3, 4];
//这一行会引发一个错误
//constList[0] = 1;
```

6.4.3 扩展运算符

扩展运算符"...",提供了一个简洁的方法来向集合中插入多个元素。可以使用"..."来向一个列表中插入另一个列表的所有元素,下面的示例扩展后输出长度为 6。

```
//数据源
var list1 = [4, 5, 6];
//使用"..."扩展了列表长度
var list2 = [1, 2, 3, ...list1];
print(list2.length);
```

如果扩展运算符右边的表达式可能为空,可以使用空感知的扩展运算符"...?"来避免异常。下面示例中 list1 为一个空变量,当 list2 进行扩展时首先会判断 list1 是否为空再进行扩展,此示例输出为 3,不会引发异常。

```
//变量为空
var list1;
//使用"...?"先判断变量是否为空再进行扩展
var list2 = [1, 2, 3, ...?list1];
print(list2.length);
```

6.5 Set 类型

Set 是没有顺序且不能重复的集合,所以不能通过索引去获取值。下面是 Set 的定义和常用方法。

```
void main() {
  //定义 Set 变量
  var set = Set();
  //输出 0
  print(set.length);

  //错误 Set 没有固定元素的定义
```

```
        //var testSet2 = Set(2);

        //添加整型类型元素
        set.add(1);
        //重复添加同一个元素无效
        set.add(1);
        //添加字符串类型元素
        set.add("a");
        //输出{1, a}
        print(set);

        //判断是否包含此元素,输出 true
        print(set.contains(1));

        //添加列表元素
        set.addAll(['b', 'c']);
        //输出{1, a, b, c}
        print(set);

        //移除某指定元素
        set.remove('b');
        //输出{1, a, c}
        print(set);
}
```

6.6 Map 类型

通常来说,Map 映射一个关联了 Key 和 Value 的对象。Key 和 Value 都可以是任意类型的对象。Key 是唯一的,但是可以多次使用相同的 Value。

常用的 Map 有如下两种定义方式,可以不指定 Key 的类型。

```
var map1 = Map();
var map2 = {"a": "this is a", "b": "this is b", "c": "this is c"};
```

创建好 Map 对象后可以调用其方法及获取属性,如输出 Map 的长度,根据 Key 返回 Value。用法如下:

```
//长度属性,输出 0
print(map1.length);
//获取值,输出 this is a
print(map2["a"]);
//如果没有 key,返回 null
print(map1["a"]);
```

```
//需要注意的是 keys 和 values 是属性而不是方法
print(map2.keys);           //返回所有 key,输出(a, b, c)
print(map2.values);         //返回所有 value 输出(this is a, this is b, this is c)
```

Map 中的 Key 及 Value 类型是可以指定的,当指定了类型后就不能使用其他类型。用法如下:

```
//key:value 的类型可以指定
var intMap = Map<int, String>();
//map 新增元素
intMap[1] = "数字 1";
//key 错误类型不正确
//intMap['a'] = "a";
intMap[2] = "数字 2";
//value 错误类型不正确
//intMap[2] = 2;
//删除元素
intMap.remove(2);
//是否存在 key 输出 true
print(intMap.containsKey(1));
```

第 7 章 运算符

Dart 支持各种类型的运算符,并且其中的一些操作符还能进行重载。完整的操作符如表 7-1 所示。

表 7-1 各种类型的运算符

描 述	运 算 符
一元后缀	expr++ expr-- () []. ?.
一元前缀	-expr ! expr ~expr ++expr --expr
乘法类型	* / % ~/
加法类型	+ -
移动位运算	<< >>
与位运算	&
异或位运算	^
或位运算	|
关系和类型测试	>= <= > < as is is!
等式	== !=
逻辑与	&&
逻辑或	||
条件	expr1 ? expr2 : expr3
级联	..
赋值	= *= /= ~/= %= += -= <<= >>= &= ^= |= ??=

使用运算符时,可以创建表达式。以下是运算符表达式的一些示例:

```
a++
a--
a + b
a = b
a == b
expr ? a : b
a is T
```

在之前的操作符表中，操作符的优先级由其所在行定义，上面行内的操作符优先级大于下面行内的操作符。例如，乘法类型操作符"％"的优先级比等价操作符"＝＝"要高，而"＝＝"操作符的优先级又比逻辑与操作符"&&"要高。这些操作符的优先级顺序将在下面的两行代码中体现出来：

```
//1.使用括号来提高可读性
if ((n % i == 0) && (d % i == 0))

//2.难以阅读,但是和上面等价
if (n % i == 0 && d % i == 0)
```

> **警告** 对于二元运算符，其左边的操作数将会决定使用的操作符的种类。例如，当使用一个 Vector 对象及一个 Point 对象时，aVector ＋ aPoint 使用的"＋"是由 Vector 所定义的。

7.1 算术运算符

Dart 支持常用的算术运算符如下：

操作符	含 义
＋	加
－	减
－expr	一元减号,也被命名为负号(使后面表达式的值反过来)
＊	乘
／	除
～／	返回一个整数值的除法
％	取余,除法下的余数

示例代码如下：

```
assert(3 + 6 == 9);
assert(3 - 6 == -3);
assert(3 * 6 == 18);
assert(7 / 2 == 3.5);     //结果是浮点型
assert(5 ~/ 2 == 2);      //结果是整型
assert(5 % 2 == 1);       //求余数
```

Dart 还支持前缀和后缀递增和递减运算符，如下所示：

操作符	含义
++var	var=var+1 表达式的值为 var+1
var++	var=var+1 表达式的值为 var
--var	var=var-1 表达式的值为 var-1
var--	var=var-1 表达式的值为 var

示例代码如下：

```
var a, b;

a = 0;
b = ++a;        //在 b 获得其值之前自增 a
assert(a == b); //1 == 1

a = 0;
b = a++;        //在 b 获得值后自增 a
assert(a != b); //1 != 0

a = 0;
b = --a;        //在 b 获得其值之前自减 a
assert(a == b); //-1 == -1

a = 0;
b = a--;        //在 b 获得值后自减 a
assert(a != b); //-1 != 0
```

7.2 关系运算符

等式和关系运算符的含义如下：

操作符	含义
==	等于
!=	不等于
>	大于
<	小于
>=	大于等于
<=	小于等于

有时需要判断两个对象是否相等，使用"=="运算符。
下面是使用每个等式和关系运算符的示例：

```
assert(2 == 2);
assert(2 != 3);
assert(3 > 2);
assert(2 < 3);
assert(3 >= 3);
assert(2 <= 3);
```

7.3 类型测试操作符

as、is 和 is! 操作符在运行时用于检查类型非常方便,含义如下:

操作符	含 义
as	类型转换
is	当对象是相应类型时返回 true
is!	当对象不是相应类型时返回 true

如果 obj 实现了 T 所定义的接口,那么 obj is T 将返回 true。

使用 as 操作符可以把一个对象转换为指定类型,前提是能够转换。转换之前用 is 判断一下更保险。思考下面这段代码:

```
if (user is User) {
  //类型检测
  user.name = 'Flutter';
}
```

如果能确定 user 是 User 的实例,则可以通过 as 直接简化代码:

```
(user as User).name = 'Flutter';
```

注意　上面两段代码并不相等。如果 user 的值为 null 或者不是一个 User 对象,第一段代码不会做任何事情,第二段代码则会报错。

7.4 赋值操作符

可以使用"="运算符赋值。要仅在变量为 null 时赋值,使用"??="运算符。代码如下:

```
//赋值给 a
a = value;
//如果 b 为空,则将值分配给 b；否则,b 保持不变
b ??= value;
```

诸如"+="之类的复合赋值运算符将操作与赋值相结合。以下是复合赋值运算符的含义:

复合赋值	等式表达式
a op b	a = a op b
a += b	a = a + b
a -= b	a = a - b

7.5 逻辑运算符

可以使用逻辑运算符反转或组合布尔表达式,逻辑运算符如下:

操作符	含义
! expr	反转表达式(将 false 更改为 true,反之亦然)
||	逻辑或
&&	逻辑与

下面是使用逻辑运算符的示例:

```
if (!expr && (test == 1 || test == 8)) {
  //...TO DO...
}
```

7.6 位运算符

通常我们指位运算为"<<"或">>"移动位运算,通过操作位的移动来达到运算的目的,而"&"、"|"、"^"、"~expr"也是操作位来达到运算的目的。具体含义如下:

操作符	含义
&	与
|	或
^	异或
~expr	一元位补码(0s 变为 1s；1s 变为 0s)
<<	左移
>>	右移

下面是使用所有位运算符的示例：

```
final value = 0x22;
final bitmask = 0x0f;

assert((value & bitmask)     == 0x02);      //与
assert((value & ~bitmask)    == 0x20);      //与非
assert((value | bitmask)     == 0x2f);      //或
assert((value ^ bitmask)     == 0x2d);      //异或
assert((value << 4)          == 0x220);     //左移
assert((value >> 4)          == 0x02);      //右移
```

7.7 条件表达式

Dart 有两种条件表达式让你简明地评估为 if-else 语句表达式。如下代码即为一种条件表达式，也可以称为三元表达式。如果条件为真，返回 expr1，否则返回 expr2：

```
condition ? expr1 : expr2
```

第二种条件表达式如下所示，如果 expr1 为非空，则返回其值；否则，计算并返回 expr2 的值：

```
expr1 ?? expr2
```

7.8 级联操作

级联用两个点 ".." 操作符允许你对同一对象执行一系列操作。类似 Java 语言里的 list.toList().toString() 处理或 JavaScript 里的 Promise 的 then 处理。级联操作主要的目的是为了简化代码。示例代码如下：

```
querySelector('#btnOK')       //获取一个 id 为 btnOK 的按钮对象
  ..text = '确定'              //使用它的成员
  ..classes.add('ButtonOKStyle')
  ..onClick.listen((e) => window.alert('确定'));
```

第一个方法调用 querySelector，返回一个按钮对象，然后再设置它的文本为"确定"，再给这个按钮添加一个样式 "'ButtonOKStyle'"，最后再监听单击事件，事件弹出一个显示"确定"的 Alert。这个例子相当于如下操作：

```
var button = querySelector('#btnOK);
button.text = '确定';
button.classes.add(''ButtonOKStyle'');
button.onClick.listen((e) => window.alert('确定'));
```

注意 严格来说，级联的"双点"符号不是运算符。这只是 Dart 语法的一部分。

第 8 章 流程控制语句

Dart 可用的流程控制语句如下：
- if 和 else；
- for 循环；
- while 和 do-while 循环；
- break 和 continue；
- switch 和 case；
- assert 断言；
- try-catch 和 throw。

8.1 if 和 else

Dart 支持 if 及 else 的多种组合。示例代码如下：

```
void main() {
  //if/else 示例
  int index = 1;
  if (index == 0) {
    print('index = 0');
  } else if (index == 1) {
    print('index = 1');
  } else {
    print('index = $ index');
  }
}
```

上面的代码输出"index = 1"，条件语句运行到第二条判断就停止了。

8.2 for 循环

标准的 for 循环使用，示例代码如下：

```
void main() {
  //定义一个数组
  var messages = [];
  //定义数组长度
  int length = 5;
  //计数器
  int i = 0;
  //使用 for 循环
  for (i; i < length; i++) {
    //向数组里添加元素
    messages.add(i.toString());
  }
  //打印数组内容
  print(messages.toString());
}
```

首先定义一个空数组 messages，然后使用 for 循环向数组里添加元素，最后输出数组内容。其中 i 为计算器，如自增 i++，自减 i--，i<length 为循环条件，当条件不满足时循环中止，大括号里为循环体添加元素。

除了常规的 for 循环外，针对可以序列化的操作数，可以使用 forEach 或 for in 语句。代码如下：

```
//forEach 迭代输出
messages.forEach((item){
    print(item);
});

//for in 语句迭代输出
for(var x in messages) {
  print(x);
}
```

上面的代码会按序列输出 0,1,2,3,4。

8.3　while 和 do-while

下面的示例是编写了一个 while 循环，定义了一个变量 temp，temp 变量在循环体里自动加 1，当条件(temp < 5)不满足时会退出循环：

```
var _temp = 0;
while(temp<5){

  print("这是一个 while 循环：" + (_temp).toString());
  temp ++;
}
```

接下来看一下 do-while 的示例，具体代码如下：

```
var _temp = 0;

  do{
    print("这是一个循环:" + (_temp).toString());
    _temp ++;
  }
  while(_temp < 5);
```

上面的两个例子都对应如下输出：

```
flutter:这是一个循环:0
flutter:这是一个循环:1
flutter:这是一个循环:2
flutter:这是一个循环:3
flutter:这是一个循环:4
```

8.4　break 和 continue

break 用来跳出整个循环，示例如下：

```
void main() {
  //数组
  var arr = [0, 1, 2, 3, 4, 5, 6];
  //for 循环
  for (var v in arr) {
    //跳出整个循环
    if(v == 2 ){
      break;
    }
    print(v);
  }
}
```

上面的代码当 v 等于 2 时循环结束，所以程序输出 0 和 1，现在把 break 改为 continue，代码如下：

```
void main() {
  //数组
  var arr = [0, 1, 2, 3, 4, 5, 6];
  //for 循环
  for (var v in arr) {
    //跳出当前循环,循环还继续向下执行
```

```
    if(v == 2 ){
      continue;
    }
    print(v);
  }
}
```

改为 continue 后,当 v 等于 2 时循环只是跳出本次循环,代码还会继续向下执行,所以输出的结果是 0,1,3,4,5,6。

8.5 switch 和 case

Dart 中 switch / case 语句使用"=="操作来比较整数、字符串或其他编译过程中的常量,从而实现分支的作用。switch / case 语句的前后操作数必须是相同的类型的对象实例。每一个非空的 case 子句最后都必须跟上 break 语句。具体示例如下:

```
String today = 'Monday';
  switch (today) {
    case 'Monday':
      print('星期一');
      break;
    case 'Tuesday':
      print('星期二');
      break;
  }
```

上面这段代码也可以改为 if /else 语句,输出的结果相同,代码输出为"星期一"。

8.6 断言 assert

Dart 语言通过使用 assert 语句来中断正常的执行流程,当 assert 判断的条件为 false 时发生中断。assert 判断的条件可以是任何可以转化为 boolean 类型的对象,即使是函数也可以。如果 assert 的判断为 true,则继续执行下面的语句。反之则会抛出一个断言错误异常 AssertionError。代码如下:

```
//确定变量的值不为 null
assert(text != null);
```

第 9 章 函　　数

9.1　函数的概念

Dart 是一个面向对象的语言，函数属于 Function 对象。函数可以像参数一样传递给其他函数，这样便于做回调处理。如下示例取两个变量的最大值：

```
int max(int a, int b) {
  return a >= b ? a:b;
}
```

即使忽略了数据类型，函数依然是可用的。代码如下：

```
//去掉变量 a,b 及返回类型,取最大值依然可用
max(a,b) {
  return a >= b ? a:b;
}
```

对于上面只包含一个表达式的函数，可以使用箭头"=>"简写。代码如下：

```
//使用箭头 =>语法简写函数
int max(int a, int b) => a >= b ? a:b;
```

这里的 => expre 语法是 { return expr; } 的简写。符号"=>"被称为箭头语法。

9.2　可选参数

函数的参数可以有选择性地进行传递。Dart 语言里可选参数分为以下两种：
- 命名参数：使用大括号{}括起来的参数；
- 位置参数：使用中括号[]括起来的参数。

9.2.1 命名参数

当定义一个函数时,使用{参数1,参数2,...}的格式来指定命名参数。如下代码中,funName 为方法名,param1 为普通参数,param2 及 param3 为可选参数。

```
void funName(String param1,{double param2,bool param3,… })
```

接下来编写一个示例,示例的函数可以设置文本的属性,如字体大小,以及字体是否加粗。实际的场景是文本的属性可能只设置了若干个,那么命名可选参数就非常实用了。代码如下:

```
void main() {
  textStyle('可选参数');
  textStyle('可选参数:',fontSize: 18.0);
  textStyle('可选参数:',fontSize: 18.0,bold: true);
}

//字体大小和是否加粗均为可选参数
void textStyle(String content,{double fontSize,bool bold}) {
  print(content + fontSize.toString() + " " + bold.toString());
}
```

上面代码中 textStyle 函数的{}里均为可选参数,在 main 函数里分别传入一个、两个或三个参数都可以正常执行。输出结果如下:

```
flutter:可选参数 null null
flutter:可选参数:18.0 null
flutter:可选参数:18.0 true
```

在 Flutter 的组件设计里处处可以看到可选参数的应用,例如容器 padding,以及文本 Style 样式的属性。代码如下:

```
import 'package:flutter/material.dart';

void main() {
  runApp(
    MaterialApp(
      title: '可选命名参数示例',
      home: MyApp(),
    ),
  );
}
```

```
class MyApp extends StatelessWidget {
  @override
  Widget build(BuildContext context) {
    return Scaffold(
      appBar: AppBar(
        title: Text('可选命名参数示例'),
      ),
      //容器组件
      body: Container(
        //内边距属性 left right top bottom 为可选参数
        padding: EdgeInsets.only(
          //左内边距值
          left: 20,
          //右内边距值
          right: 20,
          //上内边距值
          top: 30,
          //下内边距值
          bottom: 30,
        ),
        //文本组件
        child: Text(
          '可选参数',
          //文本样式 color fontSize 为可选参数
          style: TextStyle(
            //字体颜色
            color: Colors.red,
            //字体大小
            fontSize: 18.0,
          ),
        ),
      )
    );
  }
}
```

> **提示** 这里不是让你查看 Flutter 项目运行效果,可以点开 EdgeInsets.only 和 TextStyle 查看源代码,看看可选参数是怎么设置的。

命名参数通常是一种可选参数,如果想强制用户传入这个参数,可以使用 @required 注解来声明这个参数。以 Flutter 的动画容器组件 AnimatedContainer 为例,它的构造函数必须传入 duration 参数。代码如下:

```
AnimatedContainer({
  Key key,
  this.alignment,
  this.padding,
  Color color,
  Decoration decoration,
  this.foregroundDecoration,
  double width,
  double height,
  BoxConstraints constraints,
  this.margin,
  this.transform,
  this.child,
  Curve curve = Curves.linear,
  @required Duration duration,
})
```

使用@required 注解，需要导入 meta 包，可以直接导入 package:meta/meta.dart，也可以导入其他包所导出的 meta 包，例如 Flutter 的 package:flutter/material.dart。示例代码如下：

```
//导入 material.dart 后不必导入 meta.dart
//import 'package:meta/meta.dart';
import 'package:flutter/material.dart';

void main() {
  runApp(
    MaterialApp(
      title: '@required 参数示例',
      home: MyApp(),
    ),
  );
}

class MyApp extends StatelessWidget {
  @override
  Widget build(BuildContext context) {
    return Scaffold(
      appBar: AppBar(
        title: Text('@required 参数示例'),
      ),
      //动画容器组件
      body: AnimatedContainer(
        //可选参数
        margin: EdgeInsets.only(left: 10.0),
        //duration 为必传参数，不传会报异常
        duration: Duration(seconds: 2),
```

```
        //可选参数
        width: 40.0,
        //可选参数
        height: 50,
        //可选参数
        color: Colors.yellow,
      ),
    );
  }
}
```

如果去掉上述代码中的 duration 属性，编译器会引发一个异常，提示 duration 不能为空。输出内容如下所示：

```
flutter: The following assertion was thrown building MyApp(dirty):
flutter: 'package:flutter/src/widgets/implicit_animations.dart': Failed assertion: line 231
pos 15: 'duration
flutter: != null': is not true.
```

9.2.2 位置参数

在方法参数中，使用"[]"包围的参数属于可选位置参数，如下面代码所示 from 和 age 参数即为位置参数。

```
void printUserInfo(String name, [String from = '中国', int age]) {
}
```

调用包含可选位置参数的方法时，无须使用 paramName：value 的形式，因为可选位置参数是位置，如果想指定某个位置上的参数值，则前面的位置必须已经有值，即使前面的值为默认值。先看下面一个示例代码：

```
void main() {
  //可行
  printUserInfo('张三');
  //可行
  printUserInfo('张三','中国',30);
  //不可行
  //printUserInfo('张三',30);
}
//from 和 age 为可选位置参数
void printUserInfo(String name, [String from = '中国', int age]) {
  print(name + "来自" + from + "年龄" + age.toString());
}
```

这里特意使用两个不同类型的可选参数 from 和 age 作为示例。如果前后可选参数为相同类型，则会出现异常结果，并且只有在发生后才会注意到。所以这一点要特别注意。上面的代码输出如下内容：

```
flutter:张三来自中国年龄 null
flutter:张三来自中国年龄 30
```

9.3　参数默认值

如果函数参数指定了默认值，当不传入值时，函数里可以使用这个默认值；如果传入了值，则用传入的值取代默认值。你可以使用"＝"来为函数参数定义默认值，这种方法适用于命名参数和位置参数。默认值必须是编译常量。如果没有提供默认值，默认值便是 null。接下来看一个参数默认值的示例，代码如下：

```dart
void main() {
  //只传入 name
  printUserInfo('张三');
  //传入 name 及 sex
  printUserInfo('小红','女');
  //传入 name sex age
  printUserInfo('小红','女',26);
}

//参数 sex 为默认参数,当不传入 sex 参数时默认值为'男'
void printUserInfo(String name,[String sex = '男',int age]){
  //age 不传值时为 null
  if(age!= null){
    print("姓名:$ name 性别:$ sex 年龄:$ age");
  }
  print("姓名:$ name 性别:$ sex 年龄保密");
}
```

上面的示例中，name 是必传参数，sex 和 age 是可选参数。sex 参数有默认值，当 sex 参数不传时其值为'男'，当 age 不传时其值为 null。示例输出结果如下：

```
flutter:姓名:张三 性别:男 年龄保密
flutter:姓名:小红 性别:女 年龄保密
flutter:姓名:小红 性别:女 年龄:26
flutter:姓名:小红 性别:女 年龄保密
```

也可以使用 List 或者 Map 作为默认值。下面的示例中 list 和 map 为默认参数。

```dart
void main() {
  listAndMapParam();
}
//List 及 Map 默认参数
void listAndMapParam({
  List<String> list = const ['A', 'B', 'C'],
  Map<int, String> map = const {
      0: 'one',
      1: 'second',
      2: 'third'
  }}){
  print('list: $list');
  print('map: $map');
}
```

此例可以正常输出其参数值，输出结果如下：

```
flutter: list: [A, B, C]
flutter: map: {0: one, 1: second, 2: third}
```

9.4 main 函数

main 函数通常作为应用程序的入口函数。Flutter 应用程序也是从 main 函数启动的。下面的代码表示应用要启动 MyApp 类。

```dart
void main() => runApp(MyApp());
```

上面代码中的 runApp 函数启动了整个应用的根组件，这样 Flutter 应用程序的界面就从这里开始渲染了。

9.5 函数作为参数传递

函数可以作为参数传递给其他函数。代码如下：

```dart
void main(){
  //整型列表
  var listInt = [1, 2, 3];
  //把 printIntValue 作为参数
  listInt.forEach(printIntValue);
  //字符串列表
  var listString = ['A', 'B', 'C'];
  //把 printStringValue 作为参数
```

```
    listString.forEach(printStringValue);
}
//打印整型值
void printIntValue(int value) {
    print(value);
}
//打印字符串
void printStringValue(String value) {
    print(value);
}
```

上面代码中的 printIntValue 和 printStringValue 是函数,同时也是列表迭代方法 forEach 的参数。接下来打开 forEach 函数分析一下 Flutter 源码里如何实现的。代码如下:

```
/*
 * forEach 参数为一个函数 f
 * 使用 for 循环来多次调用函数 f
 * 把参数 element 回传给函数 f
 */
void forEach(void f(E element)) {
    for (E element in this) f(element);
}
```

列表迭代函数 forEach 的参数即为一个函数 f,函数体里使用 for 循环来依次调用 f 函数,这里再把参数 element 回传给函数 f,达到输出列表值的目的。

9.6 匿名函数

匿名函数即为没有名字的函数,同样具备函数的功能,目的是为了简化代码的编写。这种方式有时也被称为"lambda"或者"闭包"。你可能会将匿名函数赋值给一个变量,以便后续使用。下面的示例可以迭代输出列表 list 的值。

```
void main(){

    var list = ['I', 'love', 'study','dart'];
    //forEach 函数里的参数即为一个匿名函数
    list.forEach((item) {
        print(item);
    });

}
```

forEach 函数里的参数是需要传入一个函数的,示例中直接使用匿名函数替代有名称的函数名。匿名函数部分如下:

```
(item) {
    print(item);
}
```

从上面的示例可以看出代码简化了许多。匿名函数括号中包含 0 个或多个参数,用逗号隔开。它的完整语法格式如下:

```
([[类型] 参数 1[, …]]) {
代码块;
};
```

如果这个函数只包含一个语句,可以使用箭头符号简化它,它们的效果是等价的。代码如下:

```
//箭头函数表示方法
list.forEach(
    (item) => print(item)
);
```

在 Flutter 里处处可以看到匿名函数的用法,例如 setState 方法,如下代码所示,其改变计数器变量值的处理就是一个匿名函数。

```
//调用 State 类里的 setState 方法来更改状态值,使得计数器加 1
setState(() {
//计数器变量,单击让其加 1
_counter++;
});
```

setState 函数的源码如下所示,它需要传入一个回调函数,这个回调函数我们通常是使用匿名函数来处理的。

```
@protected
void setState(VoidCallback fn) {
    //设置状态处理
  }
```

9.7 词法作用域

Dart 是词法作用域语言,意味着变量的作用域是静态确定的,简单地通过代码的布局来确定。可以"沿着花括号向外走"来判断一个变量是否在作用域中。下面是一个模拟 Http 请求的例子,data 变量在整个 dart 文件内有效,serverUrl 变量在 main 函数的花括内有效，formData 变量在 getServerData 函数的花括内都有效。

```dart
//作用域在整个 dart 文件代码内
String data = '测试数据';

void main() {
  //请求 serverUrl 作用域在 main 函数内
  String serverUrl = 'http://127.0.0.1/getData';
  getServerData(serverUrl);
}

//获取服务器数据函数
void getServerData(String url) {
  //请求参数作用域在 getServerData 函数内
  var formData = {'id': '001'};
  //发起请求
  request(url,formData,(int statusCode){
    if(statusCode == 200){
      //此处不能读取 serverUrl 变量
      print('请求地址为:' + url);
      //此处可以读取 formData 变量
      print('请求参数为:' + formData['id']);
      //此处可以读取 data 变量
      print('成功返回数据为:' + data);
    }
  });
}

//发起请求
void request(String url,formData,Function callBack){
  print('发起 Http 请求');
  callBack(200);
}
```

上面示例输出结果如下所示：

```
flutter: 发起 Http 请求
flutter: 请求地址为:http://127.0.0.1/getData
flutter: 请求参数为:001
flutter: 成功返回数据为:测试数据
```

第2篇 面向对象编程

第 10 章 面向对象基础

Dart 是一门面向对象的编程语言,具备类和基于混入的继承。每一个对象都是一个类的实例,而所有的类都派生自 Object。本章将介绍面向对象基础知识。

10.1 面向对象概述

面向对象是相对于面向过程的一种编程方式。

面向过程的编程方式由来已久,例如 C 语言及 Basic 语言都是面向过程的编程方式。这种方式非常直观,需要一个功能,直接就写几行实现方法。比如需要操作一个人移动到某个点,直接就写代码修改一个人的坐标属性,逐格地让他移动到目标点就行了。

面向对象的编程方式操作的是一个个的对象,比如还是需要操作一个人的移动,需要先实例化那个人的一个管理类对象,然后告诉这个"人"的对象,需要移动到什么地方去,然后人就自己走过去了。至于具体是怎样走的,外部不关心,只有"人"对象本身知道。

面向对象有优点也有缺点,也存在一些争论的地方。确实,面向对象在性能上面肯定不如面向过程好,毕竟面向对象需要实例化对象,需要消耗 CPU 和内存资源,但它的优点也是很明显的,毕竟在一个大型的项目里面,面向对象易于维护和管理,条理也清晰,是一种重要的编程思想。

10.2 面向对象基本特征

面向对象有四大基本特征:
- 封装;
- 继承;
- 多态;
- 抽象。

接下来,以一家银行为例,来阐述面向对象这几大特征。

1. 封装

对于一般人来说,对银行的印象就只有一排对外办公的窗口,然后有存款和取款两种基本业务。但实际上，银行是一个结构非常复杂，功能非常众多的机构。我们并不会关心它的内部是怎样运作的，比如银行的员工是怎样数钱的，怎样记录存款，怎样开保险柜等。这些对于外部的人员来说，知道了可能会引起更多不必要的麻烦，所以银行只需要告诉你，你可以在这个窗口办理业务，可以存款和取款，这就够了。

所谓的封装，就是指把内部的实现隐藏起来，然后只暴露必要的方法让外部调用。暴露的方法我们称之为接口。

2．继承

我们知道银行有两种最基本的业务：存款和取款。但现实中，大部分的银行都不止这两种业务，还有很多其他的业务，例如投资窗口、办理对公业务的窗口等。这些业务，是在最基本的银行存取款业务的基础上添加的，所以我们可以理解成基本的银行是只有两种业务的，然后银行在保留了原有业务的基础上，再扩展了其他的业务。

如果把基本的银行看作父类（基类），包含存款和取款两个公共方法，那么后来的银行可以看作是子类，它在继承了基本银行存取款的公共方法之后，还新增了投资和对公业务两个公共方法。有些银行甚至会重写基本的存取款功能，让自己和基本银行的业务有一定的区别，这个过程就是继承。

3．多态

同样是存款业务，如果客户拿着人民币和拿着美元去银行办理存款业务，实际上银行处理的方式是不一样的。这种办理同一种业务（公共方法），由于给予的内容（传入的参数类型或者数量）不一样，而导致操作（最终实现的方法）不一样，叫作编译多态，也叫作函数的重载。

接下来，客户去了一家银行存款，客户不知道这家银行的存款业务有没有和基本银行不一样，反正客户就是把钱存进去了，然后具体业务的实现究竟是调用了基本银行存款功能，还是由这家银行新的存款功能实现，客户是不关心的。这种外部直接调用一个方法接口，然后具体实现的内容由实际处理的类来决定使用基类或者子类的方法，就叫作运行时多态。

4．抽象

有些观点并没有把抽象列为面向对象的特征，但实际上这是面向对象的一个本质的东西。

虽然银行五花八门，但我们可以找到它们的共性，如上面说的，基本的银行有存取款业务，投资银行有投资业务之类，其实就是对银行作出了一个抽象的做法。

在操作的时候，这些业务其实就是一个个的接口，客户不管面对的是什么具体的银行，只要是同一个类型的银行，都可以办理相同的业务。

10.3 类声明及构成

具有相同特性（数据元素）和行为（功能）的对象的抽象就是类，因此对象的抽象是类，类的具体化就是对象，也可以说类的实例是对象。

类是构造面向对象程序的基本单位,是抽取了同类对象的共同属性和方法所形成的对象或实体的"模板",而对象是现实世界中实体的描述,对象要创建才存在,有了对象才能对对象进行操作。类是对象的模板,对象是类的实例。关于对象的详细描述参阅"第 10 章 对象"。

Dart 语言类主要由以下几个部分构成:
- 类名;
- 成员变量;
- 成员方法;
- 构造方法。

10.3.1 类声明

Dart 语言中一个类的实现包括类声明和类体。类声明语法格式如下:

```
[abstract] class className [extends superClassName] [implements interfaceName] [with className1,className2, ...]{
   //类体
}
```

其中,class 是声明类的关键字,className 是自定义类的类名,abstract 是用来修改此类的,加上它表示此类为一个抽象类并可以省略。类名后面为继承类 extends 关键字,superClassName 为其父类名称,如果当前类不继承则可以省略。关键字 implements 用来实现某个接口,interfaceName 为接口名称,如果当前类不实现某个接口则可以省略。关键字 with 为"混入"其他类用法,className 即被混入的类名,此处可以混入多个类并用逗号","隔开,如果当前类不混入其他类则可以省略。关于 abstract、extends、implements,以及 with 的用法后面会详细介绍。

下面的代码定义了一个名称为 Person 的类。

```
//person.dart 文件
//类名为 Person 继承 Object
class Person extends Object {
   //类体
}
```

上面的代码声明了人类(Person),它继承了 Object 类。类体是类的主体,包括成员变量和方法。

10.3.2 成员变量

声明类中成员变量语法格式如下:

```
class className{
  //成员变量
  [static][const][final] type name;
}
```

其中,type 是成员变量数据类型,name 是成员变量名。数据类型前面的关键字是成员变量的修饰符,可以省略,说明如下:

- static 表示成员变量在类本身上可用,而不是在类的实例上。这就是它的意思,并没有用于其他地方;
- final 表示单一赋值,final 变量或字段必须初始化,一旦赋值,就不能改变 final 变量的值;
- const 用来定义常量,const 和 final 的区别在于,const 比 final 更加严格。final 只是要求变量在初始化后值不变,但通过 final 无法在编译时(运行之前)知道这个变量的值,而 const 所修饰的是编译时常量,在编译时就已经知道了它的值,显然它的值也是不可改变的。

提示 当 name 修改符指定为 const 或 final 时为常量。

接下来看一个类声明成员变量的例子。

```
//person.dart 文件
//类名为 Person 继承 Object
class Person extends Object {
  //成员变量
  String sex = "男";
}
```

上面代码声明了一个名称为 sex 的成员变量,并初始化了它的值。

10.3.3 成员方法

成员方法即成员函数,其定义及使用和函数是一样的。声明类体中成员方法语法格式如下:

```
class className{

  [static][type] methodName(paramType:paramName ...) [async]{
    //方法体
  }

}
```

其中,type 为方法返回值数据类型,methodName 为方法名,static 为方法修改符,表示静态方法。paramType 为方法参数类型,paramName 为方法参数名称,方法的参数可以是 0 到多个。async 表示方法是异步的,当有等待的操作时需要使用此关键字,例如数据请求处理。

下面看一个声明方法的示例,代码如下:

```dart
//person.dart
//类名为 Person 继承 Object
class Person extends Object {
  //成员变量
  String sex = "男";

  //成员方法
  String run(){
    return "人类会跑步";
  }
}
```

上面代码中 run 为 Person 类的成员方法,如果调用此方法会输出"人类会跑步"。当成员方法没有返回值时类型可设置为 void。

10.4 静态变量和静态方法

使用 static 关键词来实现类级别的变量和方法。

10.4.1 静态变量

静态变量(类变量)对类级别的状态和常数是很有用的。例如写一个 Flutter 商城项目,有些标题、标签和文本提示等内容就可以提取出来放在一个类里定义成静态字符串类型,代码如下:

```dart
//static_variable_sample/lib/string.dart 文件
class KString{
  static const String mainTitle = 'Flutter 商城';
  static const String homeTitle = '首页';
  static const String categoryTitle = '分类';
  static const String shoppingCartTitle = '购物车';
  static const String memberTitle = '会员中心';
  static const String loading = '加载中';
  static const String loadReadyText = '上拉加载';
  static const String recommendText = '商品推荐';
  static const String hotGoodsTitle = '火爆专区';
  static const String noMoreText = '没有更多了';
  static const String toBottomed = '已经到底了';
  static const String noMoreData = '暂时没有数据';
  static const String detailsPageTitle = '商品详情';
  static const String detailsPageExplain = '说明: > 急速送达 > 正品保证';
```

```dart
    static const String addToCartText = '加入购物车';
    static const String buyGoodsText = '马上购买';
    static const String cartPageTitle = '购物车';
    static const String allCheck = '全选';
    static const String allPriceTitle = '合计';
    static const String allPriceAdv = '满 10 元免配送费,预购免配送费';
    static const String orderTitle = '我的订单';
    static const String pendingPayText = '待付款';
    static const String toBeSendText = '待发货';
    static const String toBeReceivedText = '待收货';
    static const String evaluateText = '待评价';
}
```

静态变量使用非常方便,不需要实例化类的对象就可以访问。例如获取商城主标题及首页标题示例代码如下:

```dart
//static_variable_sample/lib/main.dart 文件
import 'package:flutter/material.dart';
//导入 KString 类
import 'string.dart';
//入口程序
void main() {
  runApp(
    MaterialApp(
      title: '静态变量使用示例',
      home: MyApp(),
    ),
  );
}
//主组件
class MyApp extends StatelessWidget {
  @override
  Widget build(BuildContext context) {
    return Scaffold(
      appBar: AppBar(
        //主标题
        title: Text(KString.mainTitle),
      ),
      //居中组件
      body: Center(
        //首页
        child: Text(KString.homeTitle,
          //字体大小也可以提取成静态变量
          style: TextStyle(fontSize: 28.0),
        ),
      )
    );
  }
}
```

其中，KString.mainTitle 及 KString.homeTitle 为取静态变量值。代码中首页标题的字体大小也可以提取成静态变量，因为项目中可能多处用到 2.8.0 号字。

> **提示** Flutter 项目中的文本、颜色和字体等配置项均可以使用静态变量表示。代码规范推荐使用"小驼峰"来命名。

运行此项目示例，效果如图 10-1 所示。

图 10-1 静态变量使用示例效果图

10.4.2 静态方法

静态方法（类方法）不操作实例，因此不能访问 this。调用方式如下：

```
className.methodName(paramType:param1,paramType:param2,…)
```

其中，className 为类名，methodName 为方法名，paramType 为参数类型，参数可以定义多个。

在实际项目中通常需要写一个工具类，用来放各种方法，例如产生随机数和日期处理等。静态方法的示例如下：

```dart
//static_method_sample/utils/lib/utils.dart 文件
import 'dart:io';
//工具类
class Utils{
  //静态方法,判断当前运行的平台是否为移动设备
  static bool isMobile() {
    //使用 Platform 判断平台类型
    return Platform.isAndroid || Platform.isIOS;
  }
}
```

上面代码定义了一个 Utils 工具类,添加了一个静态方法 isMobile 用来判断当前平台是否为移动设备。

接下来编写测试代码。直接使用 Utils.isMobile() 即可。代码如下:

```dart
//static_method_sample/utils/lib/main.dart 文件
//导入 Utils 类文件
import 'utils.dart';
void main(){
  print("当前设备是否为移动设备:" + Utils.isMobile().toString());
}
```

上面代码输出内容如下:

```
flutter:当前设备是否为移动设备:true
```

在 Flutter 项目里会大量使用静态方法,例如获取主题样式。示例代码如下:

```dart
//static_method_sample/theme/lib/main.dart 文件
import 'package:flutter/material.dart';

void main() {
  runApp(MyApp());
}

class MyApp extends StatelessWidget {
  @override
  Widget build(BuildContext context) {
    return MaterialApp(
      home: MyHomePage(),
    );
  }
}

class MyHomePage extends StatelessWidget {
  @override
```

```
Widget build(BuildContext context) {
  return Scaffold(
    appBar: AppBar(
      title: Text("静态方法示例"),
    ),
    body: Center(
      child: Container(
        //使用静态方法 Theme.of 获取主题的 accentColor
        color: Theme.of(context).accentColor,
        child: Text(
          '带有背景颜色的文本组件',
          //使用静态方法 Theme.of 获取主题的文本样式
          style: Theme.of(context).textTheme.title,
        ),
      ),
    ),
  );
}
```

其中，Theme.of(context).accentColor 及 Theme.of(context).textTheme.title 使用了主题类 Theme 的静态方法 of。Flutter SDK 将其设计为静态方法的原因是因为这样可以在项目的各个页面中轻易地获取到主题样式。

上面的代码运行后如图 10-2 所示。页面中间文本组件样式取自系统蓝色主题。

图 10-2 静态方法示例效果图

10.5 枚举类型

枚举类型是一种特殊的类,通常用来表示相同类型的一组常量值。每个枚举类型都用于一个 index 的 getter,用来标记元素的元素位置。第一个枚举元素的索引是 0,代码如下:

```
enum Color {
  red,
  green,
  blue
}
```

获取枚举类中所有的值,使用 values 常数:

```
List<Color> colors = Color.values;
```

因为枚举类里面的每个元素都是相同类型,可以使用 switch 语句来针对不同的值做不同的处理,示例代码如下:

```
//enum_color.dart 文件
void main(){
  //定义一个颜色变量,默认值为蓝色
  Color aColor = Color.blue;
  switch (aColor) {
    case Color.red:
      print('红色');
      break;
    case Color.green:
      print('绿色');
      break;
    default: //默认颜色
      print(aColor); //'Color.blue'
  }
}
//定义一个枚举
enum Color {
  red,
  green,
  blue
}
```

上面示例代码会输出:Color.blue。

枚举类型有以下限制:

❏ 不可以继承、混入或实现一个枚举;

❏ 不可以显式实例化一个枚举。

第 11 章 对　象

对象是面向对象程序设计的核心。所谓对象就是真实世界中的实体,对象与实体是一一对应的,也就是说现实世界中每一个实体都是一个对象,它是一种具体的概念。类的实例化可以生成对象,实例的方法是对象方法,实例变量就是对象属性。对象有以下特点:

- ❏ 对象具有属性和行为;
- ❏ 对象具有变化的状态;
- ❏ 对象具有唯一性;
- ❏ 对象都是某个类别的实例;
- ❏ 一切皆为对象,真实世界中的所有事物都可以视为对象。

一个对象的生命周期包括三个阶段:创建、使用和销毁。本章详细介绍对象的声明、初始化、方法使用,以及销毁等相关知识。

11.1　创建对象

创建对象有两个步骤:声明和实例化。

1)声明

声明对象即要创建一样类型的对象,语法格式如下:

```
type objectName;
```

其中,type 是类型,如 String 和 Person。示例代码如下:

```
Person person;
```

该语句声明了 Person 类型对象 person,此时并没有为其分配内存空间,只是一个引用。

2)实例化

实例化即为对象分配内存空间,然后调用构造方法初始化对象。示例代码如下:

```
Person person = Person();
```

最新的 Dart 版本不需要使用 new 关键字进行实例化对象。当一个引用变量没有分配内存空间，这个对象即为空对象。Dart 使用 null 关键字表示空对象。示例代码如下：

```dart
//object_null_sample.dart 文件
void main(){
  //声明 person 对象
  Person person = null;
  //实例化 person 对象
  person = Person();
  //判断对象是否为 null
  if(person != null){
    //调用成员方法
    person.run();
  }
}
//类名为 Person,继承 Object
class Person extends Object {
  //成员变量
  String sex = "男";
  //成员方法
  String run(){
    return "人类会跑步";
  }
}
```

person 对象在初始化时为 null，在使用时保险的做法是加一个不为空的判断，否则在试图调用一个空对象的成员方法时，会抛出找不到此方法的错误提示。修改上面的示例，person 对象不实例化而直接使用。代码如下：

```dart
//object_null_sample.dart 文件
void main(){
  //声明 person 对象
  Person person = null;
  //不实例化而直接调用
  person.run();
}
//类名为 Person,继承 Object
class Person extends Object {
  //成员变量
  String sex = "男";
  //成员方法
  String run(){
    return "人类会跑步";
  }
}
```

运行上面的代码会抛出 NoSuchMethodError 错误。错误提示如下：

```
[VERBOSE-2:ui_dart_state.cc(148)] Unhandled Exception: NoSuchMethodError: The method 'run'
was called on null.
Receiver: null
Tried calling: run()
```

11.2 对象成员

对象由成员方法和成员变量组成。当调用一个方法时,在一个对象上调用这个方法可以访问该对象的方法和数据。

使用一个点"."来引用实例变量或方法。示例代码如下:

```
//object_call_sample.dart 文件
void main(){
  //声明 person 对象
  Person person = null;
  //实例化 person 对象
  person = Person();
  //判断对象是否为 null
  if(person != null){
    //调用成员方法
    print("成员方法输出:" + person.run());
  }
  //调用成员变量
  print("成员方法输出:" + person.sex);
}
//类名为 Person,继承 Object
class Person extends Object {
  //成员变量
  String sex = "男";
  //成员方法
  String run(){
    return "人类会跑步";
  }
}
```

上述示例调用成功后输出如下内容:

```
flutter:成员方法输出:人类会跑步
flutter:成员变量输出:男
```

11.3 获取对象类型

要获取一个对象的类型,可以使用对象的 runtimeType 属性,它会返回一个 Type 对象。使用 is 关键字可以判断对象是否属于某个类型。示例代码如下:

```dart
//object_type_sample.dart 文件
void main(){
  //声明并实例化 person 对象
  Person person = Person();
  print("person.runtimeType:" + person.runtimeType.toString());
  //使用 is 判断是否为 Person 类
  if(person is Person){
    print("person 对象的类型是:Person");
  }else{
    print("person 对象的类型是:Animal");
  }
}
//人类 Person
class Person{
}
//动物类
class Animal{
}
```

上面代码中通过 person.runtimeType 可以直接获取该对象的类型为 Person。通过 is 关键字判断对象是否为 Person 类。示例输出结果代码如下：

```
flutter: person.runtimeType:Person
flutter: person 对象的类型是:Person
```

11.4 构造方法

构造方法是类的一种特殊方法，用来初始化类的一个新的对象，在创建对象（new 运算符）之后自动调用。Dart 中的每个类都有一个默认的构造方法，并且可以有一个以上的构造方法。

如果没有声明构造方法，Dart 会提供一个默认构造方法。默认构造方法没有参数，并且调用父类的无参构造方法。

11.4.1 声明构造方法

通过创建一个和类名一样的方法，来声明一个构造方法。最常见的构造方法形式，即生成构造方法。下面的例子定义了一个商品信息类，类名和构造方法名均为 GoodInfo。

```dart
//object_constructor_good_info.dart 文件
//商品信息
class GoodInfo{
  //商品 Id
  String goodId;
```

```
//商品数量
int amount;
//商品图片
String goodImage;
//商品价格
int goodPrice;
//商品名称
String goodName;
//商品详情
String goodDetail;

//构造方法
GoodInfo(String goodId, int amount, String goodImage, double goodPrice, String goodName, String goodDetail){
    this.goodId = goodId;
    this.amount = amount;
    this.goodImage = goodImage;
    this.goodPrice = goodPrice;
    this.goodName = goodName;
    this.goodDetail = goodDetail;
  }
}
```

上面的示例中使用构造方法 GoodInfo 传入商品信息的每个参数即可完成类的实例化。

提示 关键词 this 引用当前实例。仅当有命名冲突时使用 this。否则，Dart 的风格是省略 this。

将构造方法的参数赋值给一个实例变量，这种模式很常见，因此 Dart 用语法糖来简化操作，上面的构建方法可以简化，代码如下：

```
//object_constructor_good_info.dart 文件
//语法糖,简化构造方法
GoodInfo(this.goodId, this.amount, this.goodImage, this.goodPrice, this.goodName, this.goodDetail);
```

11.4.2　使用构造方法

可以使用"构造方法"创建一个对象。构造方法的名字格式可以是 ClassName，如下面代码，传入多个商品参数实例化商品信息类。

```
//object_constructor_good_info.dart 文件
 //调用构造方法 GoodInfo 实例化商品信息类
 GoodInfo goodInfo = GoodInfo(
     '000001',
```

```
        999,
        'http://192.168.2.168/images/1.png',
        800,
        '男士夹克外套',
        '外套男秋冬季男装连帽时尚休闲运动套装男士夹克外套男衣服 黑三件套 XL');
```

还可以使用 ClassName.identifier 方式来调用构造方法。如下面的代码使用 GoodInfo.fromJson() 构造方法创建了 GoodInfo 对象。

```
//object_constructor_good_info.dart 文件
 /*
   * ClassName.identifier 形式构造方法
   * 传入 Json 数据,实例化为 GoodInfo 对象
   */
  GoodInfo.fromJson(Map<String,dynamic> json){
    goodId = json['goodId'];
    amount = json['amount'];
    goodImage = json['goodImage'];
    goodPrice = json['goodPrice'];
    goodName = json['goodName'];
    goodDetail = json['goodDetail'];
  }
```

上面这种方式用来做数据模型非常有效。前后端数据交互往往传递的都是 Json 数据，所以通常情况还需要添加一个将当前对象转化成 Json 数据的方法，代码如下：

```
//object_constructor_good_info.dart 文件
 /*
   * 将当前对象转化成 Json 数据
   */
  Map<String,dynamic> toJson(){
    final Map<String, dynamic> data = Map<String, dynamic>();
    data['goodId'] = this.goodId;
    data['amount'] = this.amount;
    data['goodImage'] = this.goodImage;
    data['goodPrice'] = this.goodPrice;
    data['goodName'] = this.goodName;
    data['goodDetail'] = this.goodDetail;
    return data;
  }
```

GoodInfo.fromJson 构造方法调用方式代码如下：

```
//object_constructor_good_info.dart 文件
 //调用构造方法 GoodInfo.fromJson 实例化商品信息类
  GoodInfo goodInfoJson = GoodInfo.fromJson({
```

```
    'goodId':'000002',
    'amount': 666,
    'goodImage':'http://192.168.2.168/images/2.png',
    'goodPrice':688,
    'goodName':'男加厚珊瑚绒翻领开衫套装',
    'goodDetail':'南极人 睡衣男秋冬长袖可外穿法兰绒睡衣家居服男加厚珊瑚绒翻领开衫套装男经典藏青(上衣+裤子) XL'});
```

上面的代码必须传入一个 Json 对象,需要使用到键值对数据。

商品信息类的定义、构造方法定义,以及使用完整代码如下:

```
//object_constructor_good_info.dart 文件
void main(){
  //调用构造方法 GoodInfo 实例化商品信息类
  GoodInfo goodInfo = GoodInfo(
     '000001',
     999,
     'http://192.168.2.168/images/1.png',
     800,
     '男士夹克外套',
     '外套男秋冬季男装连帽时尚休闲运动套装男士夹克外套男衣服 黑三件套 XL');

  //调用构造方法 GoodInfo.fromJson 实例化商品信息类
  GoodInfo goodInfoJson = GoodInfo.fromJson({
     'goodId':'000002',
     'amount': 666,
     'goodImage':'http://192.168.2.168/images/2.png',
     'goodPrice':688,
     'goodName':'男加厚珊瑚绒翻领开衫套装',
     'goodDetail':'南极人 睡衣男秋冬长袖可外穿法兰绒睡衣家居服男加厚珊瑚绒翻领开衫套装男经典藏青(上衣+裤子) XL'});

  //打印输出 Json 数据
  print(goodInfo.toJson());
  print(goodInfoJson.toJson());

}

//商品信息
class GoodInfo{
  //商品 Id
  String goodId;
  //商品数量
  int amount;
  //商品图片
  String goodImage;
  //商品价格
```

```dart
    int goodPrice;
  //商品名称
  String goodName;
  //商品详情
  String goodDetail;

  //构造方法
//  GoodInfo(String goodId, int amount, String goodImage, double goodPrice, String goodName, String goodDetail){
//    this.goodId = goodId;
//    this.amount = amount;
//    this.goodImage = goodImage;
//    this.goodPrice = goodPrice;
//    this.goodName = goodName;
//    this.goodDetail = goodDetail;
//  }

  //语法糖,简化构造方法
  GoodInfo(this.goodId, this.amount, this.goodImage, this.goodPrice, this.goodName, this.goodDetail);

  /*
   * ClassName.identifier 形式构造方法
   * 传入Json数据实例化为GoodInfo对象
   */
  GoodInfo.fromJson(Map<String,dynamic> json){
    goodId = json['goodId'];
    amount = json['amount'];
    goodImage = json['goodImage'];
    goodPrice = json['goodPrice'];
    goodName = json['goodName'];
    goodDetail = json['goodDetail'];
  }

  /*
   * 将当前对象转化成Json数据
   */
  Map<String,dynamic> toJson(){
    final Map<String, dynamic> data = Map<String, dynamic>();
    data['goodId'] = this.goodId;
    data['amount'] = this.amount;
    data['goodImage'] = this.goodImage;
    data['goodPrice'] = this.goodPrice;
    data['goodName'] = this.goodName;
    data['goodDetail'] = this.goodDetail;
    return data;
  }

}
```

示例中调用了 toJson 方法用来生成 Json 数据。输出内容如下：

```
flutter: {goodId: 000001, amount: 999, goodImage:
http://192.168.2.168/images/1.png, goodPrice: 800, goodName: 男士夹克外套,
goodDetail: 外套男秋冬季男装连帽时尚休闲运动套装男士夹克外套男衣服 黑三件套 XL}
flutter: {goodId: 000002, amount: 666, goodImage:
http://192.168.2.168/images/2.png, goodPrice: 688, goodName: 男加厚珊瑚绒翻领开衫套装,
goodDetail: 南极人 睡衣男秋冬长袖可外穿法兰绒睡衣家居服男加厚珊瑚绒翻领开衫套装 男经典
藏青(上衣 + 裤子) XL}
```

11.4.3 命名构造方法

使用命名构造方法可以实现多个构造方法或者让代码更清晰。简单来说，因为 Dart 不支持构造方法的重载，无法像 Java 语言一样使用不同的参数来实现构造方法。示例中 Person.run 即为命名构造方法，代码如下：

```
//object_named_constructor.dart 文件
void main(){
  //调用 Person 的命名构造方法
  Person p = Person.run();
}

class Person{
  //姓名
  String name;
  //年龄
  int age;
  //默认构造方法
  Person(this.name,this.age);
  //命名构造方法
  Person.run(){
    print('命名构造方法');
  }
}
```

我们需要记住构造方法不被继承，这意味着父类的命名构造方法不会被子类继承。如果希望用父类中的命名构造方法创建子类，那么必须在子类中实现该构造方法。

11.4.4 调用父类的非默认构造方法

默认情况下，子类的构造方法会调用父类的无名、无参构造方法。父类的构造方法会在构造方法体的一开始被调用。如果初始化列表也被使用了，那么它就在父类被调用之前调用。总结执行的顺序如下：

❑ 初始化列表；

❑ 父类的无参构造方法；

❑ 子类的无参构造方法。

如果父类没有无名、无参的构造方法,那么必须手动调用父类的其中一个构造方法。在冒号":"后面,构造方法体之前(如果有的话)指定父类的构造方法。

下面的例子中,Student 类的构造方法调用了它父类 Person 的命名构造方法。

```
//object_constructor_student.dart 文件
//父类
class Person {
  //姓名
  String name;
  //年龄
  int age;
  //构造方法
  Person.fromJson(Map data) {
    print('Person construct...');
  }
}
//子类
class Student extends Person {
  //Person 没有默认构造方法,必须调用 super.fromJson(data)
  Student.fromJson(Map data) : super.fromJson(data) {
    print('Student construct...');
  }
}

main() {
  var student = Student.fromJson({});
}
```

从上面的示例可以看到先调用 Person 的构造方法,后调用 Student 构造方法,输出结果如下:

```
flutter: Person construct...
flutter: Student construct...
```

由于父类构造方法的参数在构造方法调用前被计算,参数可以是一个表达式,例如一个方法调用:

```
class Student extends Person {
  Student() : super.fromJson(getDefaultData());
  //…
}
```

警告 父类的构造方法不能访问 this,因此参数可以是静态方法,但是不能是实例方法。

11.4.5 初始化列表

调用父类构造方法的同时,也可以在构造方法体执行之前初始化实例变量。使用逗号分隔初始化器,改造商品信息类 GoodInfo 的命名构造方法 fromJson,代码如下:

```
//object_constructor_init_list.dart 文件
void main(){
  //调用构造方法 GoodInfo.fromJson 实例化商品信息类
  GoodInfo goodInfoJson = GoodInfo.fromJson({
    'goodId':'000002',
    'amount': 666,
    'goodImage':'http://192.168.2.168/images/2.png',
    'goodPrice':688,
    'goodName':'男加厚珊瑚绒翻领开衫套装',
    'goodDetail':'南极人 睡衣男秋冬长袖可外穿法兰绒睡衣家居服男加厚珊瑚绒翻领开衫套装 男经典藏青(上衣 + 裤子) XL'});

  //打印输出 Json 数据
  print(goodInfoJson.toJson());
}

//商品信息
class GoodInfo{
  //商品 Id
  String goodId;
  //商品数量
  int amount;
  //商品图片
  String goodImage;
  //商品价格
  int goodPrice;
  //商品名称
  String goodName;
  //商品详情
  String goodDetail;

  /*
   * 初始化列表在构造函数体执行前设置实例变量的值
   */
  GoodInfo.fromJson(Map<String,dynamic> json)
      //初始化列表
      : goodId = json['goodId'],
        amount = json['amount'],
        goodImage = json['goodImage'],
        goodPrice = json['goodPrice'],
        goodName = json['goodName'],
        goodDetail = json['goodDetail']{
```

```
      print('GoodInfo.fromJson 命名构造方法');
    }

    /*
     * 将当前对象转化成 Json 数据
     */
    Map<String,dynamic> toJson(){
      final Map<String, dynamic> data = Map<String, dynamic>();
      data['goodId'] = this.goodId;
      data['amount'] = this.amount;
      data['goodImage'] = this.goodImage;
      data['goodPrice'] = this.goodPrice;
      data['goodName'] = this.goodName;
      data['goodDetail'] = this.goodDetail;
      return data;
    }
}
```

警告 初始化器右边不能访问 this。

上面的示例输出内容如下：

```
flutter: GoodInfo.fromJson 命名构造方法
flutter: {goodId: 000002, amount: 666, goodImage:
http://192.168.2.168/images/2.png, goodPrice: 688, goodName: 男加厚珊瑚绒翻领开衫套装,
goodDetail: 南极人 睡衣男秋冬长袖可外穿法兰绒睡衣家居服男加厚珊瑚绒翻领开衫套装 男经典
藏青(上衣＋裤子) XL}
```

11.4.6 重定向构造方法

有时候一个构造方法的唯一目的是重定向到同一个类的另一个构造方法。一个重定向构造方法的函数体是空的，构造方法的调用在冒号":"后面。

以商品信息类为例，在新增一个商品信息时有些字段是默认值。例如：商品库存为0，商品图片没有上传从而使用默认图片，商品价格还不确定从而使用默认价格0等。这种情况在实例化商品信息类时就不需要传那么多参数。如下面示例所示，定义一个重定向构造方法 GoodInfo.redirect 传递一个商品 Id 即可。

```
//object_constructor_redirect.dart 文件
void main(){
  //实例化对象,调用重定向构造方法
  GoodInfo goodInfo = GoodInfo.redirect('000003');
  //打印输出 Json 数据
  print(goodInfo.toJson());
}
```

```dart
//商品信息
class GoodInfo{
  //商品 Id
  String goodId;
  //商品数量
  int amount;
  //商品图片
  String goodImage;
  //商品价格
  int goodPrice;
  //商品名称
  String goodName;
  //商品详情
  String goodDetail;

  //该类的主构造方法
  GoodInfo(this.goodId, this.amount, this.goodImage, this.goodPrice, this.goodName, this.goodDetail);
  //重定向构造方法,代理到主构造方法
  GoodInfo.redirect(String goodId) :
    this(goodId,0,'http://192.168.2.168/images/default.png',0,'商品名称','商品详情');

  /*
   * 将当前对象转化成 Json 数据
   */
  Map<String,dynamic> toJson(){
    final Map<String, dynamic> data = Map<String, dynamic>();
    data['goodId'] = this.goodId;
    data['amount'] = this.amount;
    data['goodImage'] = this.goodImage;
    data['goodPrice'] = this.goodPrice;
    data['goodName'] = this.goodName;
    data['goodDetail'] = this.goodDetail;
    return data;
  }
}
```

如上面代码所示。GoodInfo.redirect 为重定向构造方法,GoodInfo 为主构造方法。示例代码除了 goodId 为传入的参数,其他参数均为默认参数,输出内容如下:

```
flutter: {goodId: 000003, amount: 0, goodImage:
http://192.168.2.168/images/default.png, goodPrice: 0, goodName: 商品名称, goodDetail: 商品详情}
```

11.4.7 常量构造方法

如果类生成的对象从不改变,那么就可以让这些对象变成编译期常量。要想这样,定义一个常量构造方法并确保所有实例变量都是 final 的。下面的示例定义了一个

ImmutablePerson 常量构造方法，代码如下：

```dart
//object_constructor_const.dart 文件
void main(){
  //获取 ImmutablePerson 实例
  ImmutablePerson p = ImmutablePerson.instance;
  print('name:' + p.name);
  print('age:' + p.age.toString());
}

class ImmutablePerson {
  //静态常量
  static final ImmutablePerson instance = const ImmutablePerson('张三', 20);
  //姓名
  final String name;
  //年龄
  final int age;
  //构造方法
  const ImmutablePerson(this.name, this.age);
}
```

上面的示例输出内容如下：

```
flutter: name:张三
flutter: age:20
```

11.4.8　工厂构造方法

当要实现一个不总是创建这个类新实例的构造方法时，使用 factory 关键词。例如，一个工厂构造方法可能从缓存中返回一个实例，或者可能返回子类的一个实例。下面的代码展示了一个工厂构造方法从缓存中返回对象。

```dart
//object_constructor_factory.dart 文件
//日志类
class Logger {
  //日志名称
  final String name;

  //日志缓存_cache 用于存储 Logger 对象
  static final Map<String, Logger> _cache = <String, Logger>{};

  //工厂构造方法
  factory Logger(String name) {
    //向缓存里添加一个 Map
    return _cache.putIfAbsent(
        //key 为 name value 为 Logger 对象
```

```
    name, () => Logger._internal(name));
}

//命名构造方法,用于创建一个 Logger 对象
Logger._internal(this.name);

//输出日志
void log(String msg) {
  print(msg);
}
}
```

说明 工厂构造方法无法访问 this。

调用工厂构造方法的方式和其他构造方法一样,代码如下:

```
//object_constructor_factory.dart 文件
void main(){
  Logger logger = Logger('Dart');
  logger.log('调用工厂构造方法');
}
```

11.5 Getters 和 Setters

get()和 set()方法是专门用于读取和写入对象的属性的方法,每一个类的实例,系统都隐式地包含有 get()和 set() 方法,这和很多语言里的 VO 类相似。

例如,定义一个矩形的类,有上、下、左、右四个成员变量: top、bottom、left 和 right,使用 get 及 set 关键字分别对 right 及 bottom 进行获取和设置值。代码如下:

```
//object_getters_setters.dart 文件
class Rectangle {
  num left;
  num top;
  num width;
  num height;

  Rectangle(this.left, this.top, this.width, this.height);

  //获取 right 值
  num get right            => left + width;

  //设置 right 值,同时 left 也发生变化
  set right(num value)     => left = value - width;
```

```
  //获取 bottom 值
  num get bottom => top + height;

  //设置 bottom 值,同时 top 也发生变化
  set bottom(num value) => top = value - height;
}

main() {
  var rect = Rectangle(3, 4, 20, 15);

  print('left:' + rect.left.toString());
  print('right:' + rect.right.toString());
  rect.right = 30;
  print('更改 right 值为 30');
  print('left:' + rect.left.toString());
  print('right:' + rect.right.toString());

  print('top:' + rect.top.toString());
  print('bottom:' + rect.bottom.toString());
  rect.bottom = 50;
  print('更改 bottom 值为 50');
  print('top:' + rect.top.toString());
  print('bottom:' + rect.bottom.toString());
}
```

上面例子对应的输出为:

```
flutter: left:3
flutter: right:23
flutter: 更改 right 值为 30
flutter: left:10
flutter: right:30
flutter: top:4
flutter: bottom:19
flutter: 更改 bottom 值为 50
flutter: top:35
flutter: bottom:50
```

第 12 章 继承与多态

继承(extends)是面向对象开发方法中非常重要的一个特征,继承体现着现实世界中"一般"与"特殊"的关系。对于拥有"一般"性质的类称为"父类"或者"超类",拥有"特殊"性质的类称为"子类"。例如"动物"和"鸟",动物是一般的概念,鸟是特殊的概念,可以通过"鸟是一种特殊的动物"这句话的逻辑是否成立来判断继承关系是否成立。

12.1 Dart 中的继承

与 Java 语言类似,Dart 语言为"单继承",也就是一个类只能有一个直接的父类。如果一个类没有显式地声明父类,那么它会默认继承 Object 类。此外,Dart 语言又提供了混入(Mixin)的语法,允许子类在继承父类时混入其他类。关于混入(Mixin)的理解,请参考"第 14 章 Mixin 混入"一章。

Dart 语言中使用 extends 作为继承关键字,子类会继承父类的数据和函数。下面看一个继承的示例,代码如下:

```
//extends_basic_sample.dart 文件
void main() {
  //实例化动物类
  Animal animal = Animal();
  //实例化猫类
  Cat cat = Cat();
  //动物名称属性
  animal.name = "动物";
  //猫名称属性
  cat.name = "猫";
  //猫颜色属性就属于子类的特征
  cat.color = "黑色";
  //动物会吃方法
  animal.eat();
  //猫会吃方法
  cat.eat();
  //动物会爬树方法属于子类的特征
  cat.climb();
```

```
    }
    //动物类
    class Animal{

        //属性
        String name;

        //父类方法
        void eat(){
            print("${name}:会吃东西");
        }
    }

    //猫类继承动物类
    class Cat extends Animal{

        //子类属性
        String color;

        //子类方法
        void climb(){
            print("${color}的${name}:会爬树");
        }
    }
```

运行上面的示例,结果如下:

```
flutter:动物:会吃东西
flutter:猫:会吃东西
flutter:黑色的猫:会爬树
```

猫是一种动物,猫类继承动物类,所以动物是父类,猫是子类。动物有名字 name,动物会吃 eat,同样猫也具有这两个特征,体现了继承的特性。猫有黑猫和白猫,所以猫类里的 color 属性属于它特有特征。猫还会爬树,所以猫类里的 climb 也是它特有的特征。

12.2 方法重写

重写在面向对象中体现的现实意义是"子类与父类在同一行为上有不同的表现形式"。同 Java 语言类似,Dart 语言也支持方法的重写,子类重写父类的方法后,对象调用的即为子类的同名方法。

12.2.1 基本使用

修改猫和动物类的示例,使用@override 重写 eat 方法。示例代码如下:

```
//extends_override_method_sample.dart 文件
main() {
  //实例化动物类
  Animal animal = Animal();
  //实例化猫类
  Cat cat = Cat();
  //动物名称属性
  animal.name = "动物";
  //猫名称属性
  cat.name = "猫";
  //猫颜色属性就属于子类的特征
  cat.color = "黑色";
  //动物会吃方法
  animal.eat();
  //猫类重写了父类的 eat 方法
  cat.eat();
  //动物会爬树方法属于子类的特征
  cat.climb();
}

//动物类
class Animal{

  //属性
  String name;

  //父类方法
  void eat(){
    print("${name}:会吃东西");
  }
}

//猫类继承动物类
class Cat extends Animal{

  //子类属性
  String color;

  //子类重写父类的 eat 方法
  @override
  void eat(){
    print("${color}的${name}:会吃鱼");
  }

  //子类方法
  void climb(){
    print("${color}的${name}:会爬树");
  }
}
```

运行上面的示例,结果如下:

```
flutter: 动物:会吃东西
flutter: 黑色的猫:会吃鱼
flutter: 黑色的猫:会爬树
```

所有的动物都会吃东西,但猫会吃鱼这是猫的特性,猫类重写父类的方法就可以体现这一特征。

12.2.2　重绘 Widget 方法

在 Flutter 里重写最多的方法是 build 方法。可以重写 Widget 的 build 方法来构建一个组件,代码如下:

```
@protected Widget build(BuildContext context);
```

build 即为创建一个 Widget 的意思,返回值也是一个 Widget 对象,不管返回的是单个组件还是返回通过嵌套的方式组合的组件,都是 Widget 的实例。build 方法重写示例代码如下:

```
//extends_override_build_method.dart 文件
import 'package:flutter/material.dart';

void main() => runApp(MyApp());

//MyApp 类继承 StatelessWidget 类
class MyApp extends StatelessWidget {

  //重写 StatelessWidget 的 build 方法
  @override
  Widget build(BuildContext context) {
    //返回一个 Widget
    return MaterialApp(
      title: 'build 方法重写示例',
      home: Scaffold(
        appBar: AppBar(
          title: Text('build 方法重写示例'),
        ),
        body: Center(
          child: Text('override build'),
        ),
      ),
    );
  }
}
```

上面的代码中,MyApp 类继承 StatelessWidget 类,StatelessWidget 类为父类,MyApp 类为子类。重写 StatelessWidget 的 build 方法使得其可以返回一个 Widget 进行渲染。效

果如图 12-1 所示。

图 12-1　build 方法重写示例

12.2.3　重写高级示例

接下来再看一个重写方法的高级示例。假设想在 Flutter 的页面里画一个圆，那么就需要一支画笔 Paint 和一个画布 Canvas，通过画笔在画布上画一个圆就可以达到这个效果。

绘制圆需要调用 Canvas 的 drawCircle 方法，需要传入中心点的坐标、半径，以及画笔。代码如下：

```
canvas.drawCircle(Offset(200.0, 150.0), 150.0, _paint);
```

其中，画笔对应有填充色及没有填充色两种情况：
- PaintingStyle.fill：填充绘制；
- PaintingStyle.stroke：非填充绘制。

这里需要定义一个画圆的类 CirclePainter 继承 CustomPainter 类，同时需要重写以下两个方法：

```
paint：重写绘制内容方法
shouldRepaint：重写是否需要重绘方法
```

完整代码如下：

```dart
//extends_override_circle_painter.dart 文件
import 'package:flutter/material.dart';

void main() => runApp(MyApp());

class MyApp extends StatelessWidget {
  @override
  Widget build(BuildContext context) {
    return MaterialApp(
      title: '绘制圆示例',
      home: Scaffold(
        appBar: AppBar(
          title: Text(
            '绘制圆示例',
            style: TextStyle(color: Colors.white),
          ),
        ),
        body: Center(
          child: SizedBox(
            width: 500.0,
            height: 500.0,
            //自定义 Paint 组件
            child: CustomPaint(
              //画圆类
              painter: CirclePainter(),
              child: Center(
                child: Text(
                  '绘制圆',
                  style: const TextStyle(
                    fontSize: 38.0,
                    fontWeight: FontWeight.w600,
                    color: Colors.black,
                  ),
                ),
              ),
            ),
          ),
        ),
      ),
    );
  }
}

//继承于 CustomPainter 并且实现 CustomPainter 里面的 paint 和 shouldRepaint 方法
class CirclePainter extends CustomPainter {
  //定义画笔
```

```
Paint _paint = Paint()
  ..color = Colors.grey
  ..strokeCap = StrokeCap.square
  ..isAntiAlias = true
  ..strokeWidth = 3.0
  ..style = PaintingStyle.stroke;     //画笔样式有填充 PaintingStyle.fill 及没有填充
                                      //PaintingStyle.stroke 两种

//重写绘制内容方法
@override
void paint(Canvas canvas, Size size) {
  //绘制圆,参数为中心点、半径和画笔
  canvas.drawCircle(Offset(200.0, 150.0), 150.0, _paint);
}

//重写是否需要重绘方法
@override
bool shouldRepaint(CustomPainter oldDelegate) {
  return false;
}
}
```

上述示例代码的视图展现大致如图 12-2 所示,图中展示了非填充样式。

图 12-2　绘制圆示例

如果想画线、三角形和多边形等几何图形，只需要重写上述两个方法即可。

提示　示例中 Flutter 的组件嵌套代码不必深入理解，重点掌握重写函数的使用即可。

12.3　操作符重写

同 C++语言类似，Dart 语言支持操作符的重写，常规的四则运算和比较运算符都可以进行重写。Dart 中可重写的操作符如表 12-1 所示。

表 12-1　可重写的操作符

<	+	|	[]
>	/	^	[]=
<=	~/	&	~
>=	*	<<	==
-	%	>>	

说明　"！="不是一个可重载的运算符。表达式 e1！= e2 仅仅是！(e1 == e2) 的语法糖。

重写操作符的方法定义格式如下：

```
@override
type operator 操作符(className objectName){
    //…
}
```

重写就需要@override 修改方法体，type 为方法返回类型，operator 为操作符关键字，后面的即为操作符及方法体。

下面的示例重写了"=="和"+"的 Rectangle 类，当两个对象的 width 和 height 一致则认为它们是"相等"的，两个 Rectangle 相加则将两者的 width 和 height 进行相加后得到新的 Rectangle 对象。

```
//extends_override_operator.dart 文件
void main() {
  //初始化三个 Rectangle 对象
  Rectangle a = Rectangle(10,10);
  Rectangle b = Rectangle(5, 5);
  Rectangle c = Rectangle(10, 10);

  //判断 a 与 b 对象是否相等
```

```
    print("a == b : ${a == b}");
    //判断 a 与 c 对象是否相等
    print("a == c : ${a == c}");
    //a 与 b 相加赋给 d 对象
    Rectangle d = a + b;
    print("a.width = ${a.width} a.height = ${a.height}");
    print("d.width = ${d.width} d.height = ${d.height}");
    //判断 a 与 d 对象是否相等
    print("a == d : ${a == d}");
}

//矩形类
class Rectangle{
    //宽度属性
    int width;
    //高度属性
    int height;

    //构造方法
    Rectangle(this.width,this.height);

    //重载"=="号操作符
    @override
    bool operator == (dynamic other) {
        //判断 other 类型是否为 Rectangle 类
        if(other is! Rectangle){
            return false;
        }
        Rectangle temp = other;
        //当宽和高的数值同时相等则返回 true,否则返回 false
        return (temp.width == width && temp.height == height);
    }

    //重载"+"号操作符
    @override
    Rectangle operator + (dynamic other){
        //判断 other 类型是否为 Rectangle 类
        if(other is! Rectangle){
            return this;
        }
        Rectangle temp = other;
        //宽度等于两个对象的宽度值相加,高度等于两个对象的高度值相加
        return Rectangle( this.width + temp.width, this.height + temp.height);
    }
}
```

运行示例输出内容如下:

```
flutter: a == b : false
flutter: a == c : true
flutter: a.width = 10 a.height = 10
flutter: d.width = 15 d.height = 15
flutter: a == d : false
```

> **提示** 示例代码中用到了 dynamic 类型。dynamic 类型具有所有可能的属性和方法。Dart 语言中方法都有 dynamic 类型作为方法的返回类型，方法的参数也都有 dynamic 类型。

12.4 重写 noSuchMethod 方法

要检测或响应代码并试图使用不存在的方法或实例变量，这种情况可以重写 noSuchMethod 方法。

```
//extends_override_noSuchMethod.dart 文件
void main() {
  //实例化 Person 类
  dynamic person = Person();
  //调用一个不存在的方法
  print(person.setUserInfo('20', '张三'));
  //调用一个存在的方法
  person.someMethod();

}

class Person extends Object{

  //可调用的方法
  void someMethod(){
    print('调用此方法:someMethod');
  }

  //重写 noSuchMethod
  @override
  noSuchMethod(Invocation invocation) => '找不到此方法:方法名:${invocation.memberName} 参数:${invocation.positionalArguments}';

}
```

运行上面的示例，输出内容如下：

```
flutter:找不到此方法:方法名:Symbol("setUserInfo") 参数:[20, 张三]
flutter:调用此方法:someMethod
```

从上面的输出内容可以看到,控制台能准确地输出调用不到的方法名及参数,这样便于排除程序的错误。

12.5 多态

多态是同一个行为具有多个不同表现形式或形态。多态就是同一个接口,使用不同的实例而执行不同操作。多态性是对象多种表现形式的体现。

多态的优点如下所示:
- 消除类型之间的耦合关系;
- 可替换性;
- 可扩充性;
- 接口性;
- 灵活性;
- 简化性。

多态存在的三个必要条件:
- 继承;
- 重写;
- 父类引用指向子类对象。

下面通过一个示例理解什么是多态。父类 Animal 有 eat 和 run 方法,它有两个子类 Dog 和 Cat。这两个子类都有 eat 和 run 方法,均覆盖了其父类的方法,但具体实现方式不同。这两个方法的实现就体现了类的多态性。完整代码如下:

```
//polymorphism_sample.dart 文件
void main() {
  //子类 Dog 实例化并调用方法
  Dog d = Dog();
  d.eat();
  d.run();

  //子类 Cat 实例化并调用方法
  Cat c = Cat();
  c.eat();
  c.run();

  //声明成 Animal 类型,实例化为 Dog 类对象
  Animal animalDog = Dog();
  //调用 eat 方法体现多态性
  animalDog.eat();

  //声明成 Animal 类型,实例化为 Cat 类对象
  Animal animalCat = Cat();
```

```
  //调用 eat 方法体现多态性
  animalCat.eat();
}

//动物类
class Animal {

  //父类方法
  void eat(){
    print("动物会吃");
  }

  //父类方法
  void run(){
    print("动物会跑");
  }
}

//狗类继承动物类
class Dog extends Animal {

  //重写父类函数体现多态性
  @override
  void eat() {
    print('小狗在啃骨头');
  }

  //重写父类函数体现多态性
  @override
  void run() {
    print('小狗在遛弯');
  }

  void printInfo() {
    print('我是小狗');
  }
}

//猫类继承动物类
class Cat extends Animal {

  //重写父类方法体现多态性
  @override
  void eat() {
    print('小猫在吃鱼');
  }

  //重写父类方法体现多态性
  @override
```

```
  void run() {
    print('小猫在散步');
  }

  void printInfo() {
    print('我是小猫咪');
  }
}
```

输出内容如下。

```
flutter: 小狗在啃骨头
flutter: 小狗在遛弯
flutter: 小猫在吃鱼
flutter: 小猫在散步
flutter: 小狗在啃骨头
flutter: 小猫在吃鱼
```

从运行结果可知,当多态发生时,Dart 虚拟机运行时根据引用变量指向的实例调用它的方法,而不是根据引用变量的类型调用。

第 13 章 抽象类与接口

良好的软件系统应该具备"可复用性"和"可扩展性",能够满足用户需求的不断变更。使用抽象类和接口是实现"可复用性"和"可扩展性"重要的设计手段。

13.1 抽象类

抽象 abstract 是面向对象中的一个非常重要的概念,通常用于描述父类拥有一种行为但无法给出细节实现,而需要通过子类来实现抽象的细节。这种情况下父类被定义为抽象类,子类继承父类后实现其中的抽象方法。

同 Java 语言类似,Dart 中的抽象类也使用 abstract 来实现,不过抽象方法无须使用 abstract,直接给出定义而不给出方法体实现即可。

抽象类中可以有数据,可以有常规方法,还可以有抽象方法,但抽象类不能实例化。子类继承抽象类后必须实现其中的抽象方法。

13.1.1 抽象类定义格式

抽象类定义格式如下:

```
abstract class className{
  //成员变量
  [static] [const] [final] type name;

  //成员方法
  [type] methodName(paramType:paramName ...);
}
```

抽象类的定义和普通类基本一致,最主要的区别是抽象类的方法只能声明而不做具体实现。

提示 定义抽象类只要在类名前面加上 abstract 关键字即可。成员变量如果是常量必须初始化,成员方法没有方法体。

13.1.2 数据库操作抽象类实例

接下来写一个数据库操作的抽象类的例子。定义一个抽象类叫作 DataBaseOperate，里面定义 4 个数据库常用的操作方法"增、删、改、查"。再定义一个类命名为 DataBaseOperateImpl 继承 DataBaseOperate 类，用来实现抽象类里的方法。完整的代码如下：

```dart
//abstract_database_operate.dart 文件
void main() {

  //声明类型为 DataBaseOperate,实例化类型为 DataBaseOperateImpl
  DataBaseOperate db = DataBaseOperateImpl();
  //调用成员方法
  db.insert();
  db.delete();
  db.update();
  db.query();

}

//数据库操作抽象类
abstract class DataBaseOperate {
  void insert();        //定义插入的方法
  void delete();        //定义删除的方法
  void update();        //定义更新的方法
  void query();         //定义一个查询的方法
}

//数据库操作实现类
class DataBaseOperateImpl extends DataBaseOperate {

  //实现了插入的方法
  void insert(){
    print('实现了插入的方法');
  }

  //实现了删除的方法
  void delete(){
    print('实现了删除的方法');
  }

  //实现了更新的方法
  void update(){
    print('实现了更新的方法');
  }

  //实现了一个查询的方法
```

```
    void query(){
      print('实现了一个查询的方法');
    }

}
```

当实现类 DataBaseOperateImpl 里少实现一个方法,编译器就会报错并提示必须实现某方法。那么抽象类更像是定义了一种规范并要求子类必须实现某些方法。

上述代码输出结果为:

```
flutter:实现了插入的方法
flutter:实现了删除的方法
flutter:实现了更新的方法
flutter:实现了一个查询的方法
```

13.1.3 几何图形抽象类

不同几何图形的面积计算公式是不同的,但是它们具有的特性是相同的,如果它们具有长和宽这两个属性,那么这些图形也都具有面积计算的方法。可以定义一个抽象类,在该抽象类中含有两个属性(width 和 height)和一个抽象方法 area,具体步骤如下。

(1) 首先创建一个表示图形的抽象类 Shape,代码如下:

```
//abstract_shape.dart 文件
//图形抽象类 Shape
abstract class Shape {
  //几何图形的长
  double width;
  //几何图形的宽
  double height;

  //定义抽象方法,计算面积
  double area();
}
```

(2) 定义一个正方形类,该类继承形状类 Shape,并重写 area 抽象方法。正方形类的代码如下:

```
//abstract_shape.dart 文件
//正方形类
class Square extends Shape {

  Square(double width,double height) {
    this.width = width;
```

```
    this.height = height;
  }
  //重写父类中的抽象方法,实现计算正方形面积的功能
  @override
  double area(){
    return super.width * super.height;
  }
}
```

(3) 定义一个三角形类,该类与正方形类一样,需要继承形状类 Shape,并重写父类中的抽象方法 area。三角形类的代码实现如下:

```
//abstract_shape.dart 文件
//三角形类
class Triangle extends Shape {
  Triangle(double width,double height){
    this.width = width;
    this.height = height;
  }
  //重写父类中的抽象方法,实现计算三角形面积的功能
  @override
  double area(){
    return 0.5 * this.width * this.height;
  }
}
```

(4) 最后在主程序 main 函数里,分别创建正方形类和三角形类的对象,并调用各类中的 area() 方法,打印出不同形状的几何图形的面积。测试代码如下:

```
//abstract_shape.dart 文件
void main(){
  //创建正方形类对象
  Square square = Square(5,5);
  print("正方形的面积为: " + square.area().toString());
  //创建三角形类对象
  Triangle triangle = Triangle(2,5);
  print("三角形的面积为: " + triangle.area().toString());
}
```

在该程序中创建了三个类,分别为图形类 Shape、正方形类 Square 和三角形类 Triangle。其中图形类 Shape 是一个抽象类,创建了两个属性,分别为图形的长度和宽度。在 Shape 类的最后定义了一个抽象方法 area(),用来计算图形的面积。在这里,Shape 类只定义了计算图形面积的方法,而对于如何计算并没有任何限制。也可以这样理解,抽象类 Shape 仅定义了子类的一般形式。

正方形类 Square 继承抽象类 Shape,并实现了抽象方法 area()。三角形类 Triangle 的

实现和正方形类相同，这里不再介绍。

在测试程序的 main() 方法中，首先创建了正方形类和三角形类的实例化对象 square 和 triangle，然后分别调用 area() 方法实现了面积的计算功能。

运行该程序，输出的结果如下：

```
flutter:正方形的面积为:25.0
flutter:三角形的面积为:5.0
```

13.2 接口

和 Java 一样，Dart 也有接口，但是和 Java 还是有区别的。Dart 的接口没有用 interface 关键字定义接口，而是普通类或抽象类都可以作为接口被实现。同样使用 implements 关键字进行实现。

但是 Dart 的接口有些不同，如果实现的类是普通类，那么需要将普通类和抽象中的属性的方法全部重写一遍。

在 Dart 中只允许继承一个类，但可以实现多个接口。通过实现多个接口的方式满足多继承的设计需求。

接下来看一个接口实现的示例，代码如下：

```dart
//implements_animal.dart 文件
void main(){
  //实例化 Dog 类
  var d = Dog();
  d.name = "小狗";
  d.eat();
  d.display();
  d.swim();
  d.walk();
}

//抽象类 Animal
abstract class Animal{
  //动物名称属性
  String name;
  //显示动物名称抽象方法
  void display(){
    print("动物的名字是:${name}");
  }
  //动物进食抽象方法
  void eat();
}
```

```
//抽象类作为接口,SwimAbility 游泳能力
abstract class SwimAbility{
  void swim();
}

//普通类作为接口,WalkAbility 行走能力
class WalkAbility{
  //行走方法
  void walk(){
    //空方法
  }
}

//Dog 类继承 Animal,同时实现 Swimable 和 Walkable 接口
class Dog extends Animal implements SwimAbility, WalkAbility{
  //重写父类 Animal 方法
  @override
  void eat() {
    print(this.name + "有进食的能力");
  }

  //实现 SwimAbility 接口,并重写其 swim 方法
  @override
  void swim() {
    print(this.name + "有游泳的能力");
  }

  //实现 WalkAbility 接口,并重写其 walk 方法
  @override
  void walk() {
    print(this.name + "有行走的能力");
  }
}
```

从代码中可以看出 Dog 小狗类是具有吃、游戏和行走等能力,同时 Dog 属于动物类 Anmial。当它继承 Animal 类以后就不能再继承其他类了,所以这里使用了接口的多继承方式来实现。代码中 Dog 类同时实现了 SwimAbility 和 WalkAbility 两个接口。这样一来,Dog 小狗就具备所有的功能了。从代码中可以看出接口的类型是多样化的,如下所示:

❑ 抽象类作为接口:实现 SwimAbility 接口,具备游泳能力;
❑ 普通类作为接口:实现 WalkAbility 接口,具备行走能力。
代码输出结果如下所示:

```
flutter:小狗有进食的能力
flutter:动物的名字是:小狗
flutter:小狗有游泳的能力
flutter:小狗有行走的能力
```

提示 不管是抽象类还是实现类都可以用来实现接口,建议使用抽象类来实现接口,因为要实现实现类里面的所有方法和属性,如果不使用抽象类就显得有些乱,从而降低代码可读性。

第 14 章 Mixin 混入

在 Dart 语言中,我们经常可以看到对 with 关键字的使用,它就是 Mixin 混入,根据字面意思理解,就是混合的意思。那么 Mixin 如何使用,它的使用场景是什么？本章将详细阐述 Mixin 混入的知识。

14.1 Mixin 概念

Mixin 是面向对象程序设计语言中的类,提供了方法的实现。其他类可以访问 Mixin 类的方法而不必成为其子类。Mixin 有时被称作 Included(包含)而不是 Inherited(继承)。Mixin 为使用它的 Class 提供额外的功能,但自身却不单独使用(不能单独生成实例对象,属于抽象类)。因为有以上限制,Mixin 类通常作为功能模块使用,在需要该功能时"混入",从而不会使类的关系变得复杂。

Mixin 有利于代码复用,又避免了复杂的多继承。使用 Mixin 享有单一继承的单纯性和多重继承的共有性。接口与 Mixin 相同的地方是它们都可以多继承,不同的地方在于 Mixin 是带实现的。Mixin 也可以看作是带实现的 Interface。这种设计模式实现了依赖反转的功能。

14.2 Mixin 使用

下面以一个真实的示例,看看 Mixin 在什么时候使用,以及加入了 Mixin 的类中的代码优先执行顺序,类继承关系如图 14-1 所示。

这里有一个名为 Animal 的超类,它有三个子类(Mammal、Bird 和 Fish)。在底部,有具体的一些子类如 Dolphin 和 Bat 等。

小括号里的名称表示它们的方法,如下所示:
- Walk：表示具有此行为的类的实例可以步行(walk);
- Swim：表示具有此行为的类的实例可以游泳(swim);
- Fly：表示具有此行为的类的实例可以飞行(fly)。

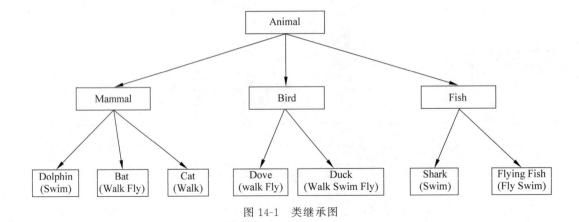

图 14-1 类继承图

有些动物有共同的行为：猫(Cat)和鸽子(Dove)都可以行走，但是猫不能飞。这些行为与此分类正交，因此无法在超类中实现这些行为。

在 Java 语言中可以借助接口(Interface)来实现相关的设计，Dart 中也可以利用隐式接口来完成相应的设计。Dart 是没有 Interface 这种东西的，但并不意味着这门语言没有接口，事实上，Dart 任何一个类都是接口，要实现任何一个类，只需要重写那个类里面的所有具体方法就可以了。

如果不同的子类在某种行为上表现得都不相同，那么使用接口来实现是一种良好的设计。但如果不同的子类在实现某种行为上有着同样的表现，那么使用接口来实现可能会造成代码的冗余，因为接口实现需要强制重写方法。

除了上述方式，也可以利用混入方式(Mixin)来完成相应的设计。对三种行为分别定义三个类描述它们，分别是 Walker、Swimmer 和 Flyer。

```
//行走类
class Walker {

  //行走方法
  void walk() {
    print("我会走路");
  }

}

//游泳类
class Swimmer {

  //游泳方法
  void swim() {
    print("我会游泳");
  }
```

```dart
}

//飞类
class Flyer{

  //飞方法
  void fly() {
    print("我会飞");
  }

}
```

如果不想让这三个类被实例化,可以使用抽象类+工厂方式定义,构造方法返回结果为空 null。代码如下:

```dart
//抽象类,行走类
abstract class Walker {

  //工厂构造方法,防止实例化
  factory Walker._() => null;

  void walk() {
    print("我会走路");
  }
}


//抽象类,游泳类
abstract class Swimmer {

  //工厂构造方法,防止实例化
  factory Swimmer._() => null;

  //游泳方法
  void swim() {
    print("我会游泳");
  }

}

//抽象类,飞类
abstract class Flyer{

  //工厂构造方法,防止实例化
  factory Flyer._() => null;

  //飞方法
```

```
  void fly() {
    print("我会飞");
  }

}
```

定义需要继承的四个父类,代码如下:

```
//动物类
abstract class Animal {

}

//哺乳动物类
abstract class Mammal extends Animal {

}

//鸟类
abstract class Bird extends Animal {

}

//鱼类
abstract class Fish extends Animal {

}
```

最后定义 Cat 和 Dove 类,使用混入的关键字是 with,它的后面可以跟随一个或多个类名。代码如下:

```
//Cat 类继承 Mammal 类,混入 Walker 类
class Cat extends Mammal with Walker {

  //输出信息方法
  void printInfo(){
    print('我是一只小猫');
  }
}

//Dove 类继承 Bird 类,混入 Walker 及 Flyer 类
class Dove extends Bird with Walker, Flyer {

  //输出信息方法
  void printInfo(){
    print('我是一只鸽子');
  }
}
```

使用时允许 Cat 和 Dove 调用 Mixin 的 walk 方法，不允许 Cat 调用未 Mixin 的 Fly 方法。完整的代码如下：

```dart
//mixin_animal.dart 文件
void main(){
  //实例化 Cat 类
  Cat cat = Cat();
  cat.printInfo();
  //具有走路功能
  cat.walk();

  //实例化 Dove 类
  Dove dove = Dove();
  dove.printInfo();
  //具有走路功能
  dove.walk();
  //具有飞的功能
  dove.fly();
}


//Cat 类继承 Mammal 类,混入 Walker 类
class Cat extends Mammal with Walker {

  //输出信息方法
  void printInfo(){
    print('我是一只小猫');
  }
}

//Dove 类继承 Bird 类,混入 Walker 及 Flyer 类
class Dove extends Bird with Walker, Flyer {

  //输出信息方法
  void printInfo(){
    print('我是一只鸽子');
  }
}

//动物类
abstract class Animal {

}

//哺乳动物类
abstract class Mammal extends Animal {

}
```

```dart
//鸟类
abstract class Bird extends Animal {

}

//鱼类
abstract class Fish extends Animal {

}

//抽象类,行走类
abstract class Walker {

  //工厂构造方法,防止实例化
  factory Walker._() => null;

  void walk() {
    print("我会走路");
  }
}

//抽象类,游泳类
abstract class Swimmer {

  //工厂构造方法,防止实例化
  factory Swimmer._() => null;

  //游泳方法
  void swim() {
    print("我会游泳");
  }

}

//抽象类,飞类
abstract class Flyer{

  //工厂构造方法,防止实例化
  factory Flyer._() => null;

  //飞方法
  void fly() {
    print("我会飞");
  }

}
```

```
////行走类
//class Walker {
//
//    //行走方法
//    void walk() {
//        print("我会走路");
//    }
//
//}
//
////游泳类
//class Swimmer {
//
//    //游泳方法
//    void swim() {
//        print("我会游泳");
//    }
//
//}
//
////飞类
//class Flyer{
//
//    //飞方法
//    void fly() {
//        print("我会飞");
//    }
//
//}
```

上述代码输出如下所示：

```
flutter: 我是一只小猫
flutter: 我会走路
flutter: 我是一只鸽子
flutter: 我会走路
flutter: 我会飞
```

14.3 重名方法处理

如果 Mixin 的类和继承类，或者混入的类之间有相同的方法，那么在调用时会产生什么样的情况，看下面的例子。

AB 类和 BA 类都使用 A 类和 B 类通过 Mixin 继承至 P 类，但顺序不同。A、B 和 P 类都有一个名为 getMessage 的方法。

```dart
//mixin_same_method.dart 文件
//A 类
class A {

  //同名方法 A
  String getMessage() => 'A';

}

//B 类
class B {

  //同名方法,返回 B
  String getMessage() => 'B';

}

//P 类
class P {

  //同名方法,返回 P
  String getMessage() => 'P';

}

//AB 类,继承 P,先混入 A 类后混入 B 类
class AB extends P with A, B {

}

//BA 类,继承 P,先混入 B 类后混入 A 类
class BA extends P with B, A {

}

void main() {
  //返回结果
  String result = '';
  //实例化 AB 类
  AB ab = AB();
  //返回结果
  result += ab.getMessage();
  //实例化 BA 类
  BA ba = BA();
  //返回结果
  result += ba.getMessage();
  print(result);
}
```

运行结果为：BA

为什么会产生这个结果？Dart 中的 Mixin 通过创建一个新类来实现,该类将 Mixin 的

实现层叠在一个超类之上以创建一个新类,它不是"在超类中",而是在超类的"顶部",因此无论如何解决查找问题都不会产生歧义。

我们先看一看 AB 类与 BA 类的定义,这段代码如下:

```
//AB类,继承P,先混入A类后混入B类
class AB extends P with A, B {

}

//BA类,继承P,先混入B类后混入A类
class BA extends P with B, A {

}
```

AB 类与 BA 类的定义在语义上等同于如下代码:

```
//AB类语义
class PA = P with A;
class PAB = PA with B;

class AB extends PAB {}

//BA类语义
class PB = P with B;
class PBA = PB with A;

class BA extends PBA {}
```

最终的继承关系如图 14-2 所示。

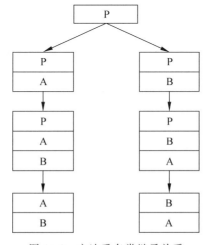

图 14-2　方法重名类继承关系

很显然，最后被继承的类重写了上面所有的 getMessage 方法，可以理解为处于 Mixin 结尾的类将前面的 getMessage 方法都覆盖（override）了。

14.4　Mixin 对象类型

Mixin 应用程序实例的类型通常是其超类的子类型，也是 Mixin 名称本身表示的类的子类型，即原始类的类型，所以这意味着下面程序示例的运行结果全部为 true。

```dart
//mixin_object_type.dart 文件
//A 类
class A {

  //同名方法,返回 A
  String getMessage() => 'A';

}

//B 类
class B {

  //同名方法,返回 B
  String getMessage() => 'B';

}

//P 类
class P {

  //同名方法,返回 P
  String getMessage() => 'P';

}

//AB 类,继承 P,先混入 A 类后混入 B 类
class AB extends P with A, B {

}

//BA 类,继承 P,先混入 B 类后混入 A 类
class BA extends P with B, A {

}

void main() {
  //实例化 AB 类
  AB ab = AB();
```

```
    print(ab is P); //true
    print(ab is A); //true
    print(ab is B); //true

    //实例化 BA 类
    BA ba = BA();
    print(ba is P); //true
    print(ba is A); //true
    print(ba is B); //true

}
```

第3篇　Dart进阶

第 15 章 异常处理

在程序设计和运行的过程中,发生错误是不可避免的。尽管 Dart 语言的设计从根本上提供了便于写出整洁、安全代码的方法,并且程序员也会尽量地减少错误的产生,但是使程序被迫停止的错误仍然不可避免。为此,Dart 提供了异常处理机制来帮助程序员检查可能出现的错误,以保证程序的可读性和可维护性。

Dart 将异常封装到一个类中,出现错误时就会抛出异常。本章将详细介绍异常处理的概念、异常处理语句,以及常见异常错误等内容。

15.1 异常概念

Dart 的异常与 Java 的异常是非常相似的。Dart 的异常是 Exception 或者 Error(包括它们的子类)的类型,甚至可以是非 Exception 或者 Error 类也可以抛出,但是不建议这么使用。

Exception 主要是程序本身可以处理的异常,例如:IOException。我们处理的异常也是以这种异常为主。

Error 是程序无法处理的错误,表示运行应用程序中较严重问题。大多数错误与代码编写者执行的操作无关,而表示代码运行时 DartVM 出现的问题。例如:内存溢出(OutOfMemoryError)等。

与 Java 不同的是,Dart 不检测异常是否是声明的,也就是说方法或者函数不需要声明要抛出哪些异常。

15.2 抛出异常

使用 throw 抛出异常,异常可以是 Exception 或者 Error 类型的,也可以是其他类型的,但是不建议这么用。另外 throw 语句在 Dart 中也是一个表达式,因此可以写成箭头函数"=>"形式。

非 Exception 或者 Error 类型是可以抛出的,但是不建议这么用。抛出异常示例如下:

```dart
//exception_throw.dart 文件
void main(){
  //调用函数,抛出异常
  testException1();
  //testException2();
  //testException3();
}

//抛出异常测试
void testException1(){
  //抛出一个异常
  throw "这是第一个异常";
}

//抛出异常测试
void testException2(){
  //抛出一个异常
  throw Exception("这是第二个异常");
}

//抛出异常测试
void testException3() => throw Exception("这是第三个异常");
```

上面的示例输出如下内容:

```
[VERBOSE-2:ui_dart_state.cc(148)] Unhandled Exception: 这是第一个异常
#0      testException1 (package:helloworld/main.dart:12:3)
#1      main (package:helloworld/main.dart:4:3)
#2      _runMainZoned.<anonymous closure>.<anonymous closure> (dart:ui/hooks.dart:199:25)
#3      _rootRun (dart:async/zone.dart:1124:13)
#4      _CustomZone.run (dart:async/zone.dart:1021:19)
#5      _runZoned (dart:async/zone.dart:1516:10)
#6      runZoned (dart:async/zone.dart:1500:12)
#7      _runMainZoned.<anonymous closure> (dart:ui/hooks.dart:190:5)
#8      _startIsolate.<anonymous closure> (dart:isolate-patch/isolate_patch.dart:300:19)
#9      _RawReceivePortImpl._handleMessage (dart:isolate-patch/isolate_patch.dart:171:12)
```

从输出内容可以看出输出了异常信息"这是一个异常",同时还指出了抛出异常的位置,输出内容如下:

```
#0      testException1 (package:helloworld/main.dart:12:3)
#1      main (package:helloworld/main.dart:4:3)
```

代码发生异常的位置在 helloworld 工程下的 main.dart 文件的第 12 行,定位非常准确,那里是抛出异常的地方。同时可以看到 main.dart 文件的第 4 行也被提示到了,那里是调用异常函数的地方。

> **提示** 在项目排查各种异常和错误时,首先要找到自己编写的代码部分,这样才能更快地定位到问题出在哪里。

从代码里可以看出也可以用"=>"来表示,代码如下:

```
void testException3() => throw Exception("这是第三个异常");
```

15.3 捕获异常

在学习本节内容之前,首先看一个现实工作中的问题。例如项目经理安排给你一个功能模块需要完成,在编写的过程中遇到了一个所用框架的坑而无法解决,这时你会把问题抛给项目经理,项目经理也没遇到过这个问题,项目经理又抛给技术总监,如果技术总监还是无法解决,那么这个问题可能就会被搁置。异常就是这样向上传递,直到有方法处理它,如果所有的方法都无法处理该异常,那么 DartVM 会终止程序运行。

15.3.1 try-catch 语句

在 Dart 中通常采用 try-catch 语句来捕获异常并处理。语法格式如下:

```
try{
    //逻辑代码块;
}
catch(e,r){
    //处理代码块;
}
```

其中,e 是异常对象,r 是 StackTrace 对象,异常的堆栈信息通常打印异常对象即可。

在以上语法中,把可能引发异常的语句封装在 try 语句块中,用以捕获可能发生的异常。

如果 try 语句块中发生异常,那么一个相应的异常对象就会被抛出,然后 catch 语句就会依据所抛出异常对象的类型进行捕获并处理。处理之后程序会跳过 try 语句块中剩余的语句,转到 catch 语句块后面的第一条语句开始执行。

如果 try 语句块中没有发生异常,那么 try 语句块正常结束后,它后面的 catch 语句块就会被跳过,程序将从 catch 语句块后的第一条语句开始执行。

下面的示例演示了 try-catch 语句的基本用法,代码如下:

```
//exception_try_catch.dart 文件
void main() {
  try{
    //调用方法
```

```
        testException();
        //e 是异常对象,r 是 StackTrace 对象,异常的堆栈信息
    } catch(e, r){
        //输出异常信息
        print(e.toString());
        //输出堆栈信息
        print(r.toString());
    }
}

//抛出异常
void testException(){
    throw FormatException("这是一个异常");
}
```

其中异常对象输出如下内容：

```
flutter: FormatException: 这是一个异常
```

StackTrace 堆栈对象 r 输出如下内容：

```
flutter: #0      testException (package:helloworld/main.dart:2:3)
#1      main (package:helloworld/main.dart:7:5)
#2      _runMainZoned.<anonymous closure>.<anonymous closure> (dart:ui/hooks.dart:199:25)
#3      _rootRun (dart:async/zone.dart:1124:13)
#4      _CustomZone.run (dart:async/zone.dart:1021:19)
#5      _runZoned (dart:async/zone.dart:1516:10)
#6      runZoned (dart:async/zone.dart:1500:12)
#7      _runMainZoned.<anonymous closure> (dart:ui/hooks.dart:190:5)
#8      _startIsolate.<anonymous closure> (dart:isolate-patch/isolate_patch.dart:300:19)
#9      _RawReceivePortImpl._handleMessage (dart:isolate-patch/isolate_patch.dart:171:12)
```

输出堆栈信息有利于分析出代码抛出异常的原因。

15.3.2 try-on-catch 语句

如果 try 代码块中有很多语句会发生异常,而且发生的异常种类又很多,那么可以使用 on 关键字。on 可以捕获到某一类的异常,但是获取不到异常对象,而 catch 可以捕获到异常对象,因此这两个关键字可以组合使用。try-on-catch 语句语法格式如下：

```
try{
    //逻辑代码块;
} on ExceptionType catch(e){
    //处理代码块;
} on ExceptionType catch(e){
    //处理代码块;
```

```
}
//多个 on-catch...

} on ExceptionType catch(e){
    //处理代码块;
}
catch(e,r){
    //处理代码块;
}
```

其中，ExceptionType 表示异常类型。可以看到可以使用多个 on-catch，当第一个异常类型匹配不到时则会匹配下一个类型。

下面的示例演示了 try-on-catch 语句的用法，代码如下：

```
//exception_try_on_catch.dart 文件
//抛出没有类型的异常
void testNoTypeException(){
  throw "这是一个没有类型的异常";
}

//抛出 Exception 类型的异常
void testException(){
  throw Exception("这是一个 Exception 类型的异常");
}

//抛出 FormatException 类型的异常
void testFormatException(){
  throw FormatException("这是一个 FormatException 类型的异常");
}

void main() {
  try{
    testNoTypeException();
   //testException();
   //testFormatException();
  } on FormatException catch(e){    //如果匹配不到 FormatException 则会继续匹配
    print(e.toString());
  } on Exception catch(e){           //匹配不到 Exception,会继续匹配
    print(e.toString()) ;
  }catch(e, r){                      //匹配所有类型的异常,e 是异常对象,r 是 StackTrace 对象,
                                     //异常的堆栈信息
    print(e);
  }
}
```

示例使用了三种异常，如下所示：

❏ 没有类型的异常；

- Exception 类型的异常；
- FormatException 类型的异常。

这三种情况分别测试，异常均可以捕获得到，输出内容如下：

```
flutter:这是一个没有类型的异常
flutter: Exception:这是一个 Exception 类型的异常
flutter: FormatException:这是一个 FormatException 类型的异常
```

15.4 重新抛出异常

在捕获中处理异常，同时允许其继续传播，使用 rethrow 关键字。rethrow 保留了异常的原始堆栈跟踪。throw 重置堆栈跟踪到最后抛出的位置。

接下来看一个重新抛出异常的示例，代码如下：

```dart
//exception_rethrow.dart 文件
void main() {
  try {
    //虽然捕获了异常,但是又重新抛出了,所以要捕获
    test();
  } catch (e) {
    print('再次捕获到异常:' + e.toString());
  }
}

//抛出异常
void testException(){
  throw FormatException("这是一个异常");
}


void test() {
  try {
    testException();
  } catch (e) {
    //捕获到异常
    print('捕获到异常:' + e.toString());
    //重新抛出了异常
    rethrow;
  }
}
```

上面代码中 test 方法里第一次捕获到一个异常，在这之后再次抛出异常，在 main 函数里再次捕获到同样的异常。输出内容如下：

```
flutter:捕获到异常:FormatException: 这是一个异常
flutter:再次捕获到异常:FormatException: 这是一个异常
```

15.5 finally 语句

finally 内部的语句,无论是否有异常,都会执行。例如网络连接、数据库连接和打开文件等操作,在使用完成后需要释放资源,可以使用 finally 代码块确保这些资源能够被释放。finally 语句语法格式如下:

```
try{
    //逻辑代码块;
} on ExceptionType catch(e){
    //处理代码块;
} on ExceptionType catch(e){
    //处理代码块;
}

//多个 on-catch...

} on ExceptionType catch(e){
    //处理代码块;
}
catch(e,r){
    //处理代码块;
} finally{
    //释放资源
}
```

finally 是在 try-catch 语句的最后面使用的。接下来看下面这个例子,代码如下:

```
//exception_finally.dart 文件
void main() {
  try{
    //调用方法
    testException();
    //e 是异常对象,r 是 StackTrace 对象,异常的堆栈信息
  } catch(e, r){
    //输出异常信息
    print(e.toString());
  } finally {
    print('释放资源');
  }
}

//抛出异常
```

```
void testException(){
  throw FormatException("这是一个异常");
}
```

上面的例子输出如下内容：

```
flutter: FormatException:这是一个异常
flutter:释放资源
```

可以看到无论 try-catch 语句如何执行，finally 部分总会执行到，这样才能最大程度保证资源的释放。

15.6 自定义异常

有些项目需要自己编写一些类库，搭一个基础的框架，这就免不了要编写一些异常类。我们可以通过实现 Exception 接口来自定义一个异常。

实现自定义异常类示例代码如下：

```
//exception_my_exception.dart 文件
void main(){
  //测试自定义异常
  try{
    testMyException();
  } catch(e){
    print(e.toString());
  }
}

//抛出异常测试
void testMyException(){
  //抛出一个异常
  throw MyException('这是一个自定义异常');
}

//实现 Exception 接口自定义一个异常
class MyException implements Exception {

  //异常信息属性
  final String msg;

  //构建方法,传入可选参数 msg
  MyException([this.msg]);

  //重写 toString 方法,输出异常信息
```

```
  @override
  String toString() => msg ?? 'MyException';
}
```

要实现自定义异常类,需要定义一个异常信息的属性 msg 用来接收异常信息,在异常类的构造方法 MyException 里提供一个可选参数 msg 用于传递异常信息,另外还要重写 toString 方法来输出异常信息。

上述示例输出内容如下所示:

```
flutter:这是一个自定义异常
```

15.7　Http 请求异常

当学会了自定义异常后就可以自己写一个简单的库,接下来看一个实际场景的异常处理。在实际项目中前后端往往需要使用 Http 进行数据交互,服务端会返回一些状态码,这些状态码反映了服务端当前处理的情况,如下所示:

- 200:(成功)服务器已成功处理了请求,通常这表示服务器提供了请求的网页;
- 404:Not Found 无法找到指定位置的资源,这也是一个常用的应答;
- 500:Internal Server Error 服务器遇到了意料不到的情况,不能完成客户的请求。

下面分步来看 Http 状态异常处理示例的过程:

步骤 1:打开 pubspec.yaml 文件添加 http 库并指定最新版本,再执行"flutter packages get"命令更新包资源。

步骤 2:定义状态枚举值,这里只列举了三种状态,实际的 Http 状态很多。代码如下:

```
enum StatusType {
  //默认状态
  DEFAULT,
  //找不到页面状态
  STATUS_404,
  //服务器内部错误状态
  STATUS_500,
}
```

步骤 3:自定义状态异常类 StatusException 需要实现 Exception 类。要实现的代码大致如下所示:

```
class StatusException implements Exception {
  //构造方法,传入状态类型及异常信息
```

```
    //StatusException ...

    //枚举值状态类型
    StatusType type;
    //异常信息
    String msg;

    //重写 toString 方法
    //输出异常信息
    //...
}
```

步骤 4：编写 Http 请求方法，发起 Http 请求，根据返回的对象判断状态并抛出异常。其中状态判断部分代码如下：

```
if(response.statusCode == 200){
    //返回 response 对象
    return response;
}else if(response.statusCode == 404){
    //抛出异常
    throw StatusException(type: StatusType.STATUS_404,msg:'找不到页面');
} else if(response.statusCode == 500){
    //抛出异常
    throw StatusException(type: StatusType.STATUS_500,msg:'服务器内部发生错误');
}else{
    //抛出异常
    throw StatusException(type: StatusType.DEFAULT,msg:'Http 请求异常');
}
```

步骤 5：将以上步骤代码串起来，完整代码如下：

```
//exception_http_status.dart
import 'dart:async';
import 'package:http/http.dart' as http;

void main(){
    //发起 Http 请求
    httpRequest();
}

//发起 Http 请求,异步处理
Future httpRequest()async{
    //try-catch 捕获异常
    try{
        //请求后台 url 路径
        var url = 'http://127.0.0.1:3000/httpException';
        //向后台发起 get 请求,response 为返回对象
```

```dart
    http.get(url).then((response) {
      print("服务端返回状态: ${response.statusCode}");
      //判断返回状态
      if(response.statusCode == 200){
        //返回 response 对象
        return response;
      }else if(response.statusCode == 404){
        //抛出异常
        throw StatusException(type: StatusType.STATUS_404,msg:'找不到页面');
      } else if(response.statusCode == 500){
        //抛出异常
        throw StatusException(type: StatusType.STATUS_500,msg:'服务器内部发生错误');
      }else{
        //抛出异常
        throw StatusException(type: StatusType.DEFAULT,msg:'Http 请求异常');
      }
    });
  }catch(e){
    //打印错误
    return print('error:::${e}');
  }
}

//状态类型
enum StatusType {
  //默认状态
  DEFAULT,
  //找不到页面状态
  STATUS_404,
  //服务器内部错误状态
  STATUS_500,
}

//自定义状态异常
class StatusException implements Exception {
  //构造方法,传入状态类型及异常信息
  StatusException({
    this.type = StatusType.DEFAULT,
    this.msg,
  });

  //枚举值状态类型
  StatusType type;
  //异常信息
  String msg;

  //输出异常信息
  String toString() {
```

```
        return msg ?? "Http 请求异常";
    }
}
```

启动后端 Node 测试程序，进入 dart_node_server 程序，执行 "npm start" 命令启动程序，然后执行前端程序。输出结果如下：

```
flutter: 服务端返回状态: 500
[VERBOSE-2:ui_dart_state.cc(148)] Unhandled Exception: 服务器内部发生错误
#0      httpRequest.<anonymous closure> (package:helloworld/main.dart:27:9)
#1      _rootRunUnary (dart:async/zone.dart:1132:38)
#2      _CustomZone.runUnary (dart:async/zone.dart:1029:19)
#3      _FutureListener.handleValue (dart:async/future_impl.dart:126:18)
#4      Future._propagateToListeners.handleValueCallback (dart:async/future_impl.dart:639:45)
#5      Future._propagateToListeners (dart:async/future_impl.dart:668:32)
#6      Future._complete (dart:async/future_impl.dart:473:7)
#7      _SyncCompleter.complete (dart:async/future_impl.dart:51:12)
#8      _AsyncAwaitCompleter.complete (dart:async-patch/async_patch.dart:28:18)
#9      _completeOnAsyncReturn (dart:async-patch/async_patch.dart:294:13)
#10     _withClient (package:http/http.dart)
<asynchronous suspension>
#11     get (package:http/http.dart:46:5)
#12     httpRequest (package:helloworld/main.dart:16:5)
<asynchronous suspension>
#13     main (package:he<…>
```

从输出结果来看，以上信息是根据 500 这个状态码来抛出异常的，程序正常抛出了异常信息，控制台也定位到异常发生的地方。其中抛出异常的代码如下：

```
throw StatusException(type: StatusType.STATUS_500,msg:'服务器内部发生错误');
```

第 16 章　集　合

集合类是 Dart 数据结构的实现,它允许以各种方式将元素分组,并定义各种使这些元素更容易操作的方法。Dart 集合类是 Dart 将一些基本的和使用频率极高的基础类进行封装和增强后再以一个类的形式提供。集合类是可以往里面保存多个对象的类,存放的是对象,不同的集合类有不同的功能和特点,适合不同的场合,用以解决一些实际问题。

16.1　集合简介

创建一个数组并给数组存储数据的时候,不知道要存储多少数据,或者是在已有数组上存储数据时发现原先的数组长度不够用,这时采用这种常规方法给数组"扩容",使得越界的数据能够存储进去。

当事先不知道要存放数据的个数,或者需要一种比数组下标存取机制更灵活的方法时,就需要用到集合类。

集合的作用如下所示：
- 在类的内部,对数据进行组织；
- 简单而快速地搜索大数量的条目；
- 有的集合接口提供了一系列排列有序的元素,并且可以在序列中间快速地插入或者删除有关的元素；
- 有的集合接口提供了映射关系,可以通过关键字(key)快速查找到对应的唯一对象,而这个关键字可以是任意类型。

Dart 集合主要有以下几种：
- List：存储一组不唯一且按插入顺序排序的对象,可以操作索引；
- Set：存储一组唯一且无序的对象；
- Map：以键值对的形式存储元素,键(key)是唯一的。

16.2　List 集合

List 是一组有序元素的集合,数据不唯一,可以重复。如图 16-1 里的数据就可以用 List 来存储。索引从 0 到 5,索引不可以重复。值并没有什么规律,值可以重复。索引和值之间是一一对应的关系。

索引	值
0	A
1	B
2	C
3	D
4	C
5	A

图 16-1　List 集合

16.2.1　常用属性

List 集合常用属性如下所示：
- length：获取 List 长度；
- reversed：List 数据反序处理；
- isEmpty：判断 List 是否为空；
- isNotEmpty：判断 List 是否不为空。

这些属性的使用请看下面的示例代码：

```dart
//list_property.dart 文件
void main(){

  List myList = ['张三','李四','王五'];
  //获取列表长度
  print(myList.length);
  //判断列表是否为空
  print(myList.isEmpty);
  //判断列表是否不为空
  print(myList.isNotEmpty);
  //对列表倒序排序
  print(myList.reversed);
  //对列表倒序排序并输出一个新的 List
  var newMyList = myList.reversed.toList();
  print(newMyList);

}
```

上述示例输出内容如下：

```
flutter: 3
flutter: false
flutter: true
flutter: (王五, 李四, 张三)
flutter: [王五, 李四, 张三]
```

16.2.2 常用方法

List 常用方法如下：
- add：增加一个元素；
- addAll：拼接数组；
- indexOf：返回元素的索引，没有则返回 -1；
- remove：根据传入具体值删除元素；
- removeAt：根据传入索引删除元素；
- insert(index,value)：根据索引位置插入元素；
- insertAll(index,list)：根据索引位置插入 List；
- toList()：其他类型转换成 List；
- join()：将 List 元素按指定元素拼接；
- split()：将字符串按指定元素拆分并转换成 List；
- map：这个方法的执行逻辑是将 List 中的每个元素调出来和 map(f)中传入的 f 函数条件进行比较，如果符合条件就会返回 true，否则就会返回 false；
- where：查找列表中满足条件的数据，条件由传入的函数参数决定。

这些方法的使用请看下面的示例代码：

```dart
//list_method.dart 文件
void main(){

  //初始 List
  List myList = ['张三','李四','王五'];
  print(myList);
  //添加元素
  myList.add('赵六');
  print(myList);
  //拼接数组
  myList.addAll(['张三','李四']);
  print(myList);
  //indexOf 查找数据,查找不到返回 -1,查找到则返回索引值
  print(myList.indexOf('小张'));
  //向指定索引位置插入数据
  myList.insert(0, '王小二');
  print(myList);
  //删除指定元素
  myList.remove('赵六');
  //删除指定索引处的元素
  myList.removeAt(1);
  print(myList);

  //将 List 元素按指定元素拼接
  var str = myList.join(' - ');
```

```
print(str);
print(str is String); //true

//将字符串按指定元素拆分并转换成 List
var list = str.split('-');
print(list);
print(list is List);

var tempList = [1,"2",3,34532,555];
//这个方法的执行逻辑是将 List 中的每个元素调出来和 map(f)中传入的 f 函数条件进行比较
//如果符合条件就会返回 true,否则就会返回 false
var testMap = tempList.map((item) => item.toString().length == 1);
print(testMap);

//查找列表中满足条件的数据,条件由传入的函数参数决定
var testWhere = tempList.where((item) => item.toString().length == 3);
print(testWhere);

}
```

示例输出内容如下：

```
flutter: [张三, 李四, 王五]
flutter: [张三, 李四, 王五, 赵六]
flutter: [张三, 李四, 王五, 赵六, 张三, 李四]
flutter: -1
flutter: [王小二, 张三, 李四, 王五, 赵六, 张三, 李四]
flutter: [王小二, 李四, 王五, 张三, 李四]
flutter: 王小二-李四-王五-张三-李四
flutter: true
flutter: [王小二, 李四, 王五, 张三, 李四]
flutter: true
flutter: (true, true, true, false, false)
flutter: (555)
```

> **注意** 这里的 map 方法和 Map 没有任何关系,执行结果和 match 更像。这个 map 方法的执行逻辑是将 List 中的每个元素调出来和 map(f)中传入的 f 函数条件进行比较,如果符合条件就会返回 true,否则就会返回 false。

16.2.3 遍历集合

遍历集合的意思就是将集合中的元素挨个取出来,进行操作或计算。List 集合遍历有三种方法：

❑ 使用 for 循环遍历,通过 list[i]的方式可以访问集合中的元素；

- 使用 for…in 循环遍历，可以直接得到集合中的每一个元素，推荐此方式；
- 使用 list.forEach 方法，可以直接得到集合中的每一个元素，也推荐此方式。

接下来通过一个示例展示遍历集合的处理方法，代码如下：

```dart
//list_for_each.dart 文件
void main(){

  List list = [1, 2, 3, 4, 5];

  print('使用 forEach 迭代每个元素');
  //遍历每个元素,此时不可 add 或 remove,否则报错但可以修改元素值
  list.forEach((element){
    element += 1;
    //直接修改 list 对应 index 的值
    list[2] = 0;
  });
  //输出列表值
  print(list);

  //使用 for 循环遍历每个元素
  print('使用 for 循环遍历每个元素');
  for(var i = 0; i<list.length; i++){
    print(list[i]);
  }

  //使用 for…in 遍历每个元素
  print('使用 for…in 遍历每个元素');
  for(var x in list){
    print(x);
  }

}
```

上面的几种方式均可以遍历并输出集合中所有元素。当需要通过索引访问集合元素时建议使用 for 循环，当只需要迭代出每个元素时建议使用 for…in 和 forEach。

上面的示例输出如下内容：

```
flutter: 使用 forEach 迭代每个元素
flutter: [1, 2, 0, 4, 5]
flutter: 使用 for 循环遍历每个元素
flutter: 1
flutter: 2
flutter: 0
flutter: 4
flutter: 5
flutter: 使用 for…in 遍历每个元素
flutter: 1
```

```
flutter: 2
flutter: 0
flutter: 4
flutter: 5
```

16.3　Set 集合

Set 表示对象的集合，其中每个对象只能出现一次。dart:core 库提供了 Set 类来实现相同的功能。图 16-2 表示了一个篮子里的水果集合。这个篮子里有一些水果，这些水果是无序的，不能通过索引进行访问，并且不能有重复的元素。

图 16-2　Set 集合

Set 集合实例化及初始数据的代码如下：

```
//set_init.dart 文件
void main(){
  //实例化 Set
  Set set = Set();
  //添加元素
  set.add('香蕉');
  set.addAll( ['苹果', '西瓜'] );
  print(set);

  //通过 from 方法初始化 Set
  Set setFrom = Set.from(['葡萄', '哈密瓜', '苹果',null]);
  print(setFrom);
}
```

从上面代码可以看出，Set 集合中允许空数据 null 存在的，但是 null 一个 Set 集合里只能有一个，否则就是重复元素了。

上面的示例输出内容如下：

```
flutter: {香蕉, 苹果, 西瓜}
flutter: {葡萄, 哈密瓜, 苹果, null}
```

16.3.1 常用属性

Set 集合常用属性如下所示：
- first：返回 Set 第一个元素；
- last：返回 Set 最后一个元素；
- length：返回 Set 的元素个数；
- isEmpty：判断 Set 是否为空；
- isNotEmpty：判断 Set 是否不为空；
- iterator：返回迭代器对象,迭代器对象用于遍历集合。

这些属性的使用请看下面的示例代码：

```dart
//set_property.dart 文件
void main(){
  Set set = Set.from(['香蕉', '苹果', '葡萄']);
  //返回第一个元素
  print(set.first);
  //返回最后一个元素
  print(set.last);
  //返回元素的数量
  print(set.length);
  //集合只有一个元素就返回元素,否则异常
  //print(set.single);
  //集合是否没有元素
  print(set.isEmpty);
  //集合是否有元素
  print(set.isNotEmpty);
  //返回集合的哈希码
  print(set.hashCode);
  //返回对象运行时的类型
  print(set.runtimeType);
  //返回集合的可迭代对象
  print(set.iterator);
}
```

示例输出内容如下：

```
flutter: 香蕉
flutter: 葡萄
flutter: 3
```

```
flutter: false
flutter: true
flutter: 145083246
flutter: _CompactLinkedHashSet<dynamic>
flutter: Instance of '_CompactIterator<dynamic>'
```

16.3.2 常用方法

Set 常用方法如下：
- add：增加一个元素；
- addAll：拼接数组，添加一些元素；
- toString：以字符串形式输出集合内容；
- join：将集合的元素用指定字符串连接，以字符串输出；
- contains：判断集合中是否包含指定的元素；
- containsAll：判断集合是否包含一些元素；
- elementAt(index)：根据索引返回集合的元素；
- remove：删除集合指定的元素；
- removeAll：删除集合的一些元素；
- clear：删除集合所有的元素。

这些方法的使用请看下面的示例代码：

```dart
//set_method.dart 文件
void main(){
  Set set = Set.from(["A", "B", "C"]);
  //添加一个值
  set.add("D");
  print(set);
  //添加一些值
  set.addAll(["E", "F"]);
  print(set);
  //以字符串输出集合
  print(set.toString());
  //将集合的值用指定字符连接,以字符串输出
  print(set.join(","));
  //集合是否包含指定值
  print(set.contains("C"));
  //集合是否包含一些值
  print(set.containsAll(["E", "F"]));
  //返回集合指定索引的值
  print(set.elementAt(1));
  //删除集合的指定值,成功则返回 true
  print(set.remove("A"));
  //删除集合的一些值
```

```
    set.removeAll(["B", "C"]);
    //删除集合的所有值
    set.clear();
}
```

示例输出内容如下:

```
flutter: {A, B, C, D}
flutter: {A, B, C, D, E, F}
flutter: {A, B, C, D, E, F}
flutter: A,B,C,D,E,F
flutter: true
flutter: true
flutter: B
flutter: true
```

16.3.3　遍历集合

Set 集合由于没有序号,所以不能使用 for 循环进行遍历,但可以使用以下两种方式进行遍历。

- 使用 for…in 循环遍历,可以直接得到集合中的每一个元素;
- 调用 Set 集合的 toList 方法会返回一个 List 对象,然后再使用 forEach 方法,这样可以直接得到集合中的每一个元素。

下面的示例演示了使用这两种方法遍历 Set 集合元素的方法,代码如下:

```
//set_for_each.dart 文件
void main(){
  //初始化集合
  Set set = Set.from(["A", "B", "C"]);

  print('使用 for…in 输出集合元素');
  //使用 for…in 输出集合元素
  for(var item in set) {
    print(item);
  }

  print('使用 toList.forEach 输出集合元素');
  //使用 toList.forEach 输出集合元素
  set.toList().forEach((value){
    print(value);
  });

}
```

示例输出内容如下:

```
flutter: 使用for…in输出集合元素
flutter: A
flutter: B
flutter: C
flutter: 使用toList.forEach输出集合元素
flutter: A
flutter: B
flutter: C
```

16.4　Map 集合

Dart 映射(Map 对象)是一个简单的键/值对。映射中的键和值可以是任何类型。映射是动态集合,就是说 Map 可以在运行时增长和缩短。映射值可以是包括 Null 在内的任何对象。图 16-3 所示是一年 12 个月的前 4 个月的集合。键是英文,表示不能重复。值是中文,表示可以重复。

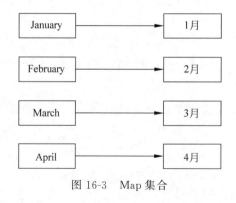

图 16-3　Map 集合

提示　Map 集合更适合通过键快速访问值的场景,例如 Mongodb 数据库就是一种典型的键值存储结构,查询效率非常高。又如,在翻阅书籍时,书的目录就相当于键,内容相当于值,当需要快速查看某个知识点时,通过目录来查找是最快的。

16.4.1　常用属性

Map 集合声明需要指定键和值的类型,初始化时可以用大括号将一对对元素括起来,如下所示:

```
Map< String, int > map = {"a":1, "b":2, "c":3};
```

Map 集合常用属性如下所示：
- hashCode：返回集合的哈希码；
- isEmpty：判断集合是否没有键值对；
- isNotEmpty：判断集合是否有键值对；
- keys：返回集合所有的键；
- values：返回集合所有的值；
- length：返回集合键值对数目；
- runtimeType：返回对象运行时类型。

如何使用这些属性，参看下面的示例代码：

```dart
//map_property.dart 文件
void main(){
  Map<String, int> map = {"a":1, "b":2, "c":3};
  //返回集合的哈希码
  print(map.hashCode);
  //集合是否没有键值对
  print(map.isEmpty);
  //集合是否有键值对
  print(map.isNotEmpty);
  //返回集合的所有键
  print(map.keys);
  //返回集合的所有值
  print(map.values);
  //返回集合键值对的数目
  print(map.length);
  //返回对象运行时的类型
  print(map.runtimeType);
}
```

示例输出内容如下：

```
flutter: 320444088
flutter: false
flutter: true
flutter: (a, b, c)
flutter: (1, 2, 3)
flutter: 3
flutter: _InternalLinkedHashMap<String, int>
```

从输出结果可以看到，map.runtimeType 输出了 Map 的定义类型<String, int>。

16.4.2 常用方法

Map 常用方法如下：
- toString：返回集合的字符串表示；

- addAll：添加其他键值对到集合中；
- containsKey：集合是否包含指定键；
- containsValue：集合是否包含指定值；
- remove：删除指定键值对；
- clear：删除所有键值对。

如何使用这些方法，参看下面的示例代码：

```dart
//map_method.dart 文件
void main(){
    Map<String, int> map = {"a":1, "b":2, "c":3};
    //返回集合的字符串表示
    print(map.toString());
    //添加其他键值对到集合中
    map.addAll({"d":4, "e":5});
    //集合是否包含指定键
    print(map.containsKey("d"));
    //集合是否包含指定值
    print(map.containsValue(5));
    //删除指定键值对
    map.remove("a");
    //删除所有键值对
    map.clear();
}
```

示例输出内容如下：

```
flutter: {a: 1, b: 2, c: 3}
flutter: true
flutter: true
```

16.4.3 遍历集合

Map 集合是可以迭代它的所有键和所有值的。可以使用以下两种方式进行遍历。
- 使用 forEach 循环遍历，可以直接得到集合中的每一对键和值；
- 使用 for…in 可以获取到第一个键，通过键可以访问其对应的值。

下面的示例演示了如何使用这两种方法遍历 Map 集合元素，代码如下：

```dart
//map_for_each.dart 文件
void main(){

    Map<String, int> map = {"a":1, "b":2, "c":3};
    print('通过forEach迭代 Map 集合');
    //按顺序迭代映射
```

```
map.forEach((key, value){
  print(key + " : " + value.toString());
});


Map< String, int > scores = {'0000001': 36};
print('通过for…in迭代Map集合');
for (var key in ['0000001', '0000002', '0000003']) {
  //查找指定键,如果不存在就添加
  scores.putIfAbsent(key, (){
    return 80;
  });
  //通过 key 访问其值
  print(scores[key]);
}
```

从示例源码中可以看出,Map集合的迭代主要是围绕键值操作的。其中,putIfAbsent方法是可以动态地向Map集合里添加键值的。

示例输出内容如下:

```
flutter: 通过forEach迭代Map集合
flutter: a : 1
flutter: b : 2
flutter: c : 3
flutter: 通过for…in迭代Map集合
flutter: 36
flutter: 80
flutter: 80
```

第 17 章 泛　型

泛型本质上是提供类型的"类型参数",也就是参数化类型。我们可以为类、接口或方法指定一个类型参数,通过这个参数限制对数据类型的操作,从而保证类型转换的绝对安全。

17.1　语法

如果你查看基本数组类型 List 的 API 文档,你会发现它的类型其实是 List＜E＞。＜...＞标记表示 List 是一个"泛型"(或带参数的)类——具有形式上的类型参数的类型。按照惯例,类型变量的名字是单个字母,例如 E、T、S、K 和 V。

以集合为例,泛型定义及使用的语法格式如下:

```
CollectionName＜T＞ identifier = CollectionName＜T＞;
```

其中,CollectionName 表明集合名称,例如 List 和 Set。T 代码集合中的数据类型,例如 String、int 或者某个类。定义好类型 T 后,集合中的每个元素必须为 T 类型。

17.2　泛型的作用

泛型通常是类型安全的要求,但它们除了让你的代码可以运行外还有诸多益处:
- 正确地指定泛型类型会写出更好的代码;
- 使用泛型可减少代码重复。

17.2.1　类型安全

如果只想让一个列表包含字符串,那么可以指定它为 List＜String＞(读作"字符串列表")。这样一来使用者的工具可以检测到将一个非字符串对象添加到该列表是错误的。参看下面一个例子:

```
//generics_type_error.dart 文件
```

```
void main(){
  //List 元素类型为 String
  var languages = List<String>();
  //类型正确
  languages.addAll(['Java', 'Kotlin', 'Dart']);
  //使用整型值会报异常
  languages.add(50);
}
```

示例中 List 类型为 String 类型，当使用了 int 类型后，编译器会抛出异常，提示不能将整型值赋值给 String 类型参数。控制台输出如下内容：

```
Compiler message:
lib/main.dart:8:17: Error: The argument type 'int' can't be assigned to the parameter type 'String'.
Try changing the type of the parameter, or casting the argument to 'String'.
    languages.add(50);
                  ^

Restarted application in 145ms.
[VERBOSE-2:ui_dart_state.cc(148)] Unhandled Exception: 'package:helloworld/main.dart':
error: lib/main.dart:8:17: Error: The argument type 'int' can't be assigned to the parameter
type 'String'.
Try changing the type of the parameter, or casting the argument to 'String'.
    languages.add(50);

#0      _runMainZoned.<anonymous closure>.<anonymous closure> (dart:ui/hooks.dart:199:25)
#1      _rootRun (dart:async/zone.dart:1124:13)
#2      _CustomZone.run (dart:async/zone.dart:1021:19)
#3      _runZoned (dart:async/zone.dart:1516:10)
#4      runZoned (dart:async/zone.dart:1500:12)
#5      _runMainZoned.<anonymous closure> (dart:ui/hooks.dart:190:5)
#6      _startIsolate.<anonymous closure> (dart:isolate-patch/isolate_patch.dart:300:19)
#7      _RawReceivePortImpl._handleMessage (dart:isolate-patch/isolate_patch.dart:171:12)
```

17.2.2 减少重复代码

使用泛型可以减少代码重复。泛型使你在多个不同类型间共享同一个接口和实现，而依然享受静态分析的优势。比如，要创建一个缓存对象的接口，代码如下：

```
//对象缓存
abstract class ObjectCache {
```

```
//获取对象
Object getByKey(String key);
//设置对象
void setByKey(String key, Object value);
}
```

你发现需要一个与此接口相对应的字符串版本,所以你创建了另一个接口,代码如下:

```
//字符串缓存
abstract class StringCache {
  //获取字符串
  String getByKey(String key);
  //设置字符串
  void setByKey(String key, String value);
}
```

之后你觉得需要一个与该接口相对应的 int 类型版本,那么又需要重新定义一个接口。泛型可以让你省去创建所有这些接口的麻烦。取而代之,你可以创建一个单一的接口并接受一个类型参数。代码如下:

```
//任意类型缓存
abstract class Cache<T> {
  //获取任意类型数据
  T getByKey(String key);
  //设置任意类型数据
  void setByKey(String key, T value);
}
```

在这段代码中,T 是替身类型。它是一个占位符,你可以将其视为开发者稍后定义的类型。

17.3 集合中使用泛型

List 和 Map 字面量是可以参数化的,定义如下:
- 参数化定义 List,要在字面量前添加<type>;
- 参数化定义 Map,要在字面量前添加<keyType, valueType>。

通过参数化定义,可以带来更安全的类型检查,并且可使用变量的自动类型推导,参看下面的示例,此示例是关于 List 和 Map 的泛型用法,代码如下:

```
//generics_list_map.dart 文件
void main(){

  //元素为 String 类型
```

```dart
  var names = <String>['张三', '李四', '王五'];
  print(names);
  //Key 和 Value 均为 String 类型
  var users = <String, String>{
    '0000001':'张三',
    '0000002':'李四',
    '0000003':'王五'
  };
  print(users);

  //Key 为 String 类型,Value 为 User 类型
  var userMap = <String,User>{
    'alex':User('alex',20),
    'kevin':User('kevin',30),
    'jennifer':User('jennifer',30),
  };
  //直接打印输出
  print(userMap);
  //输出 Map 集合的 Key 和 Value 值
  userMap.forEach((String key,User value){
    print('Key = ' + key);
    print("Value = " + "name:" + value.name + " age:" + value.age.toString());
  });

}

//用户类
class User{
  //用户姓名
  String name;
  //用户年龄
  int age;
  //构造方法
  User(this.name,this.age);
}
```

在示例代码中,泛型类型被指定为 User 类型,即不仅可以是 Dart 自带类型,也可以是自定义的类等类型。示例输出内容如下:

```
flutter: [张三, 李四, 王五]
flutter: {0000001: 张三, 0000002: 李四, 0000003: 王五}
flutter: {alex: Instance of 'User', kevin: Instance of 'User', jennifer: Instance of 'User'}
flutter: Key = alex
flutter: Value = name:alex age:20
flutter: Key = kevin
flutter: Value = name:kevin age:30
flutter: Key = jennifer
flutter: Value = name:jennifer age:30
```

可以看到输出内容 Instance of 'User' 表示此对象是 User 类的实例。

在实际项目中需要将服务端返回的 Json 数据转换成 List 集合，以商品列表为例，首先需要将一条一条商信息数据转换成 VO 类，然后再将这些 VO 类放在一个 List 集合里。参看下面的示例代码：

```dart
//generics_good_list.dart 文件
void main(){
  //服务端返回的 Json 数据
  var json = {
    //状态码
    'code':'0',
    //状态信息
    'message':'success',
    //返回数据
    'data':[
      {
        'goodId':'0000001',
        'amount': 666,
        'goodImage': 'http://192.168.2.168/images/1.png',
        'goodPrice': 15999,
        'goodName': "苹果笔记本",
        "goodDetail": "苹果 屏幕尺寸: 13.3 英寸 处理器: Intel Core i5 - 8259",
      },
      {
        'goodId':'0000002',
        'amount': 3000,
        'goodImage': 'http://192.168.2.168/images/2.png',
        'goodPrice': 5999,
        'goodName': "Dell/戴尔笔记本",
        "goodDetail": "Dell/戴尔 灵越 15(3568) Ins15E - 3525 独显 i5 游戏本超薄笔记本计算机",
      },
      {
        'goodId':'0000003',
        'amount': 999,
        'goodImage': 'http://192.168.2.168/images/3.png',
        'goodPrice': 23999,
        'goodName': "外星人笔记本",
        "goodDetail": "外星人 全新 m15 R2 九代酷睿 i7 六核 GTX1660Ti 独显 144Hz 吃鸡游戏笔记本计算机戴尔 DELL15M - R4725",
      },
    ];

    //商品信息列表
    GoodsListModel goods = GoodsListModel.fromJson(json);
    print(goods.toJson());

}
```

```dart
//商品列表数据模型
class GoodsListModel{
  //状态码
  String code;
  //状态信息
  String message;
  //商品列表数据,使用泛型
  List<GoodInfo> data;

  //构造方法
  GoodsListModel({this.code,this.message,this.data});

  //命名构造方法
  GoodsListModel.fromJson(Map<String,dynamic> json){
    code = json['code'];
    message = json['message'];
    if(json['data'] != null){
      //商品列表数据,泛型类型为 GoodInfo
      data = List<GoodInfo>();
      json['data'].forEach((v){
        data.add(GoodInfo.fromJson(v));
      });
    }
  }

  //转换成 Json 对象输出
  Map<String,dynamic> toJson(){
    final Map<String,dynamic> data = Map<String,dynamic>();
    data['code'] = this.code;
    data['message'] = this.message;
    if(this.data != null){
      data['data'] = this.data.map((v) => v.toJson()).toList();
    }
    return data;
  }

}

//商品信息 VO 类
class GoodInfo{
  //商品 Id
  String goodId;
  //商品数量
  int amount;
  //商品图片
  String goodImage;
  //商品价格
  int goodPrice;
  //商品名称
```

```
   String goodName;
   //商品详情
   String goodDetail;

   //构造方法
   GoodInfo({this.goodId, this.amount, this.goodImage, this.goodPrice, this.goodName, this.
goodDetail});

   /*
    * 初始化列表,在构造方法体执行前设置实例变量的值
    */
   GoodInfo.fromJson(Map<String,dynamic> json)
   //初始化列表
       : goodId = json['goodId'],
         amount = json['amount'],
         goodImage = json['goodImage'],
         goodPrice = json['goodPrice'],
         goodName = json['goodName'],
         goodDetail = json['goodDetail']{
   }

   /*
    * 将当前对象转化成Json数据
    */
   Map<String,dynamic> toJson(){
     final Map<String, dynamic> data = Map<String, dynamic>();
     data['goodId'] = this.goodId;
     data['amount'] = this.amount;
     data['goodImage'] = this.goodImage;
     data['goodPrice'] = this.goodPrice;
     data['goodName'] = this.goodName;
     data['goodDetail'] = this.goodDetail;
     return data;
   }
}
```

在上面的示例代码中,List<GoodInfo>使用了泛型,其中GoodInfo为泛型类型。这样做的好处是将Json数据自动转换成VO类便于页面展示使用。示例输出内容如下:

```
flutter: {code: 0, message: success, data: [{goodId: 0000001, amount: 666, goodImage: http://
192.168.2.168/images/1.png, goodPrice: 15999, goodName: 苹果笔记本, goodDetail: 苹果 屏幕尺
寸: 13.3 英寸 处理器: Intel Core i5 – 8259}, {goodId: 0000002, amount: 3000, goodImage:
http://192.168.2.168/images/2.png, goodPrice: 5999, goodName: Dell/戴尔笔记本, goodDetail:
Dell/戴尔 灵越 15(3568) Ins15E – 3525 独显 i5 游戏本超薄笔记本计算机}, {goodId: 0000003,
amount: 999, goodImage: http://192.168.2.168/images/3.png, goodPrice: 23999, goodName: 外星
人笔记本, goodDetail: 外星人 全新 m15 R2 九代酷睿 i7 六核 GTX1660Ti 独显 144Hz 吃鸡游戏笔记本
计算机戴尔 DELL15M – R4725}]}
```

17.4 构造方法中使用泛型

使用构造函数时要指定一个或多个类型,可以将类型放在类名后面的尖括号"<...>"中。示例代码如下:

```
//generics_constructor.dart 文件
void main(){

  var names = List<String>();
  names.addAll(['张三', '李四']);
  //构造方法参数必须为 String 类型
  var nameSet = Set<String>.from(names);
  print(nameSet);

}
```

在上面的示例中,Set 的构造函数参数必须为 String 类型。

17.5 判断泛型对象的类型

可以使用 is 表达式来判断泛型对象的类型,代码如下:

```
var names = new List<String>();
print(names is List<String>); //true
```

> **注意** 生产模式下不会进行类型检查,所以 List<String>可能包含非 String 对象,这种情况下,建议分别判断每个对象的类型或者处理类型转化异常。

17.6 限制泛型类型

有时候,我们希望泛型不那么泛,也就是说,希望泛型的可选类型是限制的,那么可以使用 extends 关键字实现。参看下面的示例代码:

```
//generics_check_type.dart 文件
//定义类 A
class A {

}
```

```
//定义类B继承类A
class B extends A {

}

//定义类C
class C {

}

//定义类SomeClass
class SomeClass<T extends A>{
  //...
}

main() {
  //这种情况下是可以的,因为传入的类型符合限定(自身或者子类)
  var a = SomeClass<A>();
  var b = SomeClass<B>();
  //不显式指定泛型类型,也是可以的
  var c = SomeClass();
  //这种情况下不行,因为不符合限定
  //var d = SomeClass<C>();
}
```

上面示例代码中,T是继承类A的,那么类A和类B都符合同一类型的要求,但是类C不符合类型要求,所以当其指定成SomeClass<C>类型时就出错了。

17.7 泛型方法的用法

起初Dart对泛型的支持仅限于类。一个新的语法,称为"泛型方法",允许在方法上使用类型参数,语法格式如下:

```
T first<T>(List<T> ts) {
  //做一些初始化工作或者错误检查
  T tmp = ts[0];
  //做一些额外的检查或处理
  return tmp;
}
```

这里 first<T>中的泛型参数允许你在以下几个地方使用类型参数 T:
- 在方法的返回类型中 T;
- 在参数的类型中 List<T>;
- 在局部变量的类型中 T tmp。

接下来看一个泛型方法使用的示例,代码如下:

```dart
//generics_method.dart 文件

void main(){
  print(getDataString('字符串'));
  print(getDataInt(30));
  print(getDataDynamic('dynamic'));
  //定义为 int 型,传值就传入 int 型,返回值也为 int 型
  print(getData<int>(12));
  print(getData<String>('hello'));
}

//普通方法 String 类型
String getDataString(String value){
  return value;
}

//普通方法, int 类型
int getDataInt(int value){
  return value;
}

//普通方法,不确定类型
dynamic getDataDynamic(value){
  return value;
}

//泛型方法,可以传入任意类型
T getData<T>(T value){
  return value;
}
```

代码输出内容如下:

```
flutter: 字符串
flutter: 30
flutter: dynamic
flutter: 12
flutter: hello
```

从输出内容可以看到泛型方法可以获取任意类型的数据。

17.8 泛型类的用法

把一个类设计成泛型类型,可以增强类的功能。它的格式如下:

```
class className<T>{

  //成员变量
  T variableName;

  //成员方法
  T functionName(T value ...){
    //...
    return T;
  }
}
```

可以看到类的结构和普通的类是一样的，只是涉及类型的地方统一换成了 T。

假设我们设计一个日志处理类，这个类要求能输出各种日志信息，那么采用泛型类是再合适不过的了。示例代码如下：

```
//generics_class.dart 文件

void main() {

  Log logInt = Log<int>();
  logInt.add(12);
  logInt.add(23);
  //输出 int 型数据
  logInt.printLog();

  Log logString = Log<String>();
  logString.add('这是一条日志');
  logString.add('泛型类型为 String');
  //输出 String 类型数据
  logString.printLog();

}

//日志类,类型为 T
class Log<T>{

  //定义一个列表,用来存储日志
  List list = List<T>();

  //添加数据
  void add(T value){
    //添加日志到列表里
    this.list.add(value);
  }

  //打印日志
```

```
  void printLog(){
    //循环输出日志数据
    for(var i = 0; i < this.list.length; i++){
      print(this.list[i]);
    }

  }
}
```

示例代码输出内容如下：

```
flutter: 12
flutter: 23
flutter: 这是一条日志
flutter: 泛型类型为 String
```

17.9 泛型抽象类的用法

抽象类里通常定义一些操作方法，制定一些操作规范。

假设要实现数据缓存的功能：有文件缓存和内存缓存。文件缓存和内存缓存按照接口约束实现。定义一个泛型抽象类，约束实现它的子类必须有 getByKey(key) 和 setByKey(key,value)方法，并要求 setByKey 方法的 value 类型和实例化子类所指定的类型一致。代码如下：

```
//generics_abstract_class.dart 文件
void main(){
  //实例化内存缓存对象,类型为 Map
  MemoryCache m = MemoryCache<Map>();
  m.setByKey('index', {"name":"张三","age":30});
}

//缓存抽象类
abstract class Cache<T>{
  //获取数据,类型为 T
  getByKey(String key);
  //设置数据,类型为 T
  void setByKey(String key, T value);
}

//文件缓存,实现缓存接口
class FlieCache<T> implements Cache<T>{

  //重写 getByKey 方法
  @override
```

```dart
  getByKey(String key) {
    return null;
  }

  //重写 setByKey 方法
  @override
  void setByKey(String key, T value) {
    print("我是文件缓存 把 key = ${key} value = ${value}的数据写入文件中");
  }
}

//内存缓存,实现缓存接口
class MemoryCache<T> implements Cache<T>{

  //重写 getByKey 方法
  @override
  getByKey(String key) {
    return null;
  }

  //重写 setByKey 方法
  @override
  void setByKey(String key, T value) {
    print("我是内存缓存 把 key = ${key} value = ${value} -写入内存中");
  }
}
```

从程序的设计上来看可以设计更多类型的缓存,缓存 Cache 抽象类提出了一个规范,即上述两个方法,至于存取什么类型数据,它并不关心。

上述示例输出内容如下:

```
flutter: 我是内存缓存 把 key = index value = {name: 张三, age: 30} -写入内存中
```

第 18 章 异步编程

所谓异步表示可以同时做几件事情，不需要等任何事情做完就可以做其他事情。这样可以提高程序运行的效率。本章将围绕以下几个方面阐述 Dart 异步编程的知识。

- 异步编程概念；
- Future；
- Async/Await；
- Stream；
- Isolate。

18.1 异步概念

拿做饭打个比方，我可以先把水和米放到电饭锅里面去煮，我放完水和米并盖好锅盖之后，我就可以去做其他事情了。在煮米的这个期间，我不需要等着，我可以做菜，可以听音乐，还可以和其他人聊天（无须等待），等到米煮熟了，电饭锅会自己停止程序（通知我），我就知道米煮熟了（一件事情完成）。煮饭这件事情，可以认为是一种异步。

假设在做饭的时候，我没有用电饭锅，而是老式的灶台来煮饭。在饭煮熟之前，我得一直烧火。在煮饭过程中，我不能做其他事情。在这种情况下，煮饭这件事情是一种同步过程（多数情况下叫阻塞）。

Flutter 中的异步机制涉及到的关键字有 await、async、iterator、iterable、stream 和 timer 等。比较常用的为 async 和 await。

18.1.1 单线程

编程中的代码执行，通常分为同步与异步两种。简单来说，同步就是按照代码的编写顺序，从上到下依次执行，这是我们最常接触的一种形式，但是同步执行代码的缺点也显而易见，如果其中某一行或几行代码非常耗时，那么就会阻塞，使得后面的代码不能被立刻执行。单线程的执行过程如图 18-1 所示。

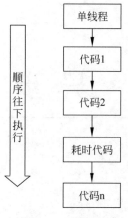

图 18-1　单线程

18.1.2　多线程

异步的出现正是为了解决这种问题,它可以使某部分耗时代码不在当前这条执行线路上立刻执行,那究竟怎么执行呢?最常见的一种方案是使用多线程,也就相当于开辟另一条执行线,然后让耗时代码在另一条执行线上运行,这样两条执行线并列执行,耗时代码自然也就不能阻塞主执行线上的代码了。多线程处理流程如图 18-2 所示。

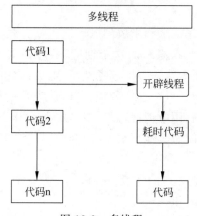

图 18-2　多线程

18.1.3　事件循环

多线程虽然好用,但是在大量并发时,仍然存在两个较大的缺陷,一个是开辟线程比较耗费资源,线程开多了机器吃不消,另一个则是线程的锁问题,多个线程操作共享内存时需要加锁,复杂情况下的锁竞争不仅会降低性能,还可能造成死锁,因此又出现了基于事件的异步模型。简单来说就是在某个单线程中存在一个事件循环和一个事件队列,事件循环不断地从事件队列中取出事件来执行,这里的事件就好比是一段代码,每当遇到耗时的事件时,事件循环不会停下来等待结果,它会跳过耗时事件,继续执行其后的事件。当不耗时的事件都完成了,再来查看耗时事件的结果,因此耗时事件不会阻塞整个事件循环,这让它后面的事件也会有机会得到执行。Node 就是事件循环的一种典型应用。事件循环的处理流程如图 18-3 所示。

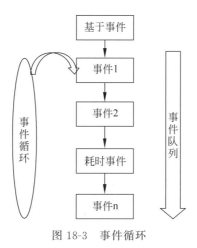

图 18-3　事件循环

我们很容易发现,这种基于事件的异步模型,只适合 I/O 密集型的耗时操作,因为 I/O 耗时操作往往是把时间浪费在等待对方传送数据或者返回结果,因此这种异步模型往往用于网络服务器并发。如果是计算密集型的操作,则应当尽可能利用处理器的多核来实现并行计算。

18.2　Future

Future 表示未来将要发生的事情,它会涉及到两个关键字 async 和 await。关键字 async 和 await 支持异步编程,可以使你用看起来像同步的方式编写异步代码,它相当于一个语法糖,同时方法的返回类型是 Future,所以通常一起使用。

18.2.1　Dart 事件循环

Dart 是基于单线程模型的语言,它也有自己的进程(或者叫线程)机制,名为 Isolate。应用的启动入口 main 方法就是一个 Isolate。对多核 CPU 的特性来说,多个 Isolate 可以显著提高运算效率,当然也要适当控制 Isolate 的数量而不应滥用。有一个很重要的点需要注意,Dart 中 Isolate 之间无法直接共享内存,不同的 Isolate 之间只能通过 Isolate API 进行通信,这里只对 Isolate 作一个简单介绍,不作深入讲解。

Dart 采用事件驱动的体系结构,该结构基于具有单个事件循环和两个队列的单线程执行模型。Dart 虽然提供调用堆栈,但是它使用事件在生产者和消费者之间传输上下文。事件循环由单个线程支持,因此根本不需要同步和锁定。

Dart 线程中有一个消息循环机制(Event Loop)和两个队列(Event Queue 和 MicroTask Queue),其中两个队列的作用如下:

- Event Queue:事件队列,它包含所有外来的事件,如 I/O、Mouse Events、Drawing Events 等,任意新增的 Event 都会放入 Event Queue 中排队等待执行,好比机场的公共排队大厅;
- MicroTask Queue:微任务队列,只在当前 Isolate 的任务队列中排队,优先级高于 Event Queue,好比机场里的某个 VIP 候机室,总是在 VIP 用户登机后,才开放公共排队入口。

Dart 事件循环执行流程如图 18-4 所示。

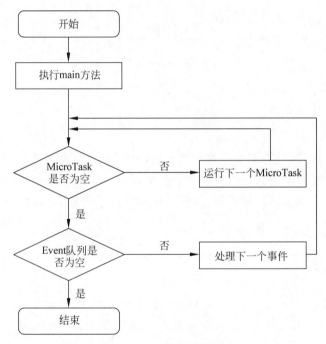

图 18-4　Dart 事件循环

先查看 MicroTask 队列是否为空,不为空则先执行 MicroTask 队列。

一个 MicroTask 执行完后,检查有没有下一个 MicroTask,直到 MicroTask 队列为空,才去执行 Event 队列。

在 Evnet 队列取出一个事件处理完后,再次返回第一步,去检查 MicroTask 队列是否为空。

我们可以看出,将任务加入到 MicroTask 中可以被尽快执行,但也需要注意,当事件循环在处理 MicroTask 队列时,Event 队列会被卡住,应用程序无法处理鼠标单击、I/O 消息等事件。

当 main 方法执行完毕并退出后,Event Loop 就会以 FIFO(先进先出)的顺序执行 MicroTask,当所有 MicroTask 执行完后它才会从 Event 队列中取事件并执行。如此反复,直到两个队列都为空。

Future 就是 Event,很多 Flutter 内置的组件是 Event,例如 Http 请求控件的 get 方法和 RefreshIndicator(下拉手势刷新控件)的 onRefresh 方法。每一个被 await 标记的方法也是一个 Event,每创建一个 Future 就会把这个 Future 扔进 Event 队列中排队并等候检查。

18.2.2 调度任务

将任务添加到 MicroTask 队列有两种方法,参看下面示例代码:

```dart
//async_micro_task.dart 文件
import 'dart:async';

void main() {
  //使用 scheduleMicrotask 方法添加
  scheduleMicrotask(myTask);

  //使用 Future 对象添加
  Future.microtask(myTask);
}

void myTask(){
  print("这是一个任务");
}
```

输出内容如下:

```
flutter: 这是一个任务
flutter: 这是一个任务
```

将任务添加到 Event 队列,示例代码如下:

```dart
//async_event_task.dart 文件
import 'dart:async';

void myTask(){
  print("这是一个任务");
}

void main() {
  //将任务传入 Future 构造方法里即可
  Future(myTask);
}
```

输出内容如下：

```
flutter: 这是一个任务
```

通过上面两个示例，我们学会了调度任务，那么现在就可以将这些处理过程放在一起来查看程序执行的过程。示例代码如下：

```dart
//async_event_and_task.dart 文件
import 'dart:async';

//测试程序执行过程
void main() {
  print("main start");

  //放入事件队列
  Future((){
    print("这是一个任务:EventTask");
  });

  //放入 MicroTask
  Future.microtask((){
    print("这是一个任务:MicroTask");
  });

  print("main stop");
}
```

示例运行结果如下：

```
flutter: main start
flutter: main stop
flutter:这是一个任务:MicroTask
flutter:这是一个任务:EventTask
```

可以看到，代码的运行顺序并不是按照我们的编写顺序来执行的，将任务添加到队列并不等于立刻执行，它们是异步执行的，当前 main 方法中的代码执行完之后，才会去执行队列中的任务，且 MicroTask 队列运行在 Event 队列之前。

18.2.3　延时任务

在程序中我们经常要用到延迟，例如延迟几秒执行一个动画，延迟几秒发起一个请求等。如需要将任务延时执行，则可使用 Future.delayed 方法，代码如下：

```dart
Future.delayed(Duration(seconds:1),(){
  print('任务延迟执行');
});
```

这段代码表示在延迟 1 秒之后将任务加入到 Event 队列。需要注意的是，这种延时方式并不是准确的，万一前面有很耗时的任务，那么你的延迟任务不一定能准时运行。

```dart
//async_delayed.dart 文件
import 'dart:async';
import 'dart:io';

void main() {
  print("main start");

  //延迟 1 秒后执行任务
  Future.delayed(Duration(seconds:1),(){
    print('延迟任务');
  });

  Future((){
    //模拟耗时 5 秒
    sleep(Duration(seconds:5));
    print("耗时 5 秒");
  });

  print("main stop");
}
```

上面这个示例输出结果如下：

```
flutter: main start
flutter: main stop
flutter: 耗时 5 秒
flutter: 延迟任务
```

从结果可以看出，delayed 方法调用在前面，但是它显然并未直接将任务加入 Event 队列，而是需要等待 1 秒之后才会去将任务加入，但在这 1 秒之间，后面的 sleep 代码直接将一个耗时任务加入到了 Event 队列，这就直接导致写在前面的 delayed 任务在 1 秒后只能被加入到耗时任务之后，只有当前面耗时任务完成后，它才有机会得到执行。这种机制使得延迟任务变得不太可靠，你无法确定延迟任务到底在延迟多久之后被执行。

18.2.4 Future 详解

Future 类是对未来结果的一个代理，即一件"将来"会发生的事情，它返回的并不是被调用的任务的返回值。将来可以从 Future 中取到一个值。参看下面的示例代码：

```dart
import 'dart:async';

//任务方法
```

```
void myTask(){
  print("这是一个任务");
}

void main() {
  //实例化 Future 对象
  Future fu = Future(myTask);
}
```

如上代码，Future 类实例 fu 并不是方法 myTask 的返回值，它只是代理了 myTask 方法，封装了该任务的执行状态。

1. 创建 Future

Future 的几种创建方法：

- Future();
- Future.microtask();
- Future.sync();
- Future.value();
- Future.delayed();
- Future.error()。

其中，sync 是同步方法，任务会被立即执行，参看一个示例代码如下：

```
//async_future_sync.dart 文件
import 'dart:async';

void main() {
  print("main start");

  //立即执行
  Future.sync((){
    print("sync task");
  });

  //最后执行
  Future((){
    print("async task");
  });

  print("main stop");
}
```

运行结果如下：

```
main start
sync task
```

```
main stop
async task
```

从运行结果来看,sync task 那段代码在运行到它时立即被执行了,最后输出 async task 表明其稍晚些被执行。

2. 注册回调

当 Future 中的任务完成后,我们往往需要一个回调方法,这个回调方法会立即执行,不会被添加到事件队列。例如前端发起 Http 数据请求,当数据返回时即触发这个注册的回调函数,进行下一步的数据处理,示例代码如下:

```
//async_future_then.dart 文件
import 'dart:async';

void main() {
  print("main start");

  Future fu = Future.value('Future 的值为 30');
  //使用 then 注册回调
  fu.then((res){
    print(res);
  });

  //链式调用,可以跟多个 then,注册多个回调
  Future((){
    print("async task");
  }).then((res){
    print("async task complete");
  }).then((res){
    print("async task after");
  });

  print("main stop");
}
```

示例运行结果如下:

```
flutter: main start
flutter: main stop
flutter: Future 的值为 30
flutter: async task
flutter: async task complete
flutter: async task after
```

从示例代码中可以看出,then 方法可以获取异步返回的结果,即可以得到 value 值。示例中还简单地使用了链式调用,即第一个 then 执行完了再执行下一个 then。

除了 then 方法,还可以使用 catchError 来处理异常,示例代码如下:

```dart
//async_future_catch.dart 文件
import 'dart:async';

void main() {
  //then catchError 用法
  Future((){
    print("async task");
  }).then((res){
    print("async task complete");
  }).catchError((e){
    print(e);
  });
}
```

通常 catchError 写在 then 的后面用于捕获异常信息。示例输出内容如下:

```
flutter: async task
flutter: async task complete
```

Future 还可以使用其静态方法 wait 等待多个任务全部完成后回调。示例代码如下:

```dart
//async_future_static_wait.dart 文件
import 'dart:async';

void main() {
  print("main start");

  //任务一
  Future task1 = Future((){
    print("task 1");
    return 1;
  });

  //任务二
  Future task2 = Future((){
    print("task 2");
    return 2;
  });

  //任务三
  Future task3 = Future((){
    print("task 3");
    return 3;
  });

  //使用 wait 方法等待三个任务完成后回调
```

```
  Future future = Future.wait([task1, task2, task3]);
  future.then((responses){
    print(responses);
  });

  print("main stop");
}
```

上面的示例中总共执行了三个任务，Future.wait 方法要等待这三个任务执行完成后才会执行其回调方法。示例运行结果如下：

```
flutter: main start
flutter: main stop
flutter: task 1
flutter: task 2
flutter: task 3
flutter: [1, 2, 3]
```

3. Async 和 Await

最新的 Dart 版本加入了 async 和 await 关键字，有了这两个关键字，我们可以更简洁地编写异步代码，而不需要调用 Future 相关的 API。

将 async 关键字作为方法声明的后缀时，具有如下意义：

- 被修饰的方法会将一个 Future 对象作为返回值；
- 该方法会同步执行其中的方法代码直到第一个 await 关键字，然后它暂停该方法其他部分的执行；
- 一旦由 await 关键字引用的 Future 任务执行完成，await 的下一行代码将立即执行。

下面的示例演示了这两个关键字的用法：

```
//async_async_wait.dart 文件
import 'dart:io';

//模拟耗时操作,调用 sleep 方法睡眠 2 秒
doTask() async{
  //等待其执行完成,耗时 2 秒
  await sleep(const Duration(seconds:2));
  return "执行了耗时操作";
}

//定义一个方法用于包装
test() async {
  //添加 await 关键字,等待异步处理
  var r = await doTask();
  //必须等待 await 关键字后面的方法 doTask 执行完成,才执行下一行代码
```

```
    print(r);
  }
void main(){
  print("main start");
  test();
  print("main end");
}
```

示例运行结果如下：

```
flutter: main start
flutter: main end
flutter: 执行了耗时操作
```

注意 async 不是并行执行，它遵循 Dart 事件循环规则来执行，并且它仅仅是一个语法糖，简化 Future API 的使用。

18.2.5 异步处理实例

在 Flutter 和纯 Dart 库中运用异步处理的情况很多，本节将以 Http 网络请求和 Flutter 列表上下拉刷新数据为例来综合运用异步处理的各种方法。

1．网络请求

在实际的项目中运用得最多的异步处理就是 Http 网络请求处理了。前端发起网络请求，需要等待服务端返回数据后才能进行下一步的处理。这里通过一个示例来综合运用 async、await，以及 Future 返回值的处理。具体步骤如下：

步骤 1：打开 pubspec.yaml 文件，添加 dio 网络请求库。具体代码如下：

```
dev_dependencies:
  flutter_test:
    sdk: flutter

  dio: ^2.0.7
```

步骤 2：导入 Dio 库，然后编写网络请求方法。方法名后面需要添加 async 关键字，表示此方法为一个异步处理的方法。其中，Dio 对象的 post 方法会向服务端发起一个 post 请求，这会有一个等待的过程，所以需要在方法前加一个 await 关键字。最后将 response 对象返回，同时方法的返回类型要定义成 Future。大致处理代码如下：

```
Future getAsyncData(url,{params}) async {
  try{
```

```
    //返回对象
    Response response;
    //实例化 Dio 对象
    Dio dio = Dio();
    //...
    response = await dio.post(url,data: params);
    //...
    return response;
  }catch(e){
    return print('error:::${e}');
  }
}
```

步骤 3：调用 getAsyncData 方法，传入请求路径 url 及请求参数 params。当 post 请求完成后根据返回的 Future 对象的 then 方法可以获取服务端返回的数据。

```
//设置请求参数
//调用 getAsyncData
Future future = getAsyncData(url,params);
future.then((value){
  //数据处理
});
```

步骤 4：启动后端 Node 测试程序，进入 dart_node_server 程序，执行"npm start"命令启动程序。Flutter 端完整的示例代码如下：

```
//async_get_async_data.dart 文件
import 'package:dio/dio.dart';
import 'dart:io';
import 'dart:async';

void main(){
  //网络请求参数
  var params = {'id':'000001'};
  //调用网络请求方法
  Future future = getAsyncData('http://192.168.2.168:3000/getAsyncData',params: params);
  //使用 Future 的 then 方法取得返回数据
  future.then((value){
    //value 即为服务端返回数据
    print(value);
  });

}

//方法后面添加 async 表示异步方法，返回值为 Future
Future getAsyncData(url,{params}) async {
```

```
    //添加 try…catch 捕获网络请求异常
    try{
      //返回对象
      Response response;
      //实例化 Dio 对象
      Dio dio = Dio();
      //设置 post 请求编码格式为 application/x-www-form-urlencoded
      dio.options.contentType = ContentType.parse('application/x-www-form-urlencoded');
      //使用 dio 发起 post 请求,使用 await 关键等待返回结果
      if(params == null){
        response = await dio.post(url);
      }else{
        response = await dio.post(url,data: params);
      }
      //当返回状态为 200 时,表示请求正常返回
      if(response.statusCode == 200){
        //返回 response 对象
        return response;
      }else{
        throw Exception('Server exception...');
      }
    }catch(e){
      return print('error:::${e}');
    }
}
```

示例正常请求后,输出结果如下:

```
flutter: {"code":"0","message":"success","data":[{"name":"张三","age":20},{"name":"李四","age":30},{"name":"王五","age":28}]}
```

提示 示例中使用 Dio 的 post 请求可以作为实际项目中的基础代码,还需要对返回的数据做进一步的处理,如把 Json 数据转换成数据模型 Model。请求的 IP 和端口 Port 根据实际计算机的配置进行修改。

2. RefreshIndicator 刷新数据

App 应用的列表数据往往很多,需要下拉刷新数据及上拉加载更多数据。Flutter 的刷新控件 RefreshIndicator 的 onRefresh 回调方法需要异步处理。

首先看看 RefreshIndicator 的属性,onRefresh 即为其刷新回调方法,代码如下:

```
RefreshIndicator(
    //刷新回调方法
    onRefresh,
    //刷新组件包裹的组件,通常为列表组件
```

```
      child,
),
```

onRefresh 方法的类型为 RefreshCallback，查看 Flutter 源码可以看到其类型定义如下：

```
typedef RefreshCallback = Future<void> Function();
```

这里 RefreshCallback 为一个异步处理的方法，所以 onRefresh 必须按照这个格式来进行处理。

接下来我们看一个列表下拉刷新和上拉加载更多数据的示例。完整示例代码如下：

```
//async_list_refresh.dart 文件
import 'package:flutter/material.dart';
import 'dart:async';

void main() => runApp(MyApp());

class MyApp extends StatelessWidget {

  @override
  Widget build(BuildContext context) {
    return MaterialApp(
      home: Scaffold(
        appBar: AppBar(
          title: Text('RefreshIndicator 示例'),
        ),
        body: DropDownRefresh(),
      ),
    );
  }
}

//创建一个有状态的组件
class DropDownRefresh extends StatefulWidget {
  @override
  _DropDownRefreshState createState() => _DropDownRefreshState();
}

class _DropDownRefreshState extends State<DropDownRefresh> {
  //列表要展示的数据
  List list = List();
  //ListView 的控制器
  ScrollController scrollController = ScrollController();
  //页数
  int page = 0;
```

```dart
//是否正在加载
bool isLoading = false;

@override
void initState() {
  super.initState();
  //初始化列表数据
  initData();
  //添加滚动监听事件
  scrollController.addListener(() {
    if (scrollController.position.pixels == scrollController.position.maxScrollExtent) {
      print('滑动到了最底部');
      //上拉加载更多数据
      getMoreData();
    }
  });
}
//初始化列表数据,加延时模仿网络请求
Future initData() async {
  //使用 Future.delayed 延迟一秒执行
  await Future.delayed(Duration(seconds: 1),(){
    //设置状态渲染列表
    setState(() {
      //初始 15 条数据
      list = List.generate(15, (i) => '初始数据 $i');
    });
  });
}

//下拉刷新方法,为 list 重新赋值
Future onRefreshData() async {
  await Future.delayed(Duration(seconds: 1), (){
    //设置状态渲染列表
    setState(() {
      //重新生成 20 条数据
      list = List.generate(20, (i) => '刷新后的数据 $i');
    });
  });
}

//根据 index 渲染某一行数据
Widget renderListItem(BuildContext context, int index){
  //当 index 显示 list.lenth 时显示列表项
  if (index < list.length) {
    return ListTile(
      title: Text(list[index]),
    );
  }
  //当索引大于等于 list.length 时,显示加载更多数据组件
```

```dart
    return showGetMoreWidget();
}

//加载更多数据时显示的组件,给用户提示
Widget showGetMoreWidget() {
  //居中显示'加载中...'
  return Center(
    child: Padding(
      padding: EdgeInsets.all(10.0),
      child: Row(
        mainAxisAlignment: MainAxisAlignment.center,
        crossAxisAlignment: CrossAxisAlignment.center,
        children: [
          Text(
            '加载中...',
            style: TextStyle(fontSize: 16.0),
          ),
          //圆形刷新提示组件
          CircularProgressIndicator(
            strokeWidth: 1.0,
          )
        ],
      ),
    ),
  );
}

//上拉加载更多
Future getMoreData() async {
  if (!isLoading) {
    setState(() {
      isLoading = true;
    });
    //延迟一秒生成更多数据
    await Future.delayed(Duration(seconds: 1),(){
      //设置状态渲染列表
      setState(() {
        //每上拉一次,重新生成5条数据,添加至现有列表
        list.addAll(List.generate(5, (i) => '第 $page 次上拉来的数据'));
        //当前页自增
        page++;
        isLoading = false;
      });
    });
  }
}

@override
Widget build(BuildContext context) {
```

```
      return Scaffold(
        appBar: AppBar(
          //标题
          title: Text(
            '下拉刷新 上拉加载更多',
            style: TextStyle(
              color: Colors.black,
              fontSize: 18.0,
            ),
          ),
          //标题居中
          centerTitle: true,
          //取消默认阴影
          elevation: 0,
          backgroundColor: Color(0xffEDEDED),
        ),
        //刷新组件
        body: RefreshIndicator(
          //刷新回调方法
          onRefresh: onRefreshData,
          //构建列表
          child: ListView.builder(
            //列表项渲染
            itemBuilder: renderListItem,
            //列表项个数
            itemCount: list.length + 1,
            controller: scrollController,
          ),
        ),
      );
    }

    @override
    void dispose() {
      super.dispose();
      scrollController.dispose();
    }
  }
```

当页面第一次打开时,用 initState 方法初始化状态,这里会调用 initData 方法初始化列表数据。处理过程如下:

```
  Future initData() async {
    //使用 Future.delayed 延迟一秒执行
    await Future.delayed(Duration(seconds: 1),(){
      //设置状态渲染列表
      setState(() {
```

```
      //初始15条数据
      list = List.generate(15, (i) => '初始数据 $i');
    });
  });
}
```

这里可以看到此方法为一个标准的异步处理方法,在方法体里使用 Future.delayed 延迟执行了一段代码,这段代码会生成 15 条数据用于初始化列表数据,设置完状态后进行渲染。

列表初始化数据完成后的效果如图 18-5 所示。

图 18-5 列表初始状态

接着我们再看看下拉刷新的处理。这里需要给 RefreshIndicator 的 onRefresh 属性设置一个回调处理方法,方法名为 onRefreshData,处理代码如下:

```
Future onRefreshData() async {
  await Future.delayed(Duration(seconds: 1), (){
    //设置状态渲染列表
    setState(() {
      //重新生成20条数据
      list = List.generate(20, (i) => '刷新后的数据 $i');
```

```
      });
    });
}
```

这里可以看到此方法同样为一个标准的异步处理方法，在方法体里使用 Future.delayed 延迟执行了一段代码，这段代码会生成 20 条数据用来重新生成列表数据，设置完状态后列表重新渲染。

列表下拉刷新后的效果如图 18-6 所示。

图 18-6 下拉刷新列表数据

我们再看看加载更多数据时的处理过程。用户在浏览数据时会不断地向上滑动列表，当滑动到底部时会加载更多数据。那么列表的最底部就是一个临界点，判断临界点的代码如下：

```
scrollController.addListener(() {
  if (scrollController.position.pixels == scrollController.position.maxScrollExtent) {
    print('滑动到了最底部');
    //上拉加载更多数据
    getMoreData();
  }
});
```

上拉加载的实现和下拉刷新的处理大同小异,不同之处在于,下拉刷新是重置列表数据,而上拉加载是向列表里追加一级数据。这部分的处理代码如下:

```
Future getMoreData() async {
  if (!isLoading) {
    setState(() {
      isLoading = true;
    });
    //延迟一秒生成更多数据
    await Future.delayed(Duration(seconds: 1),(){
      //设置状态渲染列表
      setState(() {
        //每上拉一次重新生成 5 条数据,添加至现有列表里
        list.addAll(List.generate(5, (i) => '第 $ page 次上拉来的数据'));
        //当前页自增
        page++;
        isLoading = false;
      });
    });
  }
}
```

当上拉两次后列表渲染的效果如图 18-7 所示。

图 18-7　上拉加载列表数据

> **提示** 阅读此示例之前请先补充一些 Flutter 组件基础知识、状态知识，以及列表渲染机制相关知识等。在实际项目中只需要将模拟数据换成网络请求返回的数据即可。

18.3 Stream

Stream 和 Future 都是 Dart 中异步编程的核心内容，在 18.2 节中已经详细叙述了关于 Future 的知识，本节主要介绍 Stream 相关的知识。

18.3.1 Stream 概念

Stream 是 Dart 语言中所谓异步数据序列的东西，简单理解，其实就是一个异步数据队列而已。我们知道队列的特点是先进先出的，Stream 也正是如此。

为了将 Stream 的概念可视化与简单化，可以将它想成管道 Pipe 的两端，它只允许从一端插入数据并通过管道从另外一端流出数据。

关于 Stream 相关的概念及流程如图 18-8 所示。总结有以下几点：

- 我们将这样的管道称作 Stream；
- 为了控制 Stream，我们通常可以使用 StreamController 来进行管理；
- 为了向 Stream 中插入数据，StreamController 提供了类型为 StreamSink 的属性 sink 作为入口；
- StreamController 提供 stream 属性作为数据的出口；
- StreamController.stream.listen 用来监听 Stream 上是否有数据。

图 18-8　Stream 流程

在我们刚开始学习 Flutter 的时候基本使用 StatefulWidget 和 setState((){})来刷新界面的数据，当我们熟练使用流之后就可以告别 StatefulWidget 而使用 StatelessWidget 同样达到数据刷新的效果。

> **提示** StatelessWidget 和 StatefulWidget 是 Flutter 中使用非常频繁的组件。其中 StatelessWidget 是无状态组件，而 StatefulWidget 是有状态组件。

接下来需要理解 Stream 中的数据。数据（data）是个非常抽象的概念，可以认为一切皆数据。在程序的世界里，其实只有两种东西：数据和对数据的操作。对数据的操作就是对输入的数据经过一些计算之后输出一些新数据。事件(event，如 UI 上的事件)、计算结果(value，如函数/方法的返回值)，以及从文件或网络获得的纯数据都可以认为是数据(data)。另外，Dart 中的所有事物都是对象，所以数据也一定是某种对象(object)。在本文中，可以认为事件、结果、数据和对象都是一样的，不用特意区分。

最后再看一看 Stream 和 Future 的区别。Future 表示稍后获得的一个数据，所有异步操作的返回值都用 Future 来表示，但是 Future 只能表示一次异步获得的数据，而 Stream 表示多次异步获得的数据。例如界面上的按钮可能会被用户点击多次，所以按钮上的点击事件(onClick)就是一个 Stream。简单地说，Future 将返回一个值，而 Stream 将返回多次值。

另外一点，Stream 是流式处理，比如 IO 处理的时候，一般情况是每次只会读取一部分数据(具体取决于实现)，这和一次性读取整个文件的内容相比，Stream 的好处是处理过程中内存占用较小，而 File 的 readAsString(异步读，返回 Future)或 readAsStringSync(同步读，返回 String)等方法都是一次性读取整个文件的内容，虽然获得完整内容处理起来比较方便，但是如果文件很大的话就会导致内存占用过大的问题。

18.3.2　Stream 分类

Stream 流可以分为两类：
- 单订阅流(Single Subscription)，这种流最多只能有一个监听器(listener)；
- 多订阅流(Broadcast)，这种流可以有多个监听器监听(listener)。

单订阅就是只能有一个订阅者，而广播是可以有多个订阅者，这就有点类似于消息服务的处理模式。单订阅类似于点对点，在订阅者出现之前会持有数据，在订阅者出现之后就转交给它，而广播类似于发布订阅模式，可以同时有多个订阅者，当有数据时就会传递给所有的订阅者，而不管当前是否已有订阅者存在。

18.3.3　Stream 创建方式

创建一个 Stream 有多个构造方法，其中一个是构造广播流的，这里主要看一下构造单订阅流的方法。

1. Stream < T >.periodic

该方法接收一个 Duration 对象作为参数。示例代码如下：

```
//stream_create_periodic.dart 文件
import 'dart:async';

void main(){
  //创建 Stream
  createStream();
}

createStream() async{
```

```dart
//使用periodic创建流,第一个参数为间隔时间,第二个参数为回调函数
Stream<int> stream = Stream<int>.periodic(Duration(seconds: 1), callBack);
//await for 循环从流中读取
await for(var i in stream){
    print(i);
}
}

//可以在回调函数中对值进行处理,这里直接返回了
int callBack(int value){
    return value;
}
```

打印结果如下:

```
flutter: 0
flutter: 1
flutter: 2
flutter: 3
flutter: 4
flutter: 5
flutter: 6
...
```

该方法从整数 0 开始,在指定的间隔时间内生成一个自然数列,以上设置为每一秒生成一次,callBack 函数用于对生成的整数进行处理,处理后再放入 Stream 中。这里并未处理,直接返回了。要注意,这个流是无限的,它没有任何一个约束条件使之停止,在后面会介绍如何给流设置条件。

2. Stream<T>.fromFuture

该方法从一个 Future 创建 Stream,当 Future 执行完成时,任务就会放入 Stream 中,而后从 Stream 中将任务完成的结果取出。这种用法,很像异步任务队列。示例代码如下:

```dart
//stream_create_from_future.dart 文件
import 'dart:async';

void main(){
    //创建一个 Stream
    createStream();
}

createStream() async{
    print("开始测试");
    //创建一个 Future 对象
    Future<String> future = Future((){
        return "异步任务";
```

```
    });

    //从 Future 创建 Stream
    Stream<String> stream = Stream<String>.fromFuture(future);
    //await for 循环从流中读取
    await for(var s in stream){
      print(s);
    }
    print("结束测试");
}
```

打印结果如下:

```
flutter: 开始测试
flutter: 异步任务
flutter: 结束测试
```

3. Stream<T>.fromFutures

该方法可以从多个 Future 创建 Stream,即将一系列的异步任务放入 Stream 中,每个 Future 按顺序执行,执行完成后将任务放入 Stream。示例代码如下:

```
//stream_create_from_futures.dart 文件
import 'dart:io';

void main(){
  //从多个 Future 创建 Stream
  createStreamFromFutures();
}

createStreamFromFutures() async{
  print("开始测试");

  Future<String> future1 = Future((){
    //模拟耗时 5 秒
    sleep(Duration(seconds:5));
    return "异步任务 1";
  });

  Future<String> future2 = Future((){
    return "异步任务 2";
  });

  Future<String> future3 = Future((){
    return "异步任务 3";
  });

  //将多个 Future 放入一个列表中,将该列表传入
```

```
    Stream<String> stream = Stream<String>.fromFutures([future1,future2,future3]);
    //读取 Stream
    await for(var s in stream){
      print(s);
    }

    print("结束测试");
}
```

打印结果如下:

```
flutter: 开始测试
flutter: 异步任务 1
flutter: 异步任务 2
flutter: 异步任务 3
flutter: 结束测试
```

其中,任务 1 需要执行 5 秒。

4. Stream<T>.fromIterable

该方法从一个集合创建 Stream,下面的示例从一个列表创建 Stream:

```
//stream_create_from_iterable.dart 文件
import 'dart:async';

void main(){
  //从一个集合创建 Stream
  createStream();
}

createStream() async{
  print("开始测试");
  //从集合创建 Stream
  Stream<int> stream = Stream<int>.fromIterable([1,2,3,4,5,6]);
  //读取 Stream
  await for(var s in stream){
    print(s);
  }
  print("结束测试");
}
```

打印结果输出如下:

```
flutter: 开始测试
flutter: 1
flutter: 2
flutter: 3
```

```
flutter: 4
flutter: 5
flutter: 6
flutter: 结束测试
```

18.3.4　Stream 操作方法

Stream 有一些对流的处理方法,本节将通过一些例子详细讲解这些方法的使用。

1. stream < T >. take

我们使用 Stream.periodic 方法创建了一个每隔一秒发送一次事件的无限流,如果我们想指定只发送 10 个事件则用 take 方法。示例代码如下:

```
//stream_take.dart 文件
import 'dart:async';

void main(){
  //创建 Stream
  createStream();
}

void createStream() async{
  //时间间隔为 1 秒
  Duration interval = Duration(seconds: 1);
  //每隔 1 秒发送 1 次的事件流
  Stream< int > stream = Stream.periodic(interval, (data) => data);
  //指定发送事件个数
  stream = stream.take(10);
  //输出 Stream
  await for(int i in stream ){
    print(i);
  }
}
```

这样只会打印出 0~9,不会一直打印数字,输出结果如下:

```
flutter: 0
flutter: 1
flutter: 2
flutter: 3
flutter: 4
flutter: 5
flutter: 6
flutter: 7
flutter: 8
flutter: 9
```

2. stream<T>.takeWhile

上面这种方式我们只制定了发送事件的个数,如果我们也不知道应该发送多少个事件,那么我们可以从返回的结果上限制返回值的个数,上面结果也可以用以下方式实现:

```dart
//stream_take_while.dart 文件
import 'dart:async';

void main(){
  //创建 Stream
  createStream();
}

void createStream() async {
  //时间间隔为 1 秒
  Duration interval = Duration(seconds: 1);
  //每隔 1 秒发送 1 次的事件流
  Stream<int> stream = Stream.periodic(interval, (data) => data);
  //根据返回结果限制返回值的个数
  stream = stream.takeWhile((data) {
    //返回值的限制条件
    return data < 8;
  });
  //输出 Stream
  await for (int i in stream) {
    print(i);
  }
}
```

输出结果如下:

```
flutter: 0
flutter: 1
flutter: 2
flutter: 3
flutter: 4
flutter: 5
flutter: 6
flutter: 7
```

3. stream<T>.skip(int count)

skip 可以指定跳过前面的几个事件,示例代码如下:

```dart
//stream_skip.dart 文件
import 'dart:async';

void main(){
  //创建 Stream,跳过指定个数元素
```

```
    testSkip();
}

void testSkip() async {
  //时间间隔为 1 秒
  Duration interval = Duration(seconds: 1);
  //每隔 1 秒发送 1 次的事件流
  Stream<int> stream = Stream.periodic(interval, (data) => data);
  //指定发送事件次数
  stream = stream.take(10);
  //跳过前两个元素
  stream = stream.skip(2);
  //输出 Stream
  await for (int i in stream) {
    print(i);
  }
}
```

上面的代码首先限制了发送事件的次数,然后又跳过了两个元素。这样便会跳过 0 和 1,输出 2～9,输出结果如下:

```
flutter: 2
flutter: 3
flutter: 4
flutter: 5
flutter: 6
flutter: 7
flutter: 8
flutter: 9
```

4. stream<T>.skipWhile

使用 skipWhile 方法可以指定跳过不发送事件的指定条件,示例代码如下:

```
//stream_skip_while.dart 文件
import 'dart:async';

void main(){
  //创建 Stream,按条件跳过元素
  testSkipWhile();
}

void testSkipWhile() async {
  //时间间隔为 1 秒
  Duration interval = Duration(seconds: 1);
  //每隔 1 秒发送 1 次的事件流
  Stream<int> stream = Stream.periodic(interval, (data) => data);
```

```dart
    //指定发送事件个数
    stream = stream.take(10);
    //根据条件跳过元素,条件为返回值小于 5
    stream = stream.skipWhile((data) => data < 5);
    //输出 Stream
    await for (int i in stream) {
      print(i);
    }
}
```

上面的代码首先限制了发送事件的次数,然后根据返回值的限定条件跳过了前 4 个元素。这样便会跳过 0～4,只输出 5～9,输出结果如下:

```
flutter: 5
flutter: 6
flutter: 7
flutter: 8
flutter: 9
```

5. stream<T>.toList()

此方法将流中所有的数据收集并存放在 List 中,返回 Future<List<T>>对象,toList 方法是一个异步方法,获取结果则需要使用 await 关键字,另外等待 Stream 流结束后一次性返回结果。示例代码如下:

```dart
//stream_to_list.dart 文件
import 'dart:async';

void main(){
  //创建 Stream,将流中的数据放在 List 里
  testToList();
}

void testToList() async {
  //时间间隔为 1 秒
  Duration interval = Duration(seconds: 1);
  //每隔 1 秒发送 1 次的事件流
  Stream<int> stream = Stream.periodic(interval, (data) => data);
  //指定发送事件个数
  stream = stream.take(10);
  //将流中所有的数据收集并存放在 List 中
  List<int> listData = await stream.toList();
  //输出 List 数据
  for(int i in listData){
    print(i);
  }
}
```

示例中 List 里都为 int 数据,输出内容如下:

```
flutter: 0
flutter: 1
flutter: 2
flutter: 3
flutter: 4
flutter: 5
flutter: 6
flutter: 7
flutter: 8
flutter: 9
```

6. stream＜T＞.listen( )

listen 方法是一种特定的可以用于监听数据流的方式,它和 forEach 循环的效果一致,但是返回的是 StreamSubscription＜T＞对象,查看 listen 方法源码如下:

```
StreamSubscription＜T＞ listen(void onData(T event),{Function onError, void onDone(), bool cancelOnError});
```

其中几个参数的作用如下:
- onData 是接收到数据的处理,必须要实现的方法;
- onError 流发生错误时的处理;
- onDone 流读取完成时调取;
- cancelOnError 发生错误时是否立马终止。

listen 方法的示例代码如下:

```
//stream_listen.dart 文件
import 'dart:async';

void main(){
  //创建 Stream,使用 listen 方法监听流
  testListen();
}

void testListen() async {
  //时间间隔为 1 秒
  Duration interval = Duration(seconds: 1);
  //每隔 1 秒发送 1 次的事件流
  Stream＜int＞ stream = Stream.periodic(interval, (data) => data);
  stream = stream.take(10);
  //监听流
  stream.listen((data){
    print(data);
```

```
    },onError:(error){
      print("流发生错误");
    },onDone:(){
      print("流已完成");
    }, cancelOnError: false);
}
```

示例输出结果如下:

```
flutter: 0
flutter: 1
flutter: 2
flutter: 3
flutter: 4
flutter: 5
flutter: 6
flutter: 7
flutter: 8
flutter: 9
flutter: 流已完成
```

可以看到除了值的输出外,还输出了"流已完成",这说明流读取完成时触发 onDone 回调。

7. stream<T>.forEach()

forEach 方法和 listen 方法的操作方式基本一致,也是一种监听流的方式,它只是监听了 onData,下面的示例代码也会输出 0、1、2、3、4。

```
//stream_for_each.dart 文件
import 'dart:async';

void main(){
  //创建 Stream,使用 Stream 的 forEach 迭代输出数据
  testForEach();
}

void testForEach() async {
  //时间间隔为 1 秒
  Duration interval = Duration(seconds: 1);
  //每隔 1 秒发送 1 次事件流
  Stream<int> stream = Stream.periodic(interval, (data) => data);
  stream = stream.take(5);
  //Stream 迭代输出数据
  stream.forEach((data) {
    print(data);
  });
}
```

8. stream<T>.length

lenght 用于获取等待流中所有事件完成之后统计事件的总数量,下面的示例代码会输出 5:

```dart
//stream_length.dart 文件
import 'dart:async';

void main(){
  //创建 Stream,并统计事件的总数量
  testStreamLength();
}

void testStreamLength() async {
  //时间间隔为 1 秒
  Duration interval = Duration(seconds: 1);
  //每隔 1 秒发送 1 次事件流
  Stream<int> stream = Stream.periodic(interval, (data) => data);
  stream = stream.take(5);
  //统计事件的总数量
  var allEvents = await stream.length;
  print(allEvents);
}
```

9. stream<T>.where

where 方法可以在流中添加筛选条件,过滤掉一些不想要的数据,满足条件则返回 true,不满足条件则返回 false。示例代码如下:

```dart
//stream_where.dart 文件
import 'dart:async';

void main(){
  //创建 Stream,并按指定条件筛选出数据
  testWhere();
}

void testWhere() async {
  //时间间隔为 1 秒
  Duration interval = Duration(seconds: 1);
  //每隔 1 秒发送 1 次的事件流
  Stream<int> stream = Stream.periodic(interval, (data) => data);
  //筛选条件为返回值大于 2 的所有数据
  stream = stream.where((data) => data > 2);
  //筛选条件为返回值小于 6 的所有数据
  stream = stream.where((data) => data < 6);
```

```
    //最后取上面两件条件都满足的数据
    await for(int i in stream){
      print(i);
    }
}
```

我们用上面的代码筛选出流中大于 2 而小于 6 的所有数据，这两个条件是与的关系，即两个条件都要满足的数据才行。输出内容如下：

```
flutter: 3
flutter: 4
flutter: 5
```

10. stream<S>.transform

如果我们需要进行一些流的数据转换和控制，需要使用到 transform 方法，方法的参数类型是 StreamTransformer<S,T>，S 表示转换之前的类型，T 表示转换后的输入类型。示例代码如下：

```
//stream_transform.dart 文件
import 'dart:async';

void main(){
  //创建 Stream,并测试转换方法
  testTransform();
}

void testTransform() async {
  //根据集合创建 Stream
  var stream = Stream<int>.fromIterable([123456,322445,112233]);
  //由 int 转换成 String 类型
  var st = StreamTransformer<int, String>.fromHandlers(
    //数据回调方法
    handleData: (int data, sink) {
      if (data == 112233) {
        //添加提示数据
        sink.add("密码输入正确...");
      } else {
        //添加提示数据
        sink.add("密码输入错误...");
      }
    });
  //开始转换便监听数据流
```

```
stream.transform(st).listen((String data) => print(data),onError: (error) => print("发
生错误"));
}
```

示例中程序会接收到三组数字,模拟输入了三次密码,并判断正确的密码,同时输出密码正确和密码错误提示信息。其中,handleData 为数据回调方法,可以获得到原数据,根据原数据进行判断,再向 Stream 中添加提示信息。输出结果如下:

```
flutter: 密码输入错误…
flutter: 密码输入错误…
flutter: 密码输入正确…
```

18.3.5 StreamController 使用

介绍完了 Stream 的基本概念和基本用法,以及上面直接创建流的方式,这对我们开发本身来说用途不是很大,我们在实际的开发过程中,基本使用 StreamController 来创建流。通过源码我们可以知道 Stream 的几种构造方法,最终都是通过 StreamController 进行了包装。

我们知道 Stream 有单订阅流和多订阅流之分,同样 StreamController 也可以分别进行创建。

1. 构建单订阅流

当 StreamController 实例化之后便会创建一个 Stream。下面的示例代码便创建了一个单订阅流,可以进行流的数据监听和添加数据,用完之后关闭流。

```
//stream_single.dart 文件
import 'dart:async';

void main(){
  //StreamController 里面会创建一个 Stream,我们实际操控的是 Stream
  StreamController<String> streamController = StreamController();
  //监听流数据
  streamController.stream.listen((data) => print(data));
  //添加数据
  streamController.sink.add("aaa");
  //添加数据
  streamController.add("bbb");
  //添加数据
  streamController.add("ccc");
  //关闭流
  streamController.close();
}
```

上面代码会输出如下内容：

```
flutter: aaa
flutter: bbb
flutter: ccc
```

如果我们给上面的代码再加一个 listen 会报如下异常，所以单订阅流只能有一个 listen。一般情况下我们使用单订阅流，但我们也可以将单订阅流转成多订阅流。

```
[VERBOSE-2:ui_dart_state.cc(148)] Unhandled Exception: Bad state: Stream has already been listened to.
#0    _StreamController._subscribe (dart:async/stream_controller.dart:668:7)
#1    _ControllerStream._createSubscription (dart:async/stream_controller.dart:818:19)
#2    _StreamImpl.listen (dart:async/stream_impl.dart:472:9)
#3    main (package:helloworld/main.dart:9:27)
#4    _runMainZoned.<anonymous closure>.<anonymous closure> (dart:ui/hooks.dart:199:25)
#5    _rootRun (dart:async/zone.dart:1124:13)
#6    _CustomZone.run (dart:async/zone.dart:1021:19)
#7    _runZoned (dart:async/zone.dart:1516:10)
```

2. 构建多订阅流

构建多监听器的 StreamController 有如下两种方式：

- 直接创建多订阅 Stream；
- 将单订阅流转成多订阅流。

使用 StreamController 的 broadcast 方法可以直接创建多订阅流。示例代码如下：

```dart
//stream_broadcast.dart 文件
import 'dart:async';

void main(){
  //使用 StreamController 的 broadcast 方法可以直接创建多订阅流
  StreamController<String> streamController = StreamController.broadcast();
  //第一次监听
  streamController.stream.listen((data){
    print('第一次的监听数据:' + data);
  },onError: (error){
    print(error.toString());
  });
  //第二次监听
  streamController.stream.listen((data){
      print('第二次的监听数据:' + data);
  });
  //添加数据
  streamController.add("Dart...");
}
```

我们在上面的示例代码中添加了两次监听，最后向流中添加一条数据，这样两次监听的回调方法里均输出了相同的数据。输出内容如下：

```
flutter: 第一次的监听数据:Dart...
flutter: 第二次的监听数据:Dart...
```

将单订阅流转换成多订阅流的方法是使用 stream 的 asBroadcastStream 方法。示例代码如下：

```
//stream_as_broadcast.dart 文件
import 'dart:async';

void main(){
  //实例化 StreamController 对象
  StreamController<String> streamController = StreamController();
  //将单订阅流转换成多订阅流
  Stream stream = streamController.stream.asBroadcastStream();
  //添加第一次监听
  stream.listen((data){
    print('第一次的监听数据:' + data);
  });
  //添加第二次监听
  stream.listen((data){
    print('第二次的监听数据:' + data);
  });
  streamController.sink.add("Dart...");
  //关闭流
  streamController.close();
}
```

上面的示例代码首先生成的是单订阅流，然后使用 asBroadcastStream 方法将其转换成多访问流。后面的代码同样添加了两次监听，最后向流中添加一条数据，这样两次监听的回调方法里均输出了相同的数据。输出内容如下：

```
flutter: 第一次的监听数据:Dart...
flutter: 第二次的监听数据:Dart...
```

注意 在流用完了之后记得关闭，调用 streamController.close() 方法。

3．源码相关

查看 StreamController 源码我们可以看到默认创建一个_SyncStreamController，源码如下：

```
factory StreamController(
    {
    //监听方法
    void onListen(),
    //停止方法
    void onPause(),
    //恢复方法
    void onResume(),
    //取消方法
    onCancel(),
    bool sync: false}) {
//返回异步处理
return sync
    ? new _SyncStreamController<T>(onListen, onPause, onResume, onCancel)
    : new _AsyncStreamController<T>(onListen, onPause, onResume, onCancel);
}
```

我们添加数据既使用了 streamController.sink.add() 方式,也使用了 streamController.add() 方式。实际效果是一样的,查看 sink 的源码,实际上 sink 是对 StreamController 的一种包装,最终都是调取 StreamController.add 方法,代码如下:

```
//StreamSink 包装器
class _StreamSinkWrapper<T> implements StreamSink<T> {
    //StreamController 实例
    final StreamController _target;
    _StreamSinkWrapper(this._target);
    //添加数据,实例调用 StreamController 实例添加数据
    void add(T data) {
        _target.add(data);
    }
    //添加错误信息
    void addError(Object error, [StackTrace stackTrace]) {
        _target.addError(error, stackTrace);
    }
    //关闭
    Future close() => _target.close();
    //添加流,实例调用 StreamController 实例添加流
    Future addStream(Stream<T> source) => _target.addStream(source);
    //完成
    Future get done => _target.done;
}
```

18.3.6　StreamBuilder

前面我们理解了 Stream 的原理及常用方法后,我们怎么结合 Flutter 使用呢? 在 Flutter 里面提供了一个 Widget,名叫 StreamBuilder,StreamBuilder 其实是一直记录着流

中最新的数据,当数据流发生变化时,会自动调用 build 方法重新渲染组件。

查看 Flutter 中 StreamBuilder 的源码,stream 属性需要接受一个流,我们可以传入一个 StreamController 的 Stream。builder 属性为构建器,可以根据流中的数据渲染页面,代码如下:

```
const StreamBuilder({
    Key key,
    this.initialData,
    Stream<T> stream,
    @required this.builder,
}) : assert(builder != null),
        super(key: key, stream: stream);
```

使用 StreamController 结合 StreamBuider 对 Flutter 官方的计数器进行改进,取代 setState 刷新页面,代码如下:

```
//stream_stream_builder.dart 文件
import 'dart:async';
import 'package:flutter/material.dart';

void main() => runApp(MyApp());

class MyApp extends StatelessWidget {
  @override
  Widget build(BuildContext context) {
    return MaterialApp(
      title: 'StreamBuilder 示例',
      home: MyHomePage(),
    );
  }
}

//有状态组件
class MyHomePage extends StatefulWidget {
  @override
  _MyHomePageState createState() => _MyHomePageState();
}

class _MyHomePageState extends State<MyHomePage> {
  //计数器值
  int _count = 0;
  //实例化一个 StreamController 对象
  final StreamController<int> _streamController = StreamController();

  @override
  Widget build(BuildContext context) {
```

```dart
        return Scaffold(
          appBar: AppBar(
            title: Text('StreamBuilder 示例'),
          ),
          body: Container(
            child: Center(
              //StreamBuilder 组件,数据类型为 int
              child: StreamBuilder< int >(
                //指定 stream 属性
                stream: _streamController.stream,
                //构建器,可以通过 AsyncSnapshot 获取流中的数据
                builder: (BuildContext context, AsyncSnapshot snapshot) {
                  //这里不需要_count 值,从流中取出 data 即可
                  return snapshot.data == null
                    ? Text("0",style: TextStyle(fontSize: 36.0))
                    : Text(" ${snapshot.data}", style: TextStyle(fontSize: 36.0),
                  );
                }),
            ),
          ),
          //操作按钮
          floatingActionButton: FloatingActionButton(
              child: const Icon(Icons.add),
              onPressed: (){
                //向 Stream 里添加数据
                _streamController.sink.add(++_count);
              }),
        );
      }

      @override
      void dispose() {
        //当界面销毁时关闭 Stream 流
        _streamController.close();
        super.dispose();
      }
    }
```

上面的代码通过 StreamController.sink.add 方法可以向流里添加数据,StreamBuilder 组件可以获得流中的数据,当不断地点击按钮添加数据,就可以不断地从流中读取数据进行渲染。这里我们可以看到 snapshot.data 取代了计数器状态变量_count 而达到了同样的效果。

示例代码为什么要绕一圈子取得数据,其实这样做主要是为了解耦,可以把事件触发和事件监听放在不同的地方。

示例运行的效果如图 18-9 所示。点击右下角按钮,屏幕中间数据会不断增加。

图 18-9　StreamBuilder 示例

18.3.7　响应式编程

简单来说,响应式编程是使用异步数据流的编程思想。

任何事件(如点击)、值的改变、消息、创建请求,以及任何可能改变的数据都可以被 Stream 传递和触发。

使用了响应式编程编写的应用,具有以下特征:
- 异步性;
- 由 Stream 和 listener 组成主要架构;
- 当应用中某处(事件、数值……)变化时,Stream 会收到这些变化的通知;
- 如果某个监听者监听到 Stream 的订阅,它会做出相应的处理,不管在应用的何处;
- 组件之间弱耦合。

举例来说,如果 Widget 向 Stream 传递数据,Widget 本身并不关心所传递的数据,总结如下:
- 传递后会发生的后续情况;
- 何处会使用该数据;
- 数据的使用者是谁;
- 数据会被如何使用。

Widget 只关心自己的业务逻辑。如此一来,看似应用变得无状态,但它会让应用程序具有以下优点：
- 应用中模块职责单一；
- 易于模拟数据以便测试；
- 方便组件重用；
- 应用易于重构。

18.3.8 Bloc 设计模式

Bloc(Business logic component)这一设计模式最早在 2018 年的 DartConf 大会上被提出，它由以下几个概念组成：
- 业务逻辑由一个或多个 blocs 组成；
- 业务逻辑应该尽量从展示层剥离开来，UI 只关心 UI 层面的问题；
- 使用 Stream 的高级特性，sink 作为输入，stream 作为输出；
- 保持平台独立性；
- 保持环境独立性。

事实上 Bloc 最初的构想是独立于各个平台间（Web、移动端和后端）的代码最大化地复用。设计模式如图 18-10 所示。

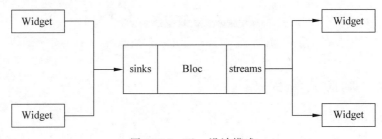

图 18-10　Bloc 设计模式

图中表达的信息有以下几点：
- Widget 通过 sink 向 Bloc 发送事件；
- Widget 通过 stream 接收 Bloc 发送事件；
- 业务逻辑由 Bloc 处理，Widget 并不关心；
- Widget 只处理用户交互与数据展示，Bloc 只处理数据。

使用该模式后，得益于业务逻辑从 UI 上的解耦，优点如下：
- 任何时候都可以改变业务逻辑，但只会对应用造成轻微影响；
- 改变 UI 时并不会对业务逻辑造成干扰；
- 使得业务逻辑易于测试。

18.3.9 Bloc 解耦

Bloc 解耦可以用于界面与逻辑处理分离。这里还是修改 Flutter 官方计数器例子，使用 Bloc 设计思想。具体步骤如下。

步骤 1：在工程的 lib 目录下添加一个 blocs 目录，然后添加一个 BlocBase 基类，里面主要添加一个 dispose 销毁的方法，目的是要求子类必须实现此方法，一般用于流的关闭处理。代码如下：

```
//stream_bloc_base.dart 文件
//定义 BlocBase 基类
abstract class BlocBase {
  //定义销毁方法,子类必须实现此方法
  void dispose();
}
```

步骤 2：在 blocs 目录再建一个 CounterBloc 类，此类继承 BlocBase 基类，所以必须实现 dispose 方法，此方法里添加流的关闭处理。CounterBloc 类主要是实例化 StreamController，提供添加流数据及获取流数据的方法。完整代码如下：

```
//stream_bloc_counter.dart 文件
import 'dart:async';
import 'bloc_base.dart';

//继承 BlocBase
class BlocCounter extends BlocBase {

  //实例化 StreamController,数据类型为 int
  final _controller = StreamController<int>();

  //获取 StreamController 的 sink,即入口可以添加数据
  get _counter => _controller.sink;

  //获取 StreamController 的 stream,即出口可以取数据
  get counter => _controller.stream;

  //增加计算器值
  void increment(int count) {
    //向流中添加数据
    _counter.add(++count);
  }

  //销毁
  void dispose() {
    //关闭流
    _controller.close();
  }
}
```

步骤 3：有了添加数据及获取数据的方法后，就可以添加按钮并通过触发事件向流里添加数据，同时添加流数据的监听来获取数据，然后进行界面渲染。完整的代码如下：

```dart
//stream_bloc_main.dart 文件
import 'package:flutter/material.dart';

import 'blocs/bloc_counter.dart';

void main() => runApp(MyApp());
class MyApp extends StatelessWidget {
  @override
  Widget build(BuildContext context) {
    return MaterialApp(
      title: 'Bloc 示例',
      home: MyHomePage(),
    );
  }
}

class MyHomePage extends StatefulWidget {
  @override
  _MyHomePageState createState() => _MyHomePageState(BlocCounter());
}

class _MyHomePageState extends State<MyHomePage> {

  //组件计数变量
  int _counter = 0;

  //计数器 BlocCounter
  final BlocCounter bloc;

  _MyHomePageState(this.bloc);

  //计数增加方法
  void _incrementCounter() {
    //调用 bloc 的方法
    bloc.increment(_counter);
  }

  @override
  void initState() {
    //监听 Bloc 里的数据
    bloc.counter.listen((_count) {
      //设置状态值
      setState(() {
        _counter = _count;
      });
    });
    super.initState();
  }
```

```
@override
Widget build(BuildContext context) {
  return Scaffold(
    appBar: AppBar(
      title: Text('Bloc 示例'),
    ),
    body: Center(
      child: Text(
        '$_counter',
        style: Theme.of(context).textTheme.display1,
      ),
    ),
    //增加按钮
    floatingActionButton: FloatingActionButton(
      //点击事件
      onPressed: _incrementCounter,
      child: Icon(Icons.add),
    ),
  );
}
```

当这几个类都编写好,一个完整的 Bloc 的应用场景就出来了。可以看到界面和数据处理进行了分离,降低了程序间的耦合。示例运行的效果如图 18-11 所示。

图 18-11　Bloc 示例

18.3.10　BlocProvider 实现

在 Flutter 的应用中往往有很多页面，通常不只有一个 Bloc，可能有数据处理 Bloc，以及事件处理 Bloc 等。由此使用一个工具类来管理这些业务逻辑组件显得尤为必要。

这里我们在上一节例子的基础上实现一个 Bloc 管理的 BlocProvider 组件。其中 BlocBase 和 BlocCounter 代码保持不变，添加 BlocProvider 组件，完整代码如下：

```dart
//stream_bloc_provider.dart 文件
import 'package:flutter/widgets.dart';

import 'bloc_base.dart';

//返回类型
Type _typeOf<T>() => T;

//BlocProvider 是一个有状态的组件，此泛型类型为 BlocBase 的子类
class BlocProvider<T extends BlocBase> extends StatefulWidget {
  BlocProvider({
    Key key,
    @required this.child,
    @required this.blocs,
  }) : super(key: key);

  //定义 child
  final Widget child;
  //定义 blocs
  final List<T> blocs;

  @override
  _BlocProviderState<T> createState() => _BlocProviderState<T>();

  /**
   * BlocProvider 的重要方法 of
   * 此泛型类型为 BlocBase 的子类
   * 返回数据为 blocs 列表
   */
  static List<T> of<T extends BlocBase>(BuildContext context) {
    final type = _typeOf<_BlocProviderInherited<T>>();
    //通过 BuildContext 可以跨组件获取对象
    //ancestorInheritedElementForWidgetOfExactType 方法获得指定类型的
    //InheritedWidget 进而获取它的共享数据.
    _BlocProviderInherited<T> provider =
        context.ancestorInheritedElementForWidgetOfExactType(type)?.widget;
    //返回所有的 blocs
    return provider?.blocs;
  }
}
```

```dart
}

class _BlocProviderState<T extends BlocBase> extends State<BlocProvider<T>> {

  //重写销毁方法
  @override
  void dispose() {
    //关闭所有的 bloc 流
    widget.blocs.map((bloc) {
      bloc.dispose();
    });
    super.dispose();
  }

  @override
  Widget build(BuildContext context){
    return _BlocProviderInherited<T>(
      blocs: widget.blocs,
      child: widget.child,
    );
  }
}

/**
 * InheritedWidget 是 Flutter 的一个功能型的 Widget 基类
 * 它能有效地将数据在当前 Widget 树向它的子 Widget 树传递
 */
class _BlocProviderInherited<T> extends InheritedWidget {
  _BlocProviderInherited({
    Key key,
    @required Widget child,
    @required this.blocs,
  }) : super(key: key, child: child);

  //所有的 bloc
  final List<T> blocs;

  //用来告诉 InheritedWidget,如果对数据进行了修改,
  //是否必须将通知传递给所有子 Widget(已注册/已订阅)
  @override
  bool updateShouldNotify(_BlocProviderInherited oldWidget) => false;
}
```

可以看到此工具类的实现相对复杂。主要需要理解以下几点:
- 由于 bloc 归 page 管理,所以 BlocProvider 设计成一个组件,把所有的 bloc 传入进来;
- 组件需要渲染一个 InheritedWidget 类型的组件,它是 Flutter 的一个功能型的

Widget 基类，它能有效地将数据从当前 Widget 树向它的子 Widget 树传递；
- BlocProvider 是一个有状态的组件，泛型类型为 BlocBase 的子类，这样可以管理所有的 bloc；
- 泛型类型需要指定为 BlocBase 的子类。

该工具类将 Page/Widget 对于 bloc 的管理变得简单，使用前我们只需要将页面和 blocs 相关联，然后调用以下代码即可。

```
final _blocs = BlocProvider.of<SomeBloc>(context);
```

其中，SomeBloc 即你要使用的 bloc，例如 CounterBloc。最后修改上面示例的 main.dart 代码如下：

```
//stream_bloc_provider_main.dart 文件
import 'package:flutter/material.dart';

import 'blocs/bloc_provider.dart';
import 'blocs/bloc_counter.dart';

void main() => runApp(MyApp());

class MyApp extends StatelessWidget {
  @override
  Widget build(BuildContext context) {
    return MaterialApp(
      title: 'BlocProvider 示例',
      //将 BlocProvider 放入顶层组件
      home: BlocProvider(
          //首页
          child: MyHomePage(),
          //所有的 bloc
          blocs: [BlocCounter()]),
    );
  }
}

class MyHomePage extends StatefulWidget {

  @override
  _MyHomePageState createState() => _MyHomePageState();
}

class _MyHomePageState extends State<MyHomePage> {
  //计数器值
  int _counter = 0;

  //增加方法
```

```dart
void _incrementCounter() {
  //通过 BlocProvider 的 of 方法获取所有 bloc
  //然后取第一个 bloc 并调用其 increment 方法向流中添加数据
  BlocProvider.of<BlocCounter>(context).first.increment(_counter);
}

@override
void initState() {
  //通过 BlocProvider 的 of 方法获取所有 bloc
  //然后取第一个 bloc 并调用其 listen 进行监听流的数据
  BlocProvider.of<BlocCounter>(context).first.counter.listen((_count) {
    //设置状态,重新渲染界面
    setState(() {
      _counter = _count;
    });
  });
  super.initState();
}

@override
Widget build(BuildContext context) {
  return Scaffold(
    appBar: AppBar(
      title: Text('BlocProvider 示例'),
    ),
    body: Center(
      //渲染流中取出的数据
      child: Text(
        '$_counter',
        style: Theme.of(context).textTheme.display1,
      ),
    ),
    //增加按钮
    floatingActionButton: FloatingActionButton(
      //点击事件
      onPressed: _incrementCounter,
      child: Icon(Icons.add),
    ),
  );
}
```

可以看到可以使用 BlocProvider.of<BlocCounter>(context) 获取 bloc,然后调用其方法添加数据和监听获取数据。这里不仅可以使用 BlocCounter 还可以使用其他继承自 BlocBase 的 bloc。

注意 这里需要将 BlocProvider 放在顶层,因为只有这样才能将数据向子 Widget 传递。

示例运行后效果如图 18-12 所示。

图 18-12　BlocProvider 示例

提示　本节的示例相当于实现了 Flutter 里一个简化版的状态管理，属于高级知识。读者只需要会使用 BlocProvider 来进行数据的添加和监听即可。

18.4　Isolate

我们通过前面所讲的异步概念知道，将非常耗时的任务添加到事件队列后，仍然会拖慢整个事件循环的处理，甚至是阻塞。可见基于事件循环的异步模型仍然是有很大缺点的，这时候我们就需要 Isolate，这个单词的中文意思是隔离。

简单来说，可以把它理解为 Dart 中的线程，但它又不同于线程，更恰当地说应该是微线程，或者说是协程。它与线程最大的区别就是不能共享内存，因此也不存在锁竞争问题，两个 Isolate 完全是两条独立的执行线，且每个 Isolate 都有自己的事件循环，它们之间只能通过发送消息进行通信，所以它的资源开销低于线程。

18.4.1　创建 Isolate

从主 Isolate 创建一个新的 Isolate 有两种方法，如下：

- spawnUri；
- spawn。

spawnUri 方法有三个必须的参数，第一个是 Uri，指定一个新 Isolate 代码文件的路径；第二个是参数列表，其类型是 List < String >；第三个是动态消息。需要注意，用于运行新 Isolate 的代码文件必须包含一个 main 函数，它是新 Isolate 的入口方法，该 main 函数中的 args 参数列表正对应 spawnUri 中的第二个参数。如不需要向新 Isolate 中传参数，该参数可传空 List。

除了可以使用 spawnUri，还可以使用 spawn 方法来创建新的 Isolate，而 spawn 方法更常用。我们通常希望将新创建的 Isolate 代码和 main Isolate 代码写在同一个文件，并且不希望出现两个 main 函数，而是将指定的耗时函数运行在新的 Isolate，这样做有利于代码的组织和代码的复用。spawn 方法有两个必须的参数，第一个是需要运行在新 Isolate 的耗时函数，第二个是动态消息，该参数通常用于传送主 Isolate 的 SendPort 对象。

首先看一个 ioslate 创建及消息通信的例子。代码如下：

```dart
//isolate_create.dart 文件
import 'dart:isolate';
import 'dart:io';

void main() {

  //主 isolate 启动
  print("main isolate start");

  //创建一个新的 isolate
  create_isolate();

  //主 isolate 停止
  print("main isolate end");

}
//创建一个新的 isolate
void create_isolate() async{

  //发送消息端口
  SendPort sendPort;

  //接收消息端口
  ReceivePort receivePort = ReceivePort();

  //创建一个新的 isolate
  //传入要执行任务方法 doWork
  //传入新 isolate 能够向主 isolate 发送的端口 receivePort.sendPort
```

```dart
    Isolate newIsolate = await Isolate.spawn(doWork, receivePort.sendPort);

    //接收消息端口监听新 isolate 发送过来的消息
    receivePort.listen((message){

      //打印接收到的所有消息
      print("main isolate listen: $message");

      //消息类型为端口
      if (message['type'] == 'port'){
        //将新 isolate 发送过来的端口赋值给 senPort
        sendPort = message['data'];
      }else{
        //当 sendPort 对象实例化后可以向新 isolate 发送消息了
        //消息类型为 message
        //消息数据为字符串
        sendPort?.send({
          'type':'message',
          'data':'main isolate message',
        });
      }
    });

}

//处理耗时任务,接收一个可以向主 isolate 发送消息的端口
void doWork(SendPort sendPort){

  //打印新 isolate 启动
  print("new isolate start");

  //接收消息端口
  ReceivePort receivePort = ReceivePort();

  //接收消息端口监听主 isolate 发送过来的消息
  receivePort.listen((message){
    print("new isolate listen:" + message['data']);
  });

  //将新 isolate 的 sendPort 发送到主 isolate 中用于通信
  sendPort.send({
    'type':'port',
    'data':receivePort.sendPort,
  });

  //模拟耗时 5 秒
  sleep(Duration(seconds:5));
```

```
//发送消息表示任务结束
sendPort.send({
  'type':'message',
  'data':'task finished',
});

//打印新 isolate 停止
print("new isolate end");
}
```

运行结果如下：

```
flutter: main isolate start
flutter: main isolate end
flutter: new isolate start
flutter: main isolate listen: {type: port, data: SendPort}
flutter: new isolate end
flutter: main isolate listen: {type: message, data: task finished}
flutter: new isolate listen:main isolate message
```

整个消息通信过程如图 18-13 所示，两个 Isolate 是通过两对 Port 对象通信，每对 Port 分别由用于接收消息的 ReceivePort 对象和用于发送消息的 SendPort 对象构成。其中 SendPort 对象不用单独创建，它已经包含在 ReceivePort 对象之中。需要注意，每对 Port 对象只能单向发消息，这就如同一根自来水管，ReceivePort 和 SendPort 分别位于水管的两头，水流只能从 SendPort 这头流向 ReceivePort 这头，因此两个 Isolate 之间的消息通信肯定是需要两根这样的水管的，这就需要两对 Port 对象。

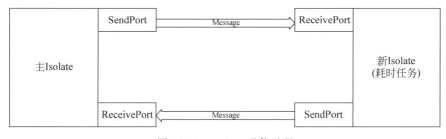

图 18-13　Isolate 通信过程

示例中我们还定义了消息收发的格式，包括消息类型及消息所携带的数据。整个消息体由使用者自己定义。格式如下所示：

```
{
  type: 消息类型
  data: 数据
}
```

通过 spawn 方法运行后会创建两个进程，一个是主 Isolate 进程，一个是新 Isolate 进程，两个进程都双向绑定了消息通信的通道，即使新的 Isolate 中的任务完成了，它的进程也不会立刻退出，因此当使用完自己创建的 Isolate 后，最好调用如下代码将 Isolate 立即杀死：

```
newIsolate.kill(priority: Isolate.immediate);
```

无论如何，在 Dart 中创建一个 Isolate 都显得有些烦琐，可惜的是 Dart 官方并未提供更高级的封装。但是，如果想在 Flutter 中创建 Isolate，则有更简便的 API，这是由 Flutter 官方进一步封装 ReceivePort 而提供的更简洁 API。

使用 compute 函数来创建新的 Isolate 并执行耗时任务，代码如下：

```
//isolate_compute.dart 文件
import 'dart:io';
import 'package:flutter/foundation.dart';

void main() {

  //主 isolate 启动
  print("main isolate start");

  //创建一个新的 isolate
  create_new_task();

  //主 isolate 停止
  print("main isolate end");

}

//创建一个新的耗时任务
create_new_task() async{
  var str = "new task finished";
  var result = await compute(doWork, str);
  print(result);
}

//开始执行
String doWork(String value){

  print("new isolate start");

  //模拟耗时 5 秒
  sleep(Duration(seconds:5));

  print("new isolate end");
  return "complete: $ value";
}
```

示例输出结果如下：

```
flutter: main isolate start
flutter: main isolate end
flutter: new isolate start
flutter: new isolate end
flutter: complete:new task finished
```

compute 函数有两个必须的参数，第一个是待执行的函数，这个函数必须是一个顶级函数，不能是类的实例方法，可以是类的静态方法；第二个参数为动态的消息类型，可以是被运行函数的参数。需要注意，使用 compute 应导入 'package:flutter/foundation.dart' 包。

18.4.2 使用场景

Isolate 虽好但也有合适的使用场景，不建议滥用 Isolate，应尽可能多地使用 Dart 中的事件循环机制去处理异步任务，这样才能更好地发挥 Dart 语言的优势。

那么应该在什么时候使用 Future，什么时候使用 Isolate 呢？一个最简单的判断方法是根据某些任务所需的平均执行时间来选择：

- 方法执行所需时间在几毫秒或十几毫秒左右的，应使用 Future；
- 如果一个任务需要几百毫秒或之上的，则建议创建单独的 Isolate。

除此之外，还有一些可以参考的场景，如下所示：

- Json 解码；
- 数据加密处理；
- 图像处理：比如剪裁；
- 网络请求：加载资源、图片。

最后再阐述一下，为什么将 Isolate 设计成内存隔离的形式？

目前移动端页面(包含 Android、iOS、Web)构建的特性包括树形结构构建布局、布局解析抽象、绘制、渲染，这一系列的复杂步骤导致必须在同一个线程完成(除了单独的渲染线程)所有任务，因为多线程操作页面 UI 元素会有并发的问题，有并发就必须要加锁，加锁就会降低执行效率，所以强制在同一线程中操作 UI 是最好的选择。

除此之外，每当有页面交互时，必定会引起布局变化而需重新绘制，这个过程会有频繁的大量的 UI 控件的创建和销毁，这就涉及耗时内存分配和回收。Dart 为了解决这个问题，就为每个 Isolate 分配各自的一块堆内存，并且独自管理此内存。这样的策略使得内存的分配和回收变得简单高效，并且不受其他 Isolate 的影响。

提示　本节属于异步编程的高级知识，在 Flutter 项目的开发过程中通常不需要使用，建议读者理解基本概念即可。

第 19 章　网络编程

自从互联网诞生以来，现在基本上所有的程序都是网络程序，很少有单机版的程序了。其中，HTTP 协议通常用于做前后端的数据交互。本章将围绕以下几个方面来阐述 Dart 的网络编程技术。

- Http 网络请求；
- HttpClient 网络请求；
- Dio 网络请求；
- Dio 文件上传；
- WebSocket。

运行本章的示例，首先需要启动后端 Node 测试程序，进入 dart_node_server 程序，执行"npm start"命令启动程序。Node 程序需要本机安装 Node 环境，Node 可到如下网址 https://nodejs.org/zh-cn/下载。后端 Node 测试程序参看本书随书源码。

19.1　Http 网络请求

在使用 Http 方式请求网络时，首先需要在 pubspec.yaml 里加入 http 库，然后在示例程序里导入 http 包。如下所示：

```
import 'package:http/http.dart' as http;
```

参看下面的完整示例代码，示例中发起了一个 http 的 get 请求，并将返回的结果信息打印到控制台里：

```
//http_sample/main.dart 文件
import 'package:flutter/material.dart';
import 'package:http/http.dart' as http;

void main() => runApp(MyApp());

class MyApp extends StatelessWidget {
```

```
@override
Widget build(BuildContext context) {
  return MaterialApp(
    title: 'http 请求示例',
    home: Scaffold(
      appBar: AppBar(
        title: Text('http 请求示例'),
      ),
      body: Center(
        child: RaisedButton(
          onPressed: () {
            //请求后台 url 路径(IP + PORT + 请求接口)
            var url = 'http://127.0.0.1:3000/getHttpData';
            //向后台发起 get 请求,response 为返回对象
            http.get(url).then((response) {
              print("状态: ${response.statusCode}");
              print("正文: ${response.body}");
            });
          },
          child: Text('发起 http 请求'),
        ),
      ),
    ),
  );
}
```

请求界面如图 19-1 所示。

图 19-1　Http 请求示例效果图

单击"发起 http 请求"按钮，程序开始请求指定的 url，如果服务器正常返回数据，则状态码为 200。控制台输出内容如下：

```
flutter:状态: 200
flutter:正文: {"code":"0","message":"success","data":[{"name":"张三"},{"name":"李四"},{"name":"王五"}]}
```

注意　服务器返回状态 200，同时返回正文。正文为后台返回的 json 数据。后端测试程序由 Node 编写，确保本地环境安装有 Node 即可。

19.2　HttpClient 网络请求

在使用 HttpClient 方式请求网络时，需要导入 io 及 convert 包，代码如下：

```
import 'dart:convert';
import 'dart:io';
```

参看下面的完整示例代码，示例中使用 HttpClient 请求了一条天气数据，并将返回的结果信息打印到控制台里。具体请求步骤看代码注释即可：

```
//http_client_sample/main.dart 文件
import 'package:flutter/material.dart';
import 'dart:convert';
import 'dart:io';

void main() => runApp(MyApp());

class MyApp extends StatelessWidget {

  //获取数据，此方法需要异步执行 async/await
  void getHttpClientData() async {
    try {
      //实例化一个 HttpClient 对象
      HttpClient httpClient = HttpClient();

      //发起请求 (IP + PORT + 请求接口)
      HttpClientRequest request = await httpClient.getUrl(
          Uri.parse("http://127.0.0.1:3000/getHttpClientData"));

      //等待服务器返回数据
      HttpClientResponse response = await request.close();

      //使用 utf8.decoder 从 response 里解析数据
```

```
      var result = await response.transform(utf8.decoder).join();
      //输出响应头
      print(result);

      //httpClient 关闭
      httpClient.close();

    } catch (e) {
      print("请求失败: $e");
    } finally {

    }
  }

  @override
  Widget build(BuildContext context) {
    return MaterialApp(
      title: 'HttpClient 请求',
      home: Scaffold(
        appBar: AppBar(
          title: Text('HttpClient 请求'),
        ),
        body: Center(
          child: RaisedButton(
            child: Text("发起 HttpClient 请求"),
            onPressed: getHttpClientData,
          ),
        ),
      ),
    );
  }
}
```

请求界面如图 19-2 所示。

单击"发起 HttpClient 请求"按钮，程序开始请求指定的 url，如果服务器正常返回数据，则状态码为 200。控制台输出内容如下：

```
flutter 200
flutter: {"code":"0","message":"success","data":[{"name":"张三","sex":"男","age":"20"},
{"name":"李四","sex":"男","age":"30"},{"name":"王五","sex":"男","age":"28"}]}
```

注意 返回的数据是 Json 格式，所以后续还需要做 Json 处理。另外还需要使用 utf8.decoder 从 response 里解析数据。

图 19-2　HttpClient 请求示例效果图

19.3　Dio 网络请求

Dio 是一个强大的 Dart Http 请求库，支持 Restful API、FormData、拦截器、请求取消、Cookie 管理、文件上传/下载、超时和自定义适配器等。

接下来是一个获取商品列表数据的示例，使用 Dio 向后台发起 Post 请求，同时传入店铺 Id 参数，服务端接收参数并返回商品列表详细数据，前端接收并解析 Json 数据，然后将 Json 数据转换成数据模型，最后使用列表渲染数据。具体步骤如下：

步骤 1：打开 pubspec.yaml 文件，添加 Dio(dio：^2.0.7)库。

步骤 2：在工程 lib 目录创建如下目录及文件：

```
├── main.dart//主程序
├── model//数据模型层
│    └── good_list_model.dart//商品列表模型
├── pages//视图层
│    └── good_list_page.dart//商品列表页面
└── service//服务层
     └── http_service.dart//http 请求服务
```

步骤 3：打开 main.dart 文件，编写应用入口程序，在 Scaffold 的 body 里添加商品列表页面组件 GoodListPage。代码如下：

```dart
//dio_sample/lib/main.dart 文件
import 'package:flutter/material.dart';
import 'pages/good_list_page.dart';

void main() => runApp(MyApp());

class MyApp extends StatelessWidget {
  @override
  Widget build(BuildContext context) {
    return MaterialApp(
      title: 'Dio 请求',
      home: Scaffold(
        appBar: AppBar(
          title: Text('Dio 请求'),
        ),
        body: GoodListPage(),
      ),
    );
  }
}
```

步骤 4：打开 http_service.dart 文件，添加 request 方法。方法传入 url 及请求参数，创建 Dio 对象，调用其 post 方法发起 Post 请求。请求返回对象为 Response，根据其状态码判断是否返回成功，statusCode 为 200 表示数据返回成功。代码如下：

```dart
//dio_sample/lib/service/http_service.dart 文件
import 'dart:io';
import 'package:dio/dio.dart';
import 'dart:async';

//Dio 请求方法封装
Future request(url, {formData}) async {
  try {
    Response response;
    Dio dio = Dio();
    dio.options.contentType = ContentType.parse('application/x-www-form-urlencoded');

    //发起 Post 请求，传入 url 及表单参数
    response = await dio.post(url, data: formData);
    //成功返回
    if (response.statusCode == 200) {
      return response;
    } else {
```

```
        throw Exception('后端接口异常,请检查测试代码和服务器运行情况...');
      }
    } catch (e) {
      return print('error:::${e}');
    }
  }
}
```

步骤5：打开good_list_model.dart文件编写商品列表数据模型,数据模型字段是根据前后端协商定义的。数据模型类里主要完成了由Josn转换成Model及由Model转换成Json两个功能。代码如下：

```
//dio_sample/lib/model/good_list_model.dart 文件
//商品列表数据模型
class GoodListModel{
  //状态码
  String code;
  //状态信息
  String message;
  //商品列表数据
  List<GoodModel> data;

  //构造方法,初始化时传入空数组[]即可
  GoodListModel(this.data);

  //通过传入Json数据转换成数据模型
  GoodListModel.fromJson(Map<String,dynamic> json){
    code = json['code'];
    message = json['message'];
    if(json['data'] != null){
      data = List<GoodModel>();
      //循环迭代Json数据并将其每一项数据转换成GoodModel
      json['data'].forEach((v){
        data.add(GoodModel.fromJson(v));
      });
    }
  }

  //将数据模型转换成Json
  Map<String,dynamic> toJson(){
    final Map<String,dynamic> data = Map<String,dynamic>();
    data['code'] = this.code;
    data['message'] = this.message;
    if(this.data != null){
      data['data'] = this.data.map((v) => v.toJson()).toList();
    }
    return data;
  }
```

```
}

//商品信息模型
class GoodModel{
  //商品图片
  String image;
  //原价
  int oriPrice;
  //现有价格
  int presentPrice;
  //商品名称
  String name;
  //商品 Id
  String goodsId;

  //构造方法
  GoodModel({this.image,this.oriPrice,this.presentPrice,this.name,this.goodsId});

  //通过传入 Json 数据转换成数据模型
  GoodModel.fromJson(Map<String,dynamic> json){
    image = json['image'];
    oriPrice = json['oriPrice'];
    presentPrice = json['presentPrice'];
    name = json['name'];
    goodsId = json['goodsId'];
  }

  //将数据模型转换成 Json
  Map<String,dynamic> toJson(){
    final Map<String,dynamic> data = new Map<String,dynamic>();
    data['image'] = this.image;
    data['oriPrice'] = this.oriPrice;
    data['presentPrice'] = this.presentPrice;
    data['name'] = this.name;
    data['goodsId'] = this.goodsId;
    return data;
  }
}
```

注意 数据模型中的字段一定要和后端返回的字段一一对应，否则会导致数据转换失败。

步骤 6：编写商品列表界面。打开 good_list_page.dart 文件，添加 GoodListPage 组件，此组件需要继承 StatefulWidget 有状态组件。在 initState 初始化状态方法里添加请求商品数据方法 getGoods，在 getGoods 方法里调用 request 方法，传入 url 及店铺 Id 参数。接着发起 Post 请求，后端返回 Json 数据，然后使用 GoodListModel.fromJson 方法将 Json 数据转换成数据模型，此时表示数据获取并转换成功。接下来一定要设置当前商品列表状态值以完成界面的刷新处理。最后在界面里添加 List 组件完成数据的渲染功能。处理细节参看如下代码：

```dart
//dio_sample/lib/pages/good_list_page.dart 文件
import 'package:flutter/material.dart';
import 'dart:convert';
import '../model/good_list_model.dart';
import '../service/http_service.dart';

//商品列表页面
class GoodListPage extends StatefulWidget {
  _GoodListPageState createState() => _GoodListPageState();
}

class _GoodListPageState extends State<GoodListPage> {
  //初始化数据模型
  GoodListModel goodsList = GoodListModel([]);
  //滚动控制
  var scrollController = ScrollController();

  @override
  void initState() {
    super.initState();
    //获取商品数据
    getGoods();
  }

  //获取商品数据
  void getGoods() async {
    //请求 url
    var url = 'http://127.0.0.1:3000/getDioData';
    //请求参数,店铺 Id
    var formData = {'shopId': '001'};

    //调用请求方法,传入 url 及表单数据
    await request(url, formData: formData).then((value) {
      //返回数据并进行 Json 解码
      var data = json.decode(value.toString());
      //打印数据
      print('商品列表数据 Json 格式:::' + data.toString());

      //设置状态刷新数据
      setState(() {
        //将返回的 Json 数据转换成 Model
        goodsList = GoodListModel.fromJson(data);
```

```
      });
    });
}

//商品列表项
Widget _ListWidget(List newList, int index) {
  return Container(
     padding: EdgeInsets.only(top: 5.0, bottom: 5.0),
     decoration: BoxDecoration(
        color: Colors.white,
        border: Border(
           bottom: BorderSide(width: 1.0, color: Colors.black12),
        )),
     //水平方向布局
     child: Row(
        children: <Widget>[
           //返回商品图片
           _goodsImage(newList, index),
           SizedBox(
              width: 10,
           ),
           //右侧使用垂直布局
           Column(
              children: <Widget>[
                 _goodsName(newList, index),
                 _goodsPrice(newList, index),
              ],
           ),
        ],
     ),
  );
}

//商品图片
Widget _goodsImage(List newList, int index) {
  return Container(
     width: 150,
     height: 150,
     child: Image.network(newList[index].image,fit: BoxFit.fitWidth,),
  );
}

//商品名称
Widget _goodsName(List newList, int index) {
  return Container(
     padding: EdgeInsets.all(5.0),
     width: 200,
     child: Text(
```

```dart
            newList[index].name,
          maxLines: 2,
          overflow: TextOverflow.ellipsis,
          style: TextStyle(fontSize: 18),
        ),
      );
    }

    //商品价格
    Widget _goodsPrice(List newList, int index) {
      return Container(
        margin: EdgeInsets.only(top: 20.0),
        width: 200,
        child: Row(
          children: <Widget>[
            Text(
              '价格：¥ ${newList[index].presentPrice}',
              style: TextStyle(color: Colors.red),
            ),
            Text(
              '¥ ${newList[index].oriPrice}',
            ),
          ],
        ),
      );
    }

    @override
    Widget build(BuildContext context) {
      //通过商品列表数组长度判断是否有数据
      if(goodsList.data.length > 0){
        return ListView.builder(
          //滚动控制器
          controller: scrollController,
          //列表长度
          itemCount: goodsList.data.length,
          //列表项构造器
          itemBuilder: (context, index) {
            //列表项,传入列表数据及索引
            return _ListWidget(goodsList.data, index);
          },
        );
      }
      //商品列表没有数据时返回空容器
      return Container();
    }
  }
```

注意 数据请求 getGoods 方法需要放在 initState 里执行，getGoods 需要使用异步处理 async/await。返回的 Json 数据转换成数据模型后一定要调用 setState 方法使界面进行刷新处理。在列表渲染之前需要判断商品列表长度是否大于 0。

Dio 获取商品数据界面如图 19-3 所示。

图 19-3　Dio 请求示例效果图

当启动 Dio 请求示例程序，后台返回的数据有商品名称、商品图片路径、商品原价，以及商品市场价等信息，返回数据如下：

flutter：商品列表数据 Json 格式：：：{code: 0, message: success, data: [{name: 苹果 屏幕尺寸: 13.3 英寸 处理器: Intel Core i5 - 8259, image: http://127.0.0.1:3000/images/goods/001/cover.jpg, presentPrice: 13999, goodsId: 001, oriPrice: 15999}, {name: 外星人 alienware 全新 m15 R2 九代酷睿 i7 六核 GTX1660Ti 独显 144Hz 吃鸡游戏笔记本计算机戴尔 DELL15M - R4725, image: http://127.0.0.1:3000/images/goods/002/cover.jpg, presentPrice: 19999, goodsId: 002, oriPrice: 23999}, {name: Dell/戴尔 灵越 15(3568) Ins15E - 3525 独显 i5 游戏本超薄笔记本计算机, image: http://127.0.0.1:3000/images/goods/003/cover.jpg, presentPrice: 6600, goodsId: 003, oriPrice: 8999}, {name: 联想 ThinkPad E480 14 英寸超薄轻薄便携官方旗舰店官网正品 IBM 全新办公用 商务大学生手提笔记本计算机 E470 新款, image: http://127.0.0.1:3000/images/goods/004/cover.jpg, presentPrice: 5699, goodsId: 004, oriPrice: 7800}, {name: 苹果 屏幕尺寸: 13.3 英寸 处理器: Intel Core i<…>

> **注意** 运行此示例,在查看前端控制台信息的同时还需要查看 Node 端控制台输出的信息,后台会打印店铺的 Id。

19.4 Dio 文件上传

在实际项目中经常要用到文件上传。例如上传图片、视频和音频等。文件上传采用的是 HTTP,Dio 库有封装文件上传的方法,隐藏了文件上传处理的细节。

接下来编写一个选择手机相册图片并上传的例子来详细说明 Dio 文件上传的用法。具体步骤如下:

步骤 1:打开 pubspec.yaml 文件,添加 dio 及 image_picker 这两个库。其中 image_picker 可以选择手机相册图片或者选择拍照图片。具体代码如下:

```
dev_dependencies:
  flutter_test:
    sdk: flutter

  image_picker: ^0.6.1+10
  dio: ^2.0.7
```

步骤 2:打开 iOS 目录下的 Info.plist 文件,添加允许 App 访问相册的权限。与之相关的权限有以下三种。

- NSCameraUsageDescription:请允许 App 访问你的相机;
- NSPhotoLibraryAddUsageDescription:请允许 App 保存图片到相册;
- NSPhotoLibraryUsageDescription:请允许 App 访问你的相册。

这三种权限对应的键值如下:

```
<key>NSCameraUsageDescription</key>
<string>请允许 App 访问你的相机</string>
<key>NSPhotoLibraryAddUsageDescription</key>
<string>请允许 App 保存图片到相册</string>
<key>NSPhotoLibraryUsageDescription</key>
<string>请允许 App 访问你的相册</string>
```

此示例里只需要访问相册的权限 NSPhotoLibraryUsageDescription 即可。Info.plist 文件的路径为 /ios/Runner/Info.plist。

> **提示** 使用 image_picker 库时 iOS 需要开启权限,而 Android 不需要。

步骤 3:编写选择相册图片的方法,使用 ImagePicker.pickImage 方法获取一个图片文件。代码如下：

```
//选择相册图片
Future _getImageFromGallery() async {
  //打开相册并选择图片
  var image = await ImagePicker.pickImage(source: ImageSource.gallery);
  //设置状态
  setState(() {
    //图片文件
    _image = image;
  });
}
```

步骤4：使用 Dio 库编写文件上传方法。创建 Form 表单数据 FormData，然后根据服务端提供的上传 url 地址发起 Post 请求。如果上传成功则返回服务器图片地址。处理代码大致如下：

```
//上传图片到服务器
_uploadImage() async {
  //创建 Form 表单数据
  FormData formData = FormData.from({"file": UploadFileInfo(_image, "imageName.png"),
  });
  //发起 Post 请求
  var response = await Dio().post("后端上传url", data: formData);
  //上传成功则返回数据
  if (response.statusCode == 200) {
    //...
  }
}
```

步骤5：编写界面包括：打开相册按钮、选择图片展示、上传图片到服务器按钮，以及服务器图片展示。完整代码如下：

```
//file_upload_sample/lib/main.dart 文件
import 'package:flutter/material.dart';
import 'dart:io';
import 'package:dio/dio.dart';
import 'package:image_picker/image_picker.dart';

void main() {
  runApp(MaterialApp(
    title: '图片上传示例',
    home: MyApp(),
  ));
}

class MyApp extends StatelessWidget {
```

```dart
    @override
    Widget build(BuildContext context) {
      return Scaffold(
        appBar: AppBar(
          title: Text('图片上传示例'),
        ),
        body: ImagePickerPage(),
      );
    }
  }

  //图片选择上传页面
  class ImagePickerPage extends StatefulWidget {
    _ImagePickerPageState createState() => _ImagePickerPageState();
  }

  class _ImagePickerPageState extends State<ImagePickerPage> {
    //记录选择的照片
    File _image;

    //当图片上传成功后,记录当前上传的图片在服务器中的位置
    String _imgServerPath;

    //选择相册图片
    Future _getImageFromGallery() async {
      //打开相册并选择图片
      var image = await ImagePicker.pickImage(source: ImageSource.gallery);
      //设置状态
      setState(() {
        //图片文件
        _image = image;
      });
    }

    //上传图片到服务器
    _uploadImage() async {
      //创建 Form 表单数据
      FormData formData = FormData.from({
        "file": UploadFileInfo(_image, "imageName.png"),
      });
      //发起 Post 请求
      var response = await Dio()
          .post("http://192.168.2.168:3000/uploadImage/", data: formData);
      print(response);
      //上传成功则返回数据
      if (response.statusCode == 200) {
        var data = response.data['data'];
        print(data[0]['path']);
        setState(() {
```

```dart
      //图片上传后的地址
      _imgServerPath = "http://192.168.2.168:3000 ${data[0]['path']}";
      print(_imgServerPath);
    });
  }
}

@override
Widget build(BuildContext context) {
  return Container(
    child: ListView(
      children: <Widget>[
        FlatButton(
          onPressed: () {
            _getImageFromGallery();
          },
          child: Text("打开相册"),
        ),
        SizedBox(height: 10),
        //展示选择的图片
        _image == null
            ? Center(child: Text("没有选择图片"),)
            : Image.file(
                _image,
                fit: BoxFit.cover,
              ),
        SizedBox(height: 10),
        FlatButton(
          onPressed: () {
            _uploadImage();
          },
          child: Text("上传图片到服务器"),
        ),
        SizedBox(height: 10),
        _imgServerPath == null
            ? Center(child: Text("没有上传图片"),)
            : Image.network(_imgServerPath),
      ],
    ),
  );
}
```

运行上述代码后,单击"打开相册"按钮会出现选择的本地图片,然后单击"上传图片到服务器"按钮会出现服务器端上传后的图片。效果如图19-4所示。

图 19-4　图片上传示例图

19.5　WebSocket

　　WebSocket 是 HTML5 开始提供的一种在单个 TCP 连接上进行全双工通信的协议。

　　WebSocket 使得客户端和服务器之间的数据交换变得更加简单，允许服务端主动向客户端推送数据。在 WebSocket API 中，浏览器和服务器只需要完成一次握手，两者之间就可以直接创建持久性的连接，并进行双向数据传输。

　　现在，很多网站为了实现推送技术，所用的技术都是 Ajax 轮询。轮询是在特定的时间间隔（如每 1 秒），由浏览器对服务器发出 Http 请求，然后由服务器返回最新的数据给客户端的浏览器。这种传统的模式带来很明显的缺点，即浏览器需要不断地向服务器发出请求，然而 Http 请求可能包含较长的头部，其中真正有效的数据可能只是很小的一部分，显然这样会浪费很多带宽等资源。

　　HTML5 定义的 WebSocket 协议，能更好地节省服务器资源和带宽，并且能够更实时地进行通信。Ajax 轮询和 WebSockets 两种数据交互如图 19-5 所示。

　　WebSocket 协议本质上是一个基于 TCP 的协议。

　　为了建立一个 WebSocket 连接，客户端浏览器首先要向服务器发起一个 Http 请求，这

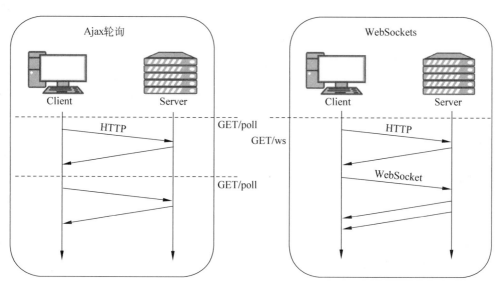

图 19-5　Ajax 轮询和 WebSockets 通信图

个请求和通常的 Http 请求不同，包含了一些附加头信息，其中附加头信息"Upgrade：WebSocket"表明这是一个申请协议升级的 Http 请求，服务器端解析这些附加的头信息，然后产生应答信息并返回给客户端，客户端和服务器端的 WebSocket 连接就建立起来了，双方就可以通过这个连接通道自由地传递信息，并且这个连接会持续存在直到客户端或者服务器端的某一方主动关闭连接。

Dart 有关 WebSocket 相关的 API 既可处理服务端也可以处理客户端。本节会详细说明在 Flutter 中如何使用 WebSocket 进行实时传递数据。

Dart 仓库中的 web_socket_channel 库可以实现 Socket 的连接、创建、收发数据和关闭等处理。其中最主要的类 IOWebSocketChannel 包装了 dart:io 包中的 WebSocket 类，可以直接创建"ws://"或"wss://"开头的连接。常规的操作代码如下：

```
//导入 web_socket_channel 库文件
import 'package:web_socket_channel/io.dart';

main() async {
  //Socket 连接
  var channel = IOWebSocketChannel.connect("ws://localhost:3000");
  //发送消息至服务器
  channel.sink.add("connected!");
  //监听并接收消息
  channel.stream.listen((message) {
    //...
  });
}
```

接下来通过编写一个网络聊天程序来展示如何使用 web_socket_channel。具体步骤如下：

步骤 1：打开 pubspec.yaml 文件添加项目所需要的第三方库。如下：

```
#日期格式化
date_format: ^1.0.6
#WebSocket 处理
web_socket_channel: ^1.0.12
```

其中，date_format 是用来做日期格式化处理的，例如收发消息的时间。

步骤 2：定义聊天程序所需要的变量如下：

- 用户 Id：String 类型，使用随机数产生 6 位数字；
- 用户名：String 类型，使用随机数产生 6 位数字再加"u_"前缀；
- 聊天消息：数组类型用于存储所有聊天消息；
- WebSocket 对象：IOWebSocketChannel 类型，用于创建、关闭 Socket，以及收发数据。

步骤 3：初始化处理，生成一个用户 Id、用户名称，以及调用创建 WebSocket 方法。生成随机数需要导入 random_string.dart 文件。处理代码如下：

```
init() async {
  //使用随机数创建 userId
  userId = randomNumeric(6);
  //使用随机数创建 userName
  userName = "u_" + randomNumeric(6);
  return await createWebsocket();
}
```

步骤 4：创建并连接 Socket 服务器，发送加入房间消息，同时监听服务器返回消息。大致处理代码如下：

```
void createWebsocket() async {
  //连接 Socket 服务器
  channel = await IOWebSocketChannel.connect('ws://192.168.2.168:3000');
  //定义加入房间消息
  //...
  //Json 编码
  //...
  //发送消息至服务器
  channel.sink.add(text);
  //监听到服务器返回消息
  channel.stream.listen((data) => listenMessage(data), onError: onError, onDone: onDone);
}
```

其中加入房间消息定义格式如下：
- type：消息类型 joinRom，表示通知服务器此用户加入聊天室；
- userId：用户 Id，发送用户 Id 至服务器；
- userName：用户名称，发送用户名称至服务器。

消息不能直接发送，需要编码后再进行发送，这里采用的是 json.encode 方法进行编码，服务端需要使用 json.parse 进行解码才可读取数据。编码处理如下：

```
String text = json.encode(message).toString();
```

步骤 5：处理服务器返回的消息，由于服务器转发过来的消息使用了 json.stringify 进行编码操作，所以客户端需要使用 jsonDecode 进行解码处理。处理代码大致如下：

```
//监听服务端返回消息
void listenMessage(data){
  //Json 解码
  var message = jsonDecode(data);
  //接收到消息,判断消息类型为公共聊天 chat_public
  if (message['type'] == 'chat_public'){
    //插入消息至消息列表
    //...
  }
}
```

由上面代码可以看出，前后端约定公共聊天消息类型为'chat_public'，所以这里做了一个消息类型的判断处理。此时便可向消息数组 message 里插入一条消息。

步骤 6：接下来处理发送消息，主要是调用 channel.sink.add 方法进行发送消息。大致处理过程如下：

```
void sendMessage(type,data){
  //定义发送消息对象
  //...
  //Json 编码
  //...
  //发送消息至服务器
  channel.sink.add(text);
}
```

其中加入房间消息定义格式如下：
- type：消息类型 chat_public，表示发送公共聊天消息至服务器；
- userId：用户 Id，发送用户 Id 至服务器；
- userName：用户名称，发送用户名称至服务器；
- msg：消息内容，未经过 json 编码的内容。

同样，这里的消息不能直接发送，需要编码后再进行发送，采用的是 json.encode 方法进行编码。编码处理如下：

```
String text = json.encode(message).toString();
```

步骤 7：当 WebSocket 的创建、连接、发送消息，以及接收消息都编写好以后，就可以编写 Flutter 界面来进行测试了，这里添加了一个输入框及消息聊天页面来展示聊天的过程。完整的代码如下：

```dart
//websocket_sample/lib/main.dart 文件
import 'package:flutter/material.dart';
import 'dart:convert';
import 'package:web_socket_channel/io.dart';
import 'random_string.dart';
import 'package:date_format/date_format.dart';

void main() {
  runApp(MaterialApp(
    title: 'WebSocket 示例',
    home: MyApp(),
  ));
}

class MyApp extends StatelessWidget {
  @override
  Widget build(BuildContext context) {
    return Scaffold(
      appBar: AppBar(
        title: Text('WebSocket 示例'),
      ),
      body: ChatPage(),
    );
  }
}

//聊天页面
class ChatPage extends StatefulWidget {
  _ChatPageState createState() => _ChatPageState();
}

class _ChatPageState extends State<ChatPage> {

  //用户 Id
  var userId = '';
  //用户名称
  var userName = '';
  //聊天消息
```

```dart
var messages = [];
//WebSocket 对象
IOWebSocketChannel channel;
//初始化
init() async {
  //使用随机数创建 userId
  userId = randomNumeric(6);
  //使用随机数创建 userName
  userName = "u_" + randomNumeric(6);
  return await createWebsocket();
}

@override
void initState() {
  super.initState();
  init();
}

@override
void dispose() {
  super.dispose();
  //当页面销毁时关闭 WebSocket
  closeWebSocket();
}

//创建并连接 Socket 服务器
void createWebsocket() async {
  //连接 Socket 服务器
  channel = await IOWebSocketChannel.connect('ws://192.168.2.168:3000');
  //定义加入房间消息
  var message = {
    'type': 'joinRoom',
    'userId': userId,
    'userName': userName,
  };
  //Json 编码
  String text = json.encode(message).toString();
  //发送消息至服务器
  channel.sink.add(text);
  //监听到服务器返回消息
  channel.stream.listen((data) => listenMessage(data),onError: onError,onDone: onDone);
}
//监听服务端返回消息
void listenMessage(data){
  //Json 解码
  var message = jsonDecode(data);
  print("receive message:" + data);
  //接收到消息,判断消息类型为公共聊天 chat_public
  if (message['type'] == 'chat_public'){
```

```dart
      //插入消息至消息列表
      setState(() {
        messages.insert(0, message);
      });
    }

  }
  //发送消息
  void sendMessage(type,data){
    //定义发送消息对象
    var message = {
      //消息类型
      "type": 'chat_public',
      'userId': userId,
      'userName': userName,
      //消息内容
      "msg": data
    };
    //Json 编码
    String text = json.encode(message).toString();
    print("send message:" + text);
    //发送消息至服务器
    channel.sink.add(text);
  }
  //监听消息错误时回调方法
  void onError(error){
    print('error: ${error}');
  }
  //当 WebSocket 断开时回调方法,此处可以做重连处理
  void onDone() {
    print('WebSocket 断开了');
  }
  //前端主动关闭 WebSocket 处理
  void closeWebSocket(){
    //关闭链接
    channel.sink.close();
    print('关闭 WebSocket');
  }

  //发送消息
  void handleSubmit(String text) {
    textEditingController.clear();
    //判断输出框内容是否为空
    if (text.length == 0 || text == '') {
      return;
    }
    //发送公共聊天消息
    sendMessage('chat_public', text);
  }
```

```dart
//文本编辑控制器
final TextEditingController textEditingController = TextEditingController();
//输入框获取焦点
FocusNode textFocusNode = FocusNode();

//创建消息输入框组件
Widget textComposerWidget() {
  return IconTheme(
    data: IconThemeData(color: Colors.blue),
    child: Container(
      margin: const EdgeInsets.symmetric(horizontal: 8.0),
      child: Row(
        children: <Widget>[
          Flexible(
            child: TextField(
              //提示内容:请输入消息
              decoration: InputDecoration.collapsed(hintText: '请输入消息'),
              //文本编辑控制器
              controller: textEditingController,
              //发送消息
              onSubmitted: handleSubmit,
              //获取焦点
              focusNode: textFocusNode,
            ),
          ),
          //发送按钮容器
          Container(
            margin: const EdgeInsets.symmetric(horizontal: 8.0),
            //发送按钮
            child: IconButton(
              icon: Icon(Icons.send),
              //按下发送消息
              onPressed: () => handleSubmit(textEditingController.text),
            ),
          )
        ],
      ),
    ),
  );
}

//根据索引创建一个带动画的消息组件
Widget messageItem(BuildContext context, int index) {
  //获取一条聊天消息
  var item = messages[index];

  return Container(
    margin: const EdgeInsets.symmetric(vertical: 10.0),
    //水平布局,左侧为消息,右侧为头像
```

```dart
      child: Row(
        crossAxisAlignment: CrossAxisAlignment.start,
        children: <Widget>[
          //左侧空余部分
          Expanded(
            child: Container(),
          ),
          //垂直排列,消息时间,消息内容
          Column(
            crossAxisAlignment: CrossAxisAlignment.start,
            children: <Widget>[
              Text(formatDate(DateTime.now(), [HH, ':', nn, ':', ss]), style: Theme.of(context).textTheme.subhead),
              Container(
                margin: const EdgeInsets.only(top: 5.0),
                child: Text(item['msg'].toString()),
              )
            ],
          ),
          //我的头像
          Container(
            margin: const EdgeInsets.only(left: 16.0),
            child: CircleAvatar(
              child: Text(item['userName'].toString()),
            ),
          ),
        ],
      ),
    );
}

@override
Widget build(BuildContext context) {
  //使用安全区域组件防止部分 iOS 设备询问不能正常显示
  return SafeArea(
    //垂直布局
    child: Column(
      children: <Widget>[
        //获取消息列表数据
        Flexible(
          //使用列表渲染消息
          child: ListView.builder(
            padding: EdgeInsets.all(8.0),
            reverse: true,
            //消息组件渲染
            itemBuilder: messageItem,
            //消息条目数
            itemCount: messages.length,
          ),
```

```
      ),
      //分隔线
      Divider(
        height: 1.0,
      ),
      //消息输入框及发送按钮
      Container(
        decoration: BoxDecoration(
          color: Theme.of(context).cardColor,
        ),
        child: textComposerWidget(),
      )
    ],
  ),
);
}
}
```

步骤8：使用"npm start"命令启动 Socket 服务端程序，然后再运行前端示例。当 Socket 正常连接后，Node 端控制会输出如下内容。

```
message.type::joinRoom
message.userId:226748
```

输入内容并单击发送，此时服务端会收到前端发来的消息，输出内容如下：

```
message.type::joinRoom
message.userId:226748
message.type::chat_public
message.type::chat_public
```

客户端会收到服务端转发过来的公共消息，控制台打印出的消息内容如下：

```
flutter: send
message:{"type":"chat_public","userId":"226748","userName":"u_777585","msg":"hello dart"}
flutter: receive
message:{"type":"chat_public","userId":"226748","userName":"u_777585","msg":"hello dart"}
flutter: send
message:{"type":"chat_public","userId":"226748","userName":"u_777585","msg":"hello flutter"}
flutter: receive
message:{"type":"chat_public","userId":"226748","userName":"u_777585","msg":"hello flutter"}
```

提示 你在做程序调试时，可以通过分析前后端输出的 Json 串来分析数据是否正确。

聊天的效果如图 19-6 所示。可以看到聊天信息列表里有用户名、消息时间，以及消息内容。

图 19-6　WebSocket 示例效果图

第 20 章 元 数 据

Dart 提供了类似于 Java 注解一样的机制 Metadata,通过使用 Metadata 可以实现与注解一样的功能,我们称它为元数据。Metadata 可以出现在库、类、typedef、参数类型、构造函数、工厂构造函数、方法、字段、参数或者变量和 import,以及 export 指令前面。可见 Metadata 使用范围之广。

20.1 元数据定义

元数据(Metadata)是描述其他数据的数据(data about other data),或者说它是用于提供某种资源的有关信息的结构数据(structured data)。元数据是描述信息资源或数据等对象的数据,其使用目的在于:识别资源,评价资源,追踪资源在使用过程中的变化,实现简单高效地管理大量网络化数据,实现信息资源的有效发现、查找、一体化组织和对使用资源的有效管理。

元数据是以@开始的修饰符,在@后面接着编译时的常量或调用一个常量构造方法。Flutter 里重写 build 方法的@override 就是使用了元数据。代码如下:

```
@override
Widget build(BuildContext context) {
  return MaterialApp(
    debugShowCheckedModeBanner: false,
    onGenerateRoute: Application.router.generator,
    theme: ThemeData(
      primaryColor: Colors.redAccent,
    ),
  );
}
```

20.2 常用元数据

Dart 内置常用的元数据有以下几个:
- @deprecated 被弃用的;

- @override 重写；
- @proxy 代理；
- @required 参数必传。

20.2.1 @deprecated

@deprecated 表示被弃用的意思，它的含义及作用如下：

- 含义：若某类或某方法加上该注解之后，表示此方法或类不再建议使用，调用时也会出现删除线，但并不代表不能用，只是说不推荐使用，因为还有更好的方法可以调用；
- 作用：因为在一个项目中，如果工程比较大，代码比较多，而在后续开发过程中，可能之前的某个方法实现得并不是很合理，这个时候就要新加一个方法，而之前的方法又不能随便删除，因为可能在别的地方还会调用它，所以加上这个注解，就方便以后开发人员的方法调用了。

接下来看一个应用场景。大家都知道手机可以支持 2G、3G、4G，甚至 5G。因为 2G 网络太慢了，被认为是不推荐的网络，这里我们定义一个手机类，假定 2G 为被弃用的方法，其他网络为推荐的方法。具体用法代码如下：

```
//metadata_deprecated.dart 文件
void main(){
  //实例化手机类
  Mobile mobile = Mobile();
  //2G 网络很慢,不推荐使用此网络
  mobile.netWork2G();
  mobile.netWork3G();
  mobile.netWork4G();
  mobile.netWork5G();
}

//定义手机类
class Mobile {

  //被弃用的方法,也可用但不推荐使用
  @deprecated
  void netWork2G(){
    print('手机使用 2G 网络');
  }

  //推荐使用的方法
  void netWork3G(){
    print('手机使用 3G 网络');
  }

  //推荐使用的方法
  void netWork4G(){
    print('手机使用 4G 网络');
```

```
  }
  //推荐使用的方法
  void netWork5G(){
    print('手机使用 5G 网络');
  }
}
```

示例输出如下内容,可以看到被弃用的方法也能正常调用。

```
flutter: 手机使用 2G 网络
flutter: 手机使用 3G 网络
flutter: 手机使用 4G 网络
flutter: 手机使用 5G 网络
```

在写代码时,IDE 会给我们提示此方法被弃用,如图 20-1 所示。方法的中间加了一根横线。

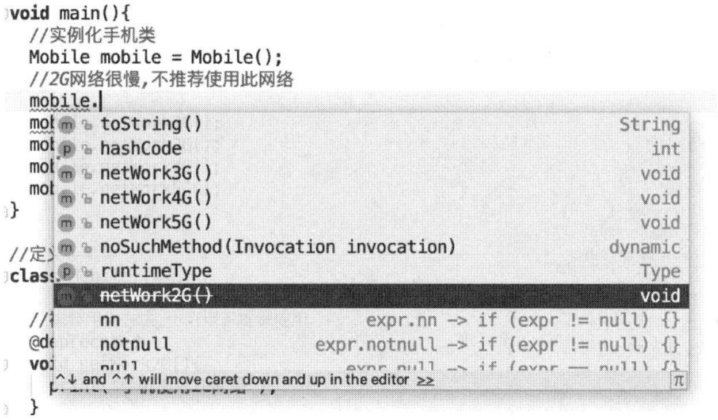

图 20-1 被弃用方法提示

20.2.2 @override

@override 是重写方法的意思,它的作用是帮助自己检查是否正确地重写了父类中已有的方法和告诉读代码的人,这是一个重写的方法。

接下来看一个子类重写父类方法的例子。代码如下:

```
//metadata_override.dart 文件
//动物类
class Animal {
  //动物会吃
  void eat(){
    print('动物会吃');
  }
```

```
  //动物会跑
  void run(){
    print('动物会跑');
  }
}
//人类
class Human extends Animal {
  void say(){
    print('人会说话');
  }

  void study(){
    print('人类也会吃');
  }

  //使用了元数据,表示重写方法
  @override
  void eat(){
    print('人类也会吃');
  }
}
void main(){
  print('实例化一个动物类');
  Animal animal = Animal();
  animal.eat();
  animal.run();

  print('实例化一个人类');
  Human human = Human();
  //重写的方法
  human.eat();
  human.run();
  human.say();
  human.study();
}
```

从代码中可以看到,Human 类是继承 Animal 类的,Human 类重写了 Animal 类的 eat 方法。eat 方法的上方使用了@override 元数据修饰。示例输出内容如下:

```
flutter: 实例化一个动物类
flutter: 动物会吃
flutter: 动物会跑
flutter: 实例话一个人类
flutter: 人类也会吃
flutter: 动物会跑
flutter: 人会说话
flutter: 人类也会吃
```

在 Flutter 使用@override 元数据最频繁的就是 build 方法了。Flutter 里一切皆为组件，项目中我们需要大量编写组件，重写 build 方法就可以重新渲染组件。示例代码如下：

```dart
//metadata_override_build.dart 文件
import 'package:flutter/material.dart';

void main() => runApp(MyApp());

//MyApp 组件继承一个没有状态的组件
class MyApp extends StatelessWidget {

  //重写 build 方法重新渲染组件
  @override
  Widget build(BuildContext context) {
    return MaterialApp(
      title: '方法重写示例',
      home: Scaffold(
        appBar: AppBar(
          title: Text('方法重写示例'),
        ),
        body: Center(
          child: Text('override build'),
        ),
      ),
    );
  }
}
```

build 方法中 return 部分即要渲染的内容。示例运行效果如图 20-2 所示。

图 20-2 方法重写示例

20.2.3 @required

@required 元数据用来标记一个参数,表示这个参数必须要传值。其作用如下:
- 告诉编译器这个参数必须要传值;
- 告诉读代码的人,这个参数必须要填写。

在 Flutter 里我们经常看到如下一段代码,它表示一个组件的构造方法,其中 key 是可选择参数,而 title 为必传参数。

```
MyHomePage({Key key, @required this.title}) : super(key: key);
```

接下来看一个 Flutter 里组件定义的例子。完整代码如下:

```
//metadata_required.dart 文件
import 'package:flutter/foundation.dart';
import 'package:flutter/material.dart';

void main() {
  runApp(MyApp());
}

//主组件
class MyApp extends StatelessWidget {

  //重写 build 方法
  @override
  Widget build(BuildContext context) {

    //定义参数变量
    final appName = '必传参数示例';

    return MaterialApp(
      //应用标题
      title: appName,
      //首页
      home: MyHomePage(
        //传入 title 参数
        title: appName,
      ),
    );
  }
}
//首页
class MyHomePage extends StatelessWidget {
  //标题
  final String title;
  //key 是可选择参数,key 为 Widget 的唯一标识,title 为标题,它是必传参数
  MyHomePage({Key key, @required this.title}) : super(key: key);
```

```
  @override
  Widget build(BuildContext context) {
    return Scaffold(
      appBar: AppBar(
        //首页标题
        title: Text(title),
      ),
      body: Center(
        child: Container(
          child: Text(
            //必传参数值
            this.title,
            //文本样式
            style: TextStyle(
              fontSize: 36.0,
            ),
          ),
        ),
      ),
    );
  }
}
```

从上面的示例代码中可以看到，MyApp 主组件给 MyHomePage 首页传入了一个标题参数。其中，MyHomePage 为一个自定义的组件，它需要传入一个 title 参数用来作标题使用，此参数为必传参数，前面加了 @required 修饰。key 是可选择参数，key 为 Widget 的唯一标识。示例运行的效果如图 20-3 所示。

图 20-3　必传参数示例

20.3 自定义元数据

可以定义自己的元数据注解。下面是定义一个接收两个参数的@Todo注解的例子。代码如下：

```dart
//custom_metadata_todo.dart 文件
library todo;

//定义元数据
class Todo {
  //名称
  final String name;
  //内容
  final String content;

  //const 构造方法
  const Todo(this.name, this.content);
}
```

然后看一下如何使用@Todo注解。代码如下：

```dart
//custom_metadata_main.dart 文件
import 'todo.dart';

void main() {

  //实例化 Test 对象
  Test test = Test();
  test.doSomething();

}

//测试类
class Test{

  //使用 Todo 元数据
  @Todo('kevin', 'make this do something')
  void doSomething() {
    print('do something');
  }

}
```

在使用@Todo进行注解时需要传入名称及内容两个参数。

提示 元数据可以出现在库、类、typedef、类型参数、构造函数、工厂构造函数、函数、字段、参数或变量声明前，以及导入和导出指令前。可以在运行期通过反射取回元数据。

20.4 元数据应用

相信大家都会遇到这样一个问题，在向服务器请求数据后，服务器往往会返回一段 Json 字符串，而我们要想更加灵活地使用数据，需要把 Json 字符串转化成对象。手写反序列化在大型项目中极不稳定，很容易导致解析失败。

这里给大家介绍的是使用 json_annotation 库自动反序列化处理。见名知意，json_annotation 就是使用注解处理 Json 的一个工具库，它有以下两个元数据：

- @JsonSerializable：实体类注解；
- @JsonKey：实体类的属性注解。

接下来我们编写一个电商中收货地址数据反序列化处理的例子。具体步骤如下：

步骤 1：打开工程目录 lib 下的 pubspec.yaml 文件，添加如下依赖库：

```
dependencies:
  flutter:
    sdk: flutter

  json_annotation: any

dev_dependencies:
  flutter_test:
    sdk: flutter

  build_runner: ^1.6.1
  json_serializable: ^3.0.0
```

这里需要添加三个依赖，它们分别是："json_annotation""build_runner"和"json_serializable"。它们的作用如下：

- json_annotation：提供 Json 注解处理的类。
- build_runner：是一个工具类，一个生成 Dart 代码文件的外部包。
- json_serializable：Json 序列化处理的类。

步骤 2：在 lib 目录下新建 entity 目录，里面存放所有的项目实体类文件。在此目录下新建一个实例类文件 address_entity.dart，这里通常需要和后端接口定义的实体类名称保持一致。源码如下：

```dart
//metadata_json_annotation_address_entity.dart 文件
import 'package:json_annotation/json_annotation.dart';

part 'address_entity.g.dart';

//地址实体类
@JsonSerializable()
class AddressEntity extends Object {

  //id
  @JsonKey(name: 'id')
  int id;

  //用户名
  @JsonKey(name: 'name')
  String name;

  //用户 id
  @JsonKey(name: 'userId')
  String userId;

  //省
  @JsonKey(name: 'province')
  String province;

  //城市
  @JsonKey(name: 'city')
  String city;

  //国家
  @JsonKey(name: 'county')
  String county;

  //详细地址
  @JsonKey(name: 'addressDetail')
  String addressDetail;

  //地区编码
  @JsonKey(name: 'areaCode')
  String areaCode;

  //电话号码
  @JsonKey(name: 'tel')
  String tel;

  //是否为默认地址
  @JsonKey(name: 'isDefault')
  bool isDefault;
```

```
    //添加地址时间
    @JsonKey(name: 'addTime')
    String addTime;

    //更新地址时间
    @JsonKey(name: 'updateTime')
    String updateTime;

    //是否删除
    @JsonKey(name: 'deleted')
    bool deleted;

    //构造方法,传入地址信息
    AddressEntity(this.id, this.name, this.userId, this.province, this.city, this.county, this.
addressDetail, this.areaCode, this.tel, this.isDefault, this.addTime, this.updateTime, this.
deleted,);

    //使用此方法将Json转换成实体对象
      factory AddressEntity.fromJson(Map<String, dynamic> srcJson) => _
$AddressEntityFromJson(srcJson);

    //使用此方法将实体对象转换成Json
    Map<String, dynamic> toJson() => _$AddressEntityToJson(this);

}
```

刚刚写完的 AddressEntity 类在运行时会报错,暂时先不用管,主要是因为还没有生成 address_entity.g.dart 辅助类。

可以看到 AddressEntity 类上面需要添加@JsonSerializable()元数据,表示此类是需要序列化的。实体类的每个字段上需要添加@JsonKey 元数据,表示此字段为实体类的一个属性。

最后两个方法 fromJson 的作用是将 Json 转换成实体对象,toJson 作用是将实体对象转换成 Json。

步骤3:使用 build_runner 工具生成 address_entity.g.dart 文件。build_runner 是 Dart 团队提供的一个生成 Dart 代码文件的外部包。

我们在当前项目根目录下运行"flutter packages pub run build_runner build"命令。控制台内容如下:

```
xuanweizideMacBook-Pro:helloworld ksj$ flutter packages pub run build_runner build
[INFO] Generating build script...
[INFO] Generating build script completed, took 410ms

[INFO] Initializing inputs
[INFO] Reading cached asset graph...
```

```
[INFO] Reading cached asset graph completed, took 79ms

[INFO] Checking for updates since last build...
[INFO] Checking for updates since last build completed, took 688ms

[INFO] Running build...
[INFO] 1.0s elapsed, 0/2 actions completed.
[INFO] Running build completed, took 1.1s

[INFO] Caching finalized dependency graph...
[INFO] Caching finalized dependency graph completed, took 28ms

[INFO] Succeeded after 1.1s with 2 outputs (4 actions)
```

可以看到最后生成文件成功。如果生成失败,可以根据控制台信息分析失败原因。或者清除之前生成的文件,尝试运行如下命令:

```
flutter packages pub run build_runner clean
```

然后再运行下面命令:

```
flutter packages pub run build_runner build --delete-conflicting-outputs
```

最后生成的文件结构如图 20-4 所示。

图 20-4 实体类生成文件

从图中可以看到新生成的文件是在原有文件名.dart 前面加了一个.g,其他部分不变。文件内容如下:

```
//metadata_json_annotation_address_entity.g.dart 文件
//GENERATED CODE - DO NOT MODIFY BY HAND

part of 'address_entity.dart';

// **************************************************************
//JsonSerializableGenerator
// ************************************************************ **
```

```dart
AddressEntity _$AddressEntityFromJson(Map<String, dynamic> json) {
  return AddressEntity(
    json['id'] as int,
    json['name'] as String,
    json['userId'] as String,
    json['province'] as String,
    json['city'] as String,
    json['county'] as String,
    json['addressDetail'] as String,
    json['areaCode'] as String,
    json['tel'] as String,
    json['isDefault'] as bool,
    json['addTime'] as String,
    json['updateTime'] as String,
    json['deleted'] as bool,
  );
}

Map<String, dynamic> _$AddressEntityToJson(AddressEntity instance) =>
    <String, dynamic>{
      'id': instance.id,
      'name': instance.name,
      'userId': instance.userId,
      'province': instance.province,
      'city': instance.city,
      'county': instance.county,
      'addressDetail': instance.addressDetail,
      'areaCode': instance.areaCode,
      'tel': instance.tel,
      'isDefault': instance.isDefault,
      'addTime': instance.addTime,
      'updateTime': instance.updateTime,
      'deleted': instance.deleted,
    };
```

此文件提供了两个重要方法，分别如下：

- _$AddressEntityFromJson：将收货地址 Json 转换成实体对象的具体实现；
- _$AddressEntityToJson：将收货地址实体对象转换成 Json 的具体实现。

从下面这段代码可以看出新生成部分代码是 address_entity.dart 的一部分。

```dart
part of 'address_entity.dart';
```

当生成文件后，再打开 address_entity.dart 文件就会发现错误消失了。

AddressEntity 类提供一个工厂构造方法 AddressEntity.fromJson，该方法实际调用生成文件的 _$AddressEntityFromJson 方法。

AddressEntity 类提供一个 AddressEntity.toJson 序列化对象的方法，实际调用生成文件的_$AddressEntityToJson方法,并将调用对象解析生成Map<String,dynamic>。

> **注意** 这段代码最上面的注释"//GENERATED CODE - DO NOT MODIFY BY HAND"。表明你可千万别手写生成文件。每次执行生成命令会覆盖此文件。

步骤4:上一步实现了Map to Dart,可是我们需要的是Json to Dart。这时候就需要Dart自带的dart:convert来帮助我们了。

dart:convert是Dart提供用于在不同数据格式之间进行转换的编码器和解码器,能够解析Json和UTF-8。

也就是说,我们需要先将Json数据使用dart:convert转成Map,这样我们就能通过Map转为Dart对象了。使用方法如下:

```dart
Map<String,dynamic> map = json.decode("jsondata");
```

知道了如何将Json字符串解析成Map以后,我们就能直接将Json转化为实体了。准备好测试数据,编写测试代码如下:

```dart
//metadata_json_annotation_address_main.dart 文件
import 'dart:convert';
import 'package:helloworld/entity/address_entity.dart';

void main(){
  //收货地址模拟数据
  var data = '''{
    "id": 1,
    "name": "张三",
    "userId": "000001",
    "province": "湖北",
    "city": "武汉",
    "county": "中国",
    "addressDetail": "某某街道",
    "areaCode": "111111",
    "tel": "88888888888",
    "isDefault": true,
    "addTime": "2019-09-25 18:32:49",
    "updateTime": "2019-09-25 18:32:49",
    "deleted": false
  }''';
  //实例化 AddressEntity 对象
  //使用 json.decode 解码,
  //使用 AddressEntity.fromJson 方法将其转换成实体对象
  AddressEntity addressEntity = AddressEntity.fromJson(json.decode(data));
  //调用实体对象的 toJson 方法输出内容
  print(addressEntity.toJson().toString());
}
```

上面的测试数据可以使用模拟数据,也可以使用服务端返回的数据。最后通过收货地址实体对象重新生成Json字符串如下:

```
flutter: {id: 1, name: 张三, userId: 000001, province: 湖北, city: 武汉, county: 中国, addressDetail: 某某街道, areaCode: 111111, tel: 88888888888, isDefault: true, addTime: 2019 – 09 – 25 18:32:49, updateTime: 2019 – 09 – 25 18:32:49, deleted: false}
```

从输出内容来看我们实现了Json与实体类的互转。

第 21 章 Dart 库

在 Flutter 项目中可以使用由其他开发者贡献给 Flutter 和 Dart 生态系统的共享软件包。这使你可以快速构建应用程序,而无须从头开始开发所有应用程序。

现有的软件包支持许多使用场景,例如,网络请求 Dio、导航路由处理 Fluro,以及获取电池电量插件库 battery 等。

在 Dart 中,库在使用时通过 import 关键字引入。library 指令可以创建一个库,每个 Dart 文件都是一个库,即使没有使用 library 指令来指定。

Dart 中的库主要有以下三种:

- 本地库:开发者自己编写的库文件;
- 系统内置库:SDK 自带的库文件;
- 第三方库:开发者发布到 Dart 仓库的共享软件包。

21.1 本地库使用

我们自定义的库。常用的导入方式如下:

```
import 'lib/mylib.dart';
```

其中,lib 为一个包名,mylib.dart 即为本地库文件。

这里我们写一个示例。首先在工程的 lib 目录下新建一个 person.dart 文件,代码如下:

```
//library_local_person.dart 文件
//定义 Person 类
class Person{

  //姓名变量
  String name;

  //年龄变量
  int age;
```

```
//构造方法
Person(this.name,this.age);

//打印信息方法
void printInfo(){
  print("${this.name} ${this.age}");
}

}
```

当编写好 Person 类的实现代码后,假设需要在 main.dart 文件里使用它,就需要导入 person.dart 文件。代码如下:

```
//library_local_main.dart 文件
//导入 person.dart 文件
import './person.dart';

void main(){
  //使用库里的 Person 类
  Person person = Person('张三', 20);
  person.printInfo();
}
```

这个示例中,person.dart 即为一个库文件,main.dart 则是库的使用者。由于这两个文件在同级目录,所以代码中使用了"./person.dart"来进行导入,"./"表示当前目录。

在 Flutter 的实际项目里我们通常不会把所有文件放在一个目录,而是通过包来进行分类。如笔者写的一个关于 Flutter 即时通讯界面实现的项目,项目结构如下:项目的源码位于 lib 目录下,包含聊天模块、好友模块、我的模块、加载页面、搜索页面、主页面、公共类,以及主程序。这样分包组织代码的好处是,结构更加清晰并有利于开发及维护。

```
├── README.md(项目说明)
├── flutter_im.iml
├── lib(源码目录)
│   ├── app.dart(主页面)
│   ├── chat(聊天模块)
│   │   ├── message_data.dart(消息数据)
│   │   ├── message_item.dart(消息项)
│   │   └── message_page.dart(消息页面)
│   ├── common(公共类)
│   │   ├── im_item.dart(IM 列表项)
│   │   └── touch_callback.dart(触摸回调封装)
│   ├── contacts(好友模块)
│   │   ├── contact_header.dart(好友列表头部)
│   │   ├── contact_item.dart(好友列表项)
│   │   ├── contact_sider_list.dart(好友列表)
```

```
│   │       ├── contact_vo.dart(好友 vo 类)
│   │       └── contacts.dart(好友页面)
│   ├── loading.dart(加载页面)
│   ├── main.dart(主程序)
│   ├── personal(我的模块)
│   │       └── personal.dart(我的页面)
│   └── search.dart(搜索页面)
├── pubspec.lock
└── pubspec.yaml(项目配置文件)
```

如果在 app.dart(主界面)中想导入其他几个模块的内容,可以使用相对路径,代码如下:

```
//导入聊天页面
import './chat/message_page.dart';
//导入联系人页面
import './contacts/contacts.dart';
//导入个人中心页面
import './personal/personal.dart';
```

假设 personal 包下的 personal.dart 文件需要导入 comman 包下的文件,可以使用"../"来跳到上一级目录再进行访问。personal/personal.dart 引入代码如下:

```
//跳至上级目录后,再找到 common 目录下的文件
import '../common/touch_callback.dart';
import '../common/im_item.dart';
```

如果需要跳多级目录就使用多个"../"符号。如:"../../../lib/xxx.dart"。

21.2 系统内置库使用

系统内置库即 SDK 自带的一些库文件。如数学相关的库、IO 处理的库,以及数据转换的库等。导入方式如下:

```
import 'dart:math';      //数据库
import 'dart:io';        //IO 操作库
import 'dart:convert';   //数据转换库
```

在 Flutter 里,我们大量使用其 SDK 的 Material 库。如下面示例所示,导入并使用了 Material 风格的组件。MaterialApp 即为一个 Material 风格的组件,代码如下:

```
//library_system.dart 文件
//导入 material 库
```

```
import 'package:flutter/material.dart';

void main() => runApp(MyApp());

class MyApp extends StatelessWidget {
  @override
  Widget build(BuildContext context) {
    //Material 风格组件
    return MaterialApp(
      title: 'Welcome to Flutter',
      home: Scaffold(
        appBar: AppBar(
          title: Text('Welcome to Flutter'),
        ),
        body: Center(
          child: Text('Hello World'),
        ),
      ),
    );
  }
}
```

21.3 第三方库使用

第三方库是指由全球众多开发者贡献的软件包,可以打开 https://pub.dev/ 网址找到你想要的第三方库,如图 21-1 所示。

从图中我们可以看到,第三方库会根据使用的范围分为以下几类:
- Flutter:用于 Flutter 移动端或桌面程序库;
- WEB:用于网页开发使用的库;
- OTHER:用于其他方面,如服务端应用开发。

这几个分类都支持的库表明其支持的平台最多,例如 Dio 库(网络请求库)可以用来开发手机、桌面,以及 Web 程序。shared_preferences 库(本地存储)只能用于 Flutter 的应用,不能用于 Web 程序,并且值得注意的是当你需要将其应用于开发桌面程序时,需要将其扩展至 Windows 或 macOS 系统。

提示 当我们自己开发一个第三方库时尽量做到适配更多的平台,例如要做一个绘图的 2D 库,可以使用 Canvas 的 API 来进行绘制而不是采用插件原生的 API 来进行绘制。

接下来通过使用及分析 shared_preferences 这个库来详细说明第三方库的使用方法。在使用之前首先补充一下 Key-Value 存储的知识。

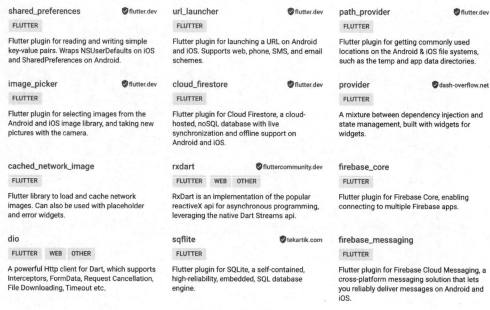

图 21-1　常用第三方库

21.3.1　Key-Value 存储介绍

应用开发时会有很多的数据存储需求，这个时候就需要用到持久化存储技术，与 iOS、Android 一样，Flutter 中也有很多种持久化存储方式，例如 Key-Value 存储、文件存储和数据库存储等，但其实质都是通过平台对应的模块来实现的。

Key-Value（键值对）存储主要是平台提供特定的 API 来供我们操作，其本质依然是将数据存储到特定文件中，只不过这些工作由平台帮我们做，例如 iOS 平台中的 NSUserDefaults、Android 平台的 SharedPreferences 等。

21.3.2　shared_preferences 使用

Flutter 中可以使用 shared_preferences 插件来实现 Key-Value 存储。主要存储数据类型包括 bool、int、double、String 和 List 等。

接下来通过一个示例来描述其使用过程，具体步骤如下：

步骤 1：引入插件，在 pubspec.yaml 文件中添加 shared_preferences 插件，代码如下：

```yaml
dependencies:
  flutter:
    sdk: flutter
  cupertino_icons: ^0.1.2
  # shared_preferences 插件
  shared_preferences: ^0.5.3+4
```

用命令行执行"flutter packages get"即可将插件下载到本地。

步骤2：插件引入到项目后，在使用的Dart文件中导入shared_preferences.dart文件，代码如下：

```
import 'package:shared_preferences/shared_preferences.dart';
```

导入文件后需要实例化其对象，获取SharedPreferences的实例方法是一个异步方法，所以在使用时需要注意使用await获取其真实对象，如下：

```
Future<SharedPreferences> _prefs = SharedPreferences.getInstance();
```

步骤3：编写保存数据及获取数据方法，代码如下：

```
//保存数据
void saveMethodName() async {
    SharedPreferences prefs = await _prefs;
    prefs.setString("strKey", "strValue");
    ...
}

//获取数据
void initFromCache() async {
    SharedPreferences prefs = await _prefs;
    String strValue = prefs.getString("strKey");
    ...
}
```

步骤4：编写UI组件，调用保存数据方法及在页面初始化时获取数据方法。完整代码如下：

```
//shared_preferences_main.dart文件
import 'package:flutter/material.dart';
import 'package:shared_preferences/shared_preferences.dart';

void main() => runApp(MyApp());

class MyApp extends StatelessWidget {
  @override
  Widget build(BuildContext context) {
    return MaterialApp(
      title: 'SharedPreferences 第三方库',
      theme: ThemeData(
        primarySwatch: Colors.blue,
      ),
      home: PersistentDemo(),
    );
```

```dart
    }
}

//本地存储,使用 Key-Value 存储
class PersistentDemo extends StatefulWidget {
  @override
  State<StatefulWidget> createState() => PersistentDemoState();
}

class PersistentDemoState extends State<PersistentDemo> {
  //实例化本地存储对象
  Future<SharedPreferences> _prefs = SharedPreferences.getInstance();
  //昵称及选择语言的值
  var controller = TextEditingController();
  bool value_dart = false;
  bool value_js = false;
  bool value_java = false;

  @override
  void initState() {
    super.initState();
    initFromCache();
  }

  @override
  void dispose() {
    super.dispose();
    controller = null;
  }

  //从缓存中获取信息填充
  void initFromCache() async {
    final SharedPreferences prefs = await _prefs;
    //根据键 key 获取本地存储的值 value
    final value_nickname = prefs.getString("key_nickname");
    final value_dart = prefs.getBool("key_dart");
    final value_js = prefs.getBool("key_js");
    final value_java = prefs.getBool("key_java");

    //获取缓存中的值后,使用 setState 更新界面信息
    setState(() {
      controller.text = (value_nickname == null ? "" : value_nickname);
      this.value_dart = (value_dart == null ? false : value_dart);
      this.value_js = (value_js == null ? false : value_js);
      this.value_java = (value_java == null ? false : value_java);
    });
  }

  //保存界面的输入信息
```

```dart
void saveInfo(String value_nickname) async {
  final SharedPreferences prefs = await _prefs;
  prefs.setString("key_nickname", value_nickname);
  prefs.setBool("key_dart", value_dart);
  prefs.setBool("key_js", value_js);
  prefs.setBool("key_java", value_java);
}

@override
Widget build(BuildContext context) {
  return Scaffold(
      appBar: AppBar(
        title: Text('SharedPreferences 第三方库'),
      ),
      body: Container(
        padding: EdgeInsets.all(15),
        child: Column(
          crossAxisAlignment: CrossAxisAlignment.center,
          children: <Widget>[
            TextField(
              controller: controller,
              decoration: InputDecoration(
                labelText: '昵称:',
                hintText: '请输入名称',
              ),
            ),
            Text('你喜欢的编程语言'),
            Row(
              mainAxisAlignment: MainAxisAlignment.spaceBetween,
              children: <Widget>[
                Text('Dart'),
                Switch(
                  value: value_dart,
                  onChanged: (isChanged) {
                    //设置状态改变要存储的值
                    setState(() {
                      this.value_dart = isChanged;
                    });
                  },
                )
              ],
            ),
            Row(
              mainAxisAlignment: MainAxisAlignment.spaceBetween,
              children: <Widget>[
                Text('JavaScript'),
                Switch(
                  value: value_js,
                  onChanged: (isChanged) {
```

```
              setState(() {
                this.value_js = isChanged;
              });
            },
          )
        ],
      ),
      Row(
        mainAxisAlignment: MainAxisAlignment.spaceBetween,
        children: <Widget>[
          Text('Java'),
          Switch(
            value: value_java,
            onChanged: (isChanged) {
              setState(() {
                this.value_java = isChanged;
              });
            },
          )
        ],
      ),
      MaterialButton(
        child: Text('保存'),
        onPressed: () {
          saveInfo(controller.text);
        },
      ),
    ],
  ),
  )
  );
 }
}
```

运行示例后，首先输入名字"kevin"，然后选中"Dart"，最后单击"保存"按钮。当我们再次运行示例后发现页面显示的是上一次保存的页面内容。示例的运行效果如图 21-2 所示。

21.3.3　shared-preferences 实现原理

shared_preferences 插件的实现原理很简单，通过源码分析，我们发现主要是通过 Channel 与原生平台进行交互，通过 iOS 平台的 NSUserDefaults、Android 平台的 SharedPreferences 来实现具体数据存取操作。

1. iOS 平台

iOS 平台插件关键代码如下：

图 21-2　SharedPreferences 第三方库

```objc
+ (void)registerWithRegistrar:(NSObject<FlutterPluginRegistrar> *)registrar {
  FlutterMethodChannel * channel =
      [FlutterMethodChannel methodChannelWithName:CHANNEL_NAME binaryMessenger:registrar.messenger];
  [channel setMethodCallHandler:^(FlutterMethodCall * call, FlutterResult result) {
    NSString * method = [call method];
    NSDictionary * arguments = [call arguments];

    if ([method isEqualToString:@"getAll"]) {
      result(getAllPrefs());
    } else if ([method isEqualToString:@"setBool"]) {
      NSString * key = arguments[@"key"];
      NSNumber * value = arguments[@"value"];
      [[NSUserDefaults standardUserDefaults] setBool:value.boolValue forKey:key];
      result(@YES);
    } else if ([method isEqualToString:@"setInt"]) {
      NSString * key = arguments[@"key"];
      NSNumber * value = arguments[@"value"];
      //int type in Dart can come to native side in a variety of forms
      //It is best to store it as is and send it back when needed.
```

```
            //Platform channel will handle the conversion.
            [[NSUserDefaults standardUserDefaults] setValue:value forKey:key];
            result(@YES);
        } else {
            .
            .
            .
        }
    }];
}
```

由源码可以看出,Flutter 中代码通过传输指定的 method 和对应的存储数据给 iOS 平台,iOS 端接收到指令数据后会根据方法名来判断执行操作的类型,然后进行对应的操作,例如 setBool 就会通过[NSUserDefaults standardUserDefaults]对象来保存 bool 类型的数据。

2. Android 平台

Android 平台插件关键代码如下:

```
private final android.content.SharedPreferences preferences;

public static void registerWith(PluginRegistry.Registrar registrar) {
    MethodChannel channel = new MethodChannel(registrar.messenger(), CHANNEL_NAME);
    SharedPreferencesPlugin instance = new SharedPreferencesPlugin(registrar.context());
    channel.setMethodCallHandler(instance);
}

private SharedPreferencesPlugin(Context context) {
    preferences = context.getSharedPreferences(SHARED_PREFERENCES_NAME, Context.MODE_PRIVATE);
}

@Override
public void onMethodCall(MethodCall call, MethodChannel.Result result) {
    String key = call.argument("key");
    boolean status = false;
    try {
        switch (call.method) {
            case "setBool":
                status = preferences.edit().putBoolean(key, (boolean) call.argument("value")).commit();
                break;
            case "setDouble":
                float floatValue = ((Number) call.argument("value")).floatValue();
                status = preferences.edit().putFloat(key, floatValue).commit();
                break;
```

```
          .
          .
          .
      }
      result.success(status);
    } catch (IOException e) {
      result.error("IOException encountered", call.method, e);
    }
  }
}
```

Android 端和 iOS 端类似，会通过一个 SharedPreferences 实例对象 preferences 来根据指定 method 做对应的存取数据操作。

假设想把库扩展至 macOS 和 Windows 上只需要实现对应平台的插件即可，这样就能支持更多的平台了。

21.4　库重名与冲突解决

当引入的两个库中有相同名称标识符的时候，在 Java 语言里通常通过写上完整的包名路径来指定使用的具体标识符，甚至不用 import 库都可以，但是在 Dart 语言里是必须 import 库的。当冲突的时候，可以使用 as 关键字来指定库的前缀。

接下来通过一个例子来说明如何处理库的重名情况。首先添加一个 person1.dart 文件。定义一个 Person 类。代码如下：

```
//library_same_name_person1.dart 文件
//定义 Person 类
class Person{

  //姓名变量
  String name;

  //年龄变量
  int age;

  //构造方法
  Person(this.name,this.age);

  //打印信息方法
  void printInfo(){
    print("${this.name} ${this.age}");
  }

}
```

再添加一个 person2.dart 文件。同样定义一个 Person 类。代码如下：

```dart
//library_same_name_person2.dart 文件
//定义 Person 类
class Person{

  //姓名变量
  String name;

  //年龄变量
  int age;

  //构造方法
  Person(this.name,this.age);

  //打印信息方法
  void printInfo(){
    print("${this.name} ${this.age}");
  }

}
```

在 main.dart 文件里同时导入这两个文件,就会有相同的 Person 类,编译不知道要用哪个 Person 类所以就会报错。这里可以指定一个前缀 lib 来解决这些问题。代码如下:

```dart
//library_same_name_main.dart 文件
//导入 person1.dart
import 'Person1.dart';
//导入 person2.dart 重命名为 lib
import 'Person2.dart' as lib;

void main(List<String> args) {

  //直接使用 Person 类
  Person p1 = Person('张三', 20);
  p1.printInfo();

  //使用 lib.Person 类
  lib.Person p2 = lib.Person('李四', 20);
  p2.printInfo();

}
```

示例运行后可以正常输出如下内容:

```
flutter: 张三 20
flutter: 李四 20
```

21.5 显示或隐藏成员

如果只需要导入库的一部分，有以下两种模式：
- show 关键字可以显示某个成员（屏蔽其他成员）；
- hide 关键字可以隐藏某个成员（显示其他成员）。

接下来通过一个示例来说明如何显示和隐藏库文件里的成员。首先新建一个 person.dart 文件，定义 Person 和 Student 两个类。代码如下：

```dart
//library_show_hide_person.dart 文件
//定义 Person 类
class Person{

  //姓名变量
  String name;

  //年龄变量
  int age;

  //构造方法
  Person(this.name,this.age);

  //打印信息方法
  void printInfo(){
    print("Person:${this.name} ${this.age}");
  }

}
//定义 Student 类
class Student{

  //姓名变量
  String name;

  //年龄变量
  int age;

  //构造方法
  Student(this.name,this.age);

  //打印信息方法
  void printInfo(){
    print("Student:${this.name} ${this.age}");
  }
```

```
  //学习方法
  void study(){
    print('Student:study');
  }

}
```

这里假设 Person 和 Student 两个类都使用,可以使用 show 关键字。按如下方式导入:

```
import 'person.dart' show Student, Person;
```

如果不想使用 Person 类,使用其他的类,可以使用 hide 关键字。按如下方式导入:

```
import 'person.dart' hide Person;
```

最后编写 main.dart 的完整测试代码,如下:

```
//library_show_hide_main.dart 文件
//可以使用 Student 和 Person 类
import 'person.dart' show Student, Person;
//不能使用 Person 类
//import 'person.dart' hide Person;

void main(List<String> args) {

  //使用 Person 类
  Person person = Person('张三', 20);
  person.printInfo();

  //使用 Student 类
  Student student = Student('李四', 20);
  student.printInfo();
  student.study();

}
```

示例输出如下内容:

```
flutter: Person:张三 20
flutter: Student:李四 20
flutter: Student:study
```

21.6 库的命名与拆分

如果需要自定义一个库则需要给库起一个名字。要显式声明库,请使用库语句。声明库的语法如下:

```
library library_name;
```

有的时候一个库可能太大,不能方便地保存在一个文件当中,这时 Dart 允许我们把一个库拆分成一个或者多个较小的 part 组件。或者当我们想让某一些库共享它们的私有对象的时候,我们需要使用关键字 part。

接下来通过一个工具库的示例来展示库的拆分及使用。步骤如下:

步骤 1:新建一个 util.dart 文件,将库名命名为 util。代码如下:

```
//library_part_util.dart 文件
//库命名为 util
library util;
//导入 math 库
import 'dart:math';

//日志工具为 util 库的一部分
part 'logger.dart';
//计算工具为 util 库的一部分
part 'calculator.dart';
```

从上面的代码可以看出,库的名称是 util,此库导入了 math 库,这样它所包含的部分也可以共享这个库,而不需要重新导入了,它包含了如下两个部分工具:

- logger.dart:日志工具;
- calculator.dart:数学计算工具。

步骤 2:新建 logger.dart 文件,添加如下代码:

```
//library_part_logger.dart 文件
//日志工具为 util 库的一部分
part of util;

//日志类
class Logger {

  //应用名称
  String _app_name;

  //构造方法
  Logger(this._app_name){}

  //错误
  void error(error) {
    print('[' + _app_name + '] ERROR: ' + error);
  }

  //调试
```

```
  void debug(msg) {
    print('[' + _app_name + '] DEBUG: ' + msg);
  }

  //警告
  void warn(msg) {
    print('[' + _app_name + '] WARN: ' + msg);
  }

  //失败
  void failure(error) {
    var log = '[' + _app_name + '] FAILURE: ' + error;
    print(log);
    throw (log);
  }
}
```

这里我们不用关心日志工具是如何实现的,其中 part of util 这行代码表明日志工具属于工具库 util 的一部分。

步骤 3:新建 calculator.dart 文件,添加如下代码:

```
//library_part_calculator.dart 文件
//计算工具为 util 库的一部分
part of util;

//加法
int add(int firstNumber, int secondNumber) {
  print("Calculator 库里的 add 方法");
  return firstNumber + secondNumber;
}

//减法
int sub(int firstNumber, int secondNumber) {
  print("Calculator 库里的 sub 方法");
  return firstNumber - secondNumber;
}

//生成随机数
int random(int no) {
  print("Calculator 库里的 random 方法");
  return Random().nextInt(no);
}
```

这里实现了几个数学的计算方法,其中 part of util 这行代码表明日志工具属于工具库 util 的一部分。

> **提示** 由于 util 库中已经导入了 math 库,所以这里不需要再次导入。作为 util 库的一部分它们之间共享导入的库的内容。

步骤 4:打开 main.dart 文件,导入 util 库测试并使用。代码如下:

```
//library_part_main.dart 文件
import './util.dart';

void main() {

  //使用日志工具
  Logger logger = Logger('Demo');
  logger.debug('这是调试信息');
  logger.error('这是错误信息');
  logger.warn('这是警告信息');
  //logger.failure('这是失败信息');

  //使用计算工具
  print(add(12, 34));
  print(sub(30,20));
}
```

从示例代码可以看出,只要导入 util.dart 文件了,日志和计算工具库不需要再次导入,这样就达到了库的拆分的目的。示例输出内容如下所示:

```
flutter: [Demo] DEBUG: 这是调试信息
flutter: [Demo] ERROR: 这是错误信息
flutter: [Demo] WARN: 这是警告信息
flutter: Calculator 库里的 add 方法
flutter: 46
flutter: Calculator 库里的 sub 方法
flutter: 10
```

如果取消注释下面的代码会引发异常。

```
logger.failure('这是失败信息');
```

输出内容如下:

```
flutter: [Demo] DEBUG: 这是调试信息
flutter: [Demo] ERROR: 这是错误信息
flutter: [Demo] WARN: 这是警告信息
flutter: [Demo] FAILURE: 这是失败信息
[VERBOSE-2:ui_dart_state.cc(157)] Unhandled Exception: [Demo] FAILURE: 这是失败信息
```

```
#0      Logger.failure (package:helloworld/logger.dart:33:5)
#1      main (package:helloworld/main.dart:11:10)
#2      _runMainZoned.<anonymous closure>.<anonymous closure> (dart:ui/hooks.dart:239:25)
#3      _rootRun (dart:async/zone.dart:1126:13)
#4      _CustomZone.run (dart:async/zone.dart:1023:19)
#5      _runZoned (dart:async/zone.dart:1518:10)
#6      runZoned (dart:async/zone.dart:1502:12)
#7      _runMainZoned.<anonymous closure> (dart:ui/hooks.dart:231:5)
#8      _startIsolate.<anonymous closure> (dart:isolate-patch/isolate_patch.dart:307:19)
#9      _RawReceivePortImpl._handleMessage (dart:isolate-patch/isolate_patch.dart:174:12)
```

这个异常是 util 库里的日志工具的 failure 方法所抛出的。从这里我们得到一个启示，一个好的库不仅仅功能要完善，还要编写详细的注释和处理好各种异常。

21.7 导出库

当设计的库比较庞大时，你可以使用 export 关键字导出指定的部分。如下面代码所示，库 math 导出了 random.dart 及 point.dart 部分。

```
//库命名
library math;
//导出库
export 'random.dart';
//导出库
export 'point.dart';
```

你也可以将导出部分组合成一个新库。

```
//库命名
library math;
//导出库并且显示 Random
export 'random.dart' show Random;
//导出库并且隐藏 Sin
export 'point.dart' hide Sin;
```

以笔者参与的一个 Flutter 的库项目为例。这个库名称为 flutter-webrtc，实现了 Flutter 的音视频插件，库里包含摄像头的获取、媒体连接、音视频录制和媒体状态信息等内容。工具目录如图 21-3 所示。

可以看到此库所包含的内容很多，这里就可以使用 export 功能来导出必要的部分。此库的导出部分放在 webrtc.dart 文件里。代码如下：

```
flutter-webrtc  library root, ~/Desktop/yu/flutter-webrtc
  .dart_tool
  .github
  .idea
  android
  example
  ios
  lib
    web
      enums.dart
      get_user_media.dart
      media_recorder.dart
      media_stream.dart
      media_stream_track.dart
      rtc_data_channel.dart
      rtc_dtmf_sender.dart
      rtc_ice_candidate.dart
      rtc_peerconnection.dart
      rtc_peerconnection_factory.dart
      rtc_session_description.dart
      rtc_stats_report.dart
      rtc_video_view.dart
      utils.dart
    webrtc.dart
```

图 21-3　flutter-webrtc 工程结构图

```
export 'get_user_media.dart'
    if (dart.library.js) 'web/get_user_media.dart';
export 'media_stream_track.dart'
    if (dart.library.js) 'web/media_stream_track.dart';
export 'media_stream.dart'
    if (dart.library.js) 'web/media_stream.dart';
export 'rtc_data_channel.dart'
    if (dart.library.js) 'web/rtc_data_channel.dart';
export 'rtc_video_view.dart'
    if (dart.library.js) 'web/rtc_video_view.dart';
export 'rtc_ice_candidate.dart'
    if (dart.library.js) 'web/rtc_ice_candidate.dart';
export 'rtc_session_description.dart'
    if (dart.library.js) 'web/rtc_session_description.dart';
export 'rtc_peerconnection.dart'
    if (dart.library.js) 'web/rtc_peerconnection.dart';
export 'rtc_peerconnection_factory.dart'
    if (dart.library.js) 'web/rtc_peerconnection_factory.dart';
export 'rtc_stats_report.dart';
export 'media_recorder.dart'
    if (dart.library.js) 'web/media_recorder.dart';
export 'utils.dart'
    if (dart.library.js) 'web/utils.dart';
export 'enums.dart';
```

从代码中可以看出，在 export 里还可以加 if 判断，这里是用来判断应用所处的平台，如 Web 平台就使用 web 包下的库文件。

> 提示　flutter-webrtc 库部分只需要了解库的导出知识即可，如果想进一步研究音视频插件，可以打开如下网址 https://github.com/cloudwebrtc/flutter-webrtc。

第4篇 商城项目实战

第 22 章 项目简介

通过前面几章的学习,我们掌握了 Dart 及 Flutter 相关的知识,从本章开始将通过一个大型案例把这些知识贯穿起来,以便掌握 Flutter 商业级项目的开发过程。

22.1 功能介绍

商城项目是一种常见的功能较多、内容较为复杂的 App,如淘宝、京东、当当、唯品会等。商城 App 通常包括以下几大核心模块:

- 登录注册;
- 首页;
- 分类;
- 购物车;
- 商品列表;
- 商品详情;
- 订单列表;
- 订单详情;
- 填写订单;
- 收货地址;
- 个人收藏;
- 我的订单。

商城的项目可大可小,如可以添加团购、分销和 IM 等功能,应用的主要功能效果如图 22-1 所示。

图 22-1 App 效果图

22.2 总体架构

商城项目使用一个完整的技术栈,包括前端、后端、后台管理,以及数据库。各个部分所使用的功能如下:

❑ 前端 App:使用 Flutter 技术,可打包成 Android 及 iOS 程序;

- 后端接口：使用流行的 SpringBoot（Java 版）技术，提供前端数据接口、后台管理数据接口，并与数据库交互以存取数据；
- 后台管理：使用 Vue 技术编写后台管理页面；
- 数据库：使用 MySQL 数据库存储商城项目中产生的数据。

其中前端、后端与后台管理之间采用 Http+Json 的数据交互方式，技术总体架构如图 22-2 所示。

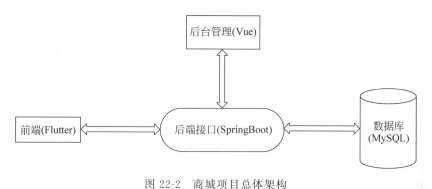

图 22-2　商城项目总体架构

> 提示　项目前端参考 FlutterMall 开源项目 https://github.com/youxinLu/mall，后端参考 LiteMall 开源项目 https://github.com/linlinjava/litemall。

接下来详细说明各个部分所使用的技术栈及其在项目中的使用情况。

22.2.1　前端 Flutter

前端 Flutter 的工程名为 shop-flutter，使用的第三方库如下：

- flutter_swipe：用于首页及商品详情图片轮播；
- dio：网络框架，用于 Flutter 与 SpringBoot 数据交互处理；
- flutter_spinkit：旋转组件，用于数据等待提示；
- event_bus：用于不同模块或页面事件通知；
- flutter_html：html 框架，用于商品详情展示 html 数据；
- flutter_screenutil：用于各个界面屏幕适配；
- fluttertoast：用于界面中间弹出轻量提示文本；
- shared_preferences：用于本地数据缓存，如存取 token；
- fluro：企业级路由，用于页面导航及参数传递；
- city_pickers：用于收货地址城市选择；
- flutter_easyrefresh：上拉加载和下拉刷新，如商品列表数据的刷新；
- flutter_webview_plugin：用于加载网页的插件；
- cached_network_image：图片缓存组件。

> **提示** 针对 Flutter 端使用的技术栈,在项目开发之前可以先大致查询一下各个库或组件的使用情况,运行一个简单的示例进行测试,然后再使用到项目中去。这样可以避免由于页面及功能复杂而不好排查的情况。

22.2.2 后端接口 SpringBoot

后端接口采用 SpringBoot 开发,工程名为 shop-server,主要是给前端 Flutter 及后台管理 Vue 提供数据接口,同时与数据库 MySQL 交互以存取数据,项目结构如图 22-3 所示。

其中各个模块的作用如下:
- shop-admin-api:提供后台管理数据接口;
- shop-client-api:提供前端数据接口;
- shop-db:后端与数据库交互处理;
- shop-core:后端接口核心包;
- shop-all:后端接口程序启动入口;
- storage:文件及图片存储目录;
- logs:日志目录。

图 22-3 后端接口项目结构

后端接口 SpringBoot 这里不过多介绍,我们只要掌握如何启动并使用它即可。

22.2.3 后台管理 Vue

后台管理 Vue 的工程名为 shop-admin,主要使用 Vue 技术开发,用来提供商品管理、广告管理、订单管理和地区管理等功能,实现了响应式的布局,界面展示华丽。后台管理页面如图 22-4 所示。

图 22-4 后台管理页面

项目采用的主要技术栈如下：
- vue：Vue 框架；
- vuex：Vuex 是一个专为 Vue.js 应用程序开发的状态管理模式；
- vue-router：Vue 路由导航工具；
- vue-count-to：数字滚动插件；
- element-ui：饿了么开发的 UI 框架；
- font-awesome：字体图标库；
- js-cookie：cookie 插件；
- nprogress：进度条组件；
- screenfull：使用全屏功能的库。

后台管理 Vue 这里不过多介绍，我们只要掌握如何启动并使用它即可。

22.2.4 数据库 MySQL

数据库部分用来存取商城产生的数据，如商品数据、订单数据、用户数据和地址数据等，由后台 SpringBoot 来进行操作，使用的是 MySQL 数据库。

数据库表的设计是根据需求分析来制定的，商城项目中涉及的表及其作用如下：
- shop_ad：广告表，用于存储轮播图数据；
- shop_address：收货地址表，用于存储收货地址信息，包含姓名、电话和详细地址等信息；
- shop_admin：管理员表，用于存储管理员数据，包含管理员名称、密码、角色和最后登录时间等数据；
- shop_cart：购物车表，用于存储购物车中的商品信息及是否选中的状态；
- shop_category：分类表，用于存储商品分类类目名称、图片，以及父类 Id 数据；
- shop_collect：收藏表，用于记录被收藏的商品 Id 及用户 Id；
- shop_goods：商品表，用于记录商品 Id、名称和详情等基本数据；
- shop_goods_attribute：商品属性表，用于记录商品的属性，如衣服的面料、款式等数据；
- shop_goods_product：商品货品表，用于记录商品货品的价格、数量和图片等数据；
- shop_goods_specification：商品规格表，用于记录商品的规格数据，如衣服的大码、均码以及小码数据；
- shop_issue：常见问题表，记录商品详情中的常见问题，如使用什么快递等；
- shop_log：日志表，用于记录管理员登录及操作相关数据；
- shop_order：订单表，用于记录订单状态、收货人信息和订单总价等数据；
- shop_order_goods：订单商品表，用于记录订单中商品的基本信息；
- shop_permission：权限表，用于记录角色对应的权限数据；
- shop_region：地区表，用于记录地区信息，如省、市和区等；

- shop_role：角色表，用于记录系统角色信息；
- shop_storage：文件存储表，用于记录用户文件存储数据，如用户从后台上传的商品图片数据；
- shop_system：系统配置表，用于记录系统配置数据；
- shop_user：普通用户表，用于记录商城普通用户数据，如用户名称、密码和头像等数据。

数据库表的设计这里不做深入描述，只需要建好数据库，导入 sql 语句即可。

22.3 后端及数据库准备

在开发商城 App 之前首先需要将后端 SpringBoot 和后台管理 Vue 运行起来，同时连上 MySQL 数据库。

22.3.1 MySQL 安装

步骤 1：安装 MySQL 数据库，设置好 root 的密码（笔者设置为 12345678）。打开 MySQL 图形管理工具 Navicat Premium，新建一个连接，填写好连接名称、主机、用户名和密码后单击"连接测试"按钮，确保连接成功，如图 22-5 所示。

图 22-5　新建连接 MySQL

打开连接后,新建一个数据库并命名为 shop,选择好字符集和排序规则后单击"确定"按钮,如图 22-6 所示。

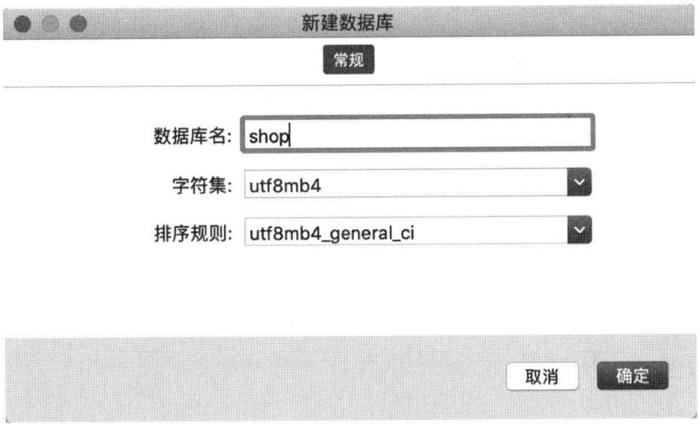

图 22-6 新建数据库

步骤 2:新建好数据库后,选中数据库名称并单击右键运行 shop-server/shop-db/sql/shop.sql 文件,这样便会生成数据库的表及数据,如图 22-7 所示。

图 22-7 数据库表

22.3.2 JDK 安装

数据库安装好并导入数据后,接下来准备后端 SpringBoot 的环境。shop-server 工程使用的是 Java 版的 SpringBoot,所以需要安装 JDK,JDK 的安装这里不过多说明,在控制台输入"java-version"命令核实是否安装成功。注意 JDK 需要 1.8 或以上版本。输出内容如下:

```
xuanweizideMacBook-Pro:~ ksj$ java -version
java version "1.8.0_111"
```

22.3.3 Maven 安装

安装好 JDK 后可以安装 Maven 工具。以 Windows 环境为例,具体步骤如下。

步骤 1:首先下载 Maven,地址为 https://maven.apache.org/download.cgi,然后将其解压至当前目录如下:

```
D:\Program Files\apache-maven-3.6.0
```

步骤 2:设置环境变量。打开环境变量设置界面,新建环境变量 MAVEN_HOME,赋值如下:

```
D:\Program Files\apache-maven-3.6.0
```

然后编辑环境变量 Path,追加 %MAVEN_HOME%\bin\。至此 Maven 已经完成了安装,我们可以通过 DOS 命令检查是否安装成功。

步骤 3:配置仓库镜像。由于网络问题可以配置阿里的镜像。打开 maven/conf/settings.xml 文件,修改内容如下:

```
<mirrors>
    <!-- 阿里云仓库 -->
    <mirror>
        <id>alimaven</id>
        <mirrorOf>*</mirrorOf>
        <name>aliyun maven</name>
        <url>http://maven.aliyun.com/nexus/content/repositories/central/</url>
    </mirror>
    <!-- 中央仓库 1 -->
    <mirror>
        <id>repo1</id>
        <mirrorOf>central</mirrorOf>
        <name>Human Readable Name for this Mirror.</name>
```

```
        <url>http://repo1.maven.org/maven2/</url>
    </mirror>
    <!-- 中央仓库2 -->
    <mirror>
        <id>repo2</id>
        <mirrorOf>central</mirrorOf>
        <name>Human Readable Name for this Mirror.</name>
        <url>http://repo2.maven.org/maven2/</url>
    </mirror>
</mirrors>
```

macOS 环境下下载对应的 Maven 版本后解压,使用命令 vim ~/.bash_profile 打开环境变量配置文件,添加如下内容后,执行 source ~/.bash_profile 即可。

```
export M2_HOME=/Users/ksj/Desktop/software/maven/
export PATH=$PATH:$M2_HOME/bin
```

22.3.4 IntelliJ IDEA 启动项目

IntelliJ IDEA 是一个 Java 开发工具,关于它的下载及安装这里不过多说明。导入 shop-server 工程后,IDEA 工具会自动下载所依赖的各种包,请耐心等待直至下载完成。

打开数据库连接配置文件,确保数据库名、用户名和密码正确,如图 22-8 中箭头所示。

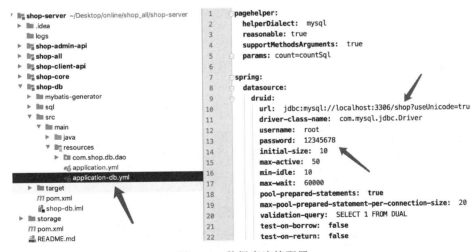

图 22-8 数据库连接配置

当准备工作完成后,单击 IDEA 工具的调试按钮启动 SpringBoot 项目。当看到输出如下信息则表明后端接口启动成功。

```
com.shop.Application - Started Application in 9.696 seconds (JVM running for 11.287)
```

22.3.5　Node 安装

由于后台管理 shop-admin 项目需要在 Node 环境下运行，所以需要安装 Node。关于它的安装这里不过多说明。控制台下执行"node -v"命令查看是否安装成功。输出内容如下：

```
xuanweizideMacBook-Pro:~ ksj$ node -v
v13.2.0
```

进入到 shop-admin 目录，依次执行如下命令：

```
//安装 cnpm
npm install -g cnpm --registry=https://registry.npm.taobao.org
//安装依赖库
cnpm install
//启动程序
cnpm run dev
```

这里使用的是 taobao 的仓库，这样下载资源更快。如果不需要则可以直接执行如下命令：

```
npm i
npm run dev
```

当控制台输出如下内容则表明 shop-admin 启动成功：

```
Your application is running here: http://localhost:3000
```

后台管理启动成功后，首先进入后台登录界面，如图 22-9 所示。

图 22-9　管理员登录界面

至此后端及数据库准备完毕，可以开发前端 App 项目了。

第 23 章 项目框架搭建

由于项目模块较多,可以交给多名开发人员开发。这里就涉及一个工程化的问题,搭建一个良好的基础框架可以让不同开发人员分工协作。同时约定好命名规范和编码规范,这样就可以减少开发中带来的一些不必要的错误。

本章带领大家一步步把项目的框架搭建起来,包括添加资源和插件等内容。

23.1 新建项目

新建项目是项目开发的第一步,包括工程创建、资源添加、库添加和项目配置等内容。具体步骤如下:

步骤 1:新建工程参考第 3 章。基本的步骤是一样的,只是需要注意项目命名为 shop_flutter,另外再填写一个项目描述即可。

步骤 2:准备好项目中使用的各种图标、背景图片、加载图片,以及字体等资源。图片放入 images 目录而字体放入 fonts 目录下即可,如图 23-1 所示。

图 23-1 项目图片及字体资源

> **提示** 图标可以从图标库下载:http://www.iconfont.cn。

打开项目配置文件 pubspec.yaml,在 assets 资源节点下添加项目中用到的所有位图资源。在 fonts 下添加字体文件。代码如下:

```
assets:
  - images/no_data.png
  - images/pw.png
  - images/loading.png
  - images/network_error.png
  - images/avatar.jpeg

fonts:
```

```
      - family: ShopIcon
        fonts:
          - asset: fonts/iconfont.ttf
```

提示 图片的配置要与 images 文件夹下的文件名保持一致。命名遵循一定的规则,例如 avatar 就表示头像图标。

位图及字体资源添加完后,单击配置文件右上角"Packages get"按钮更新配置。

步骤 3:再次打开 pubspec.yaml 文件,添加项目所需的第三方库。配置项如下:

```
dependencies:
  flutter:
    sdk: flutter
  #轮播组件
  flutter_swiper: ^1.1.6
  #json 处理
  json_annotation: ^2.3.0
  #网络请求
  dio: ^2.1.7
  #加载等待组件
  flutter_spinkit: ^3.1.0
  #事件处理
  event_bus: ^1.1.0
  #html 显示
  flutter_html: ^0.9.6
  #屏幕适配
  flutter_screenutil: ^0.5.3
  #提示组件
  fluttertoast: ^3.1.0
  #本地存储
  shared_preferences: ^0.5.3+4
  #路由导航
  fluro: "^1.5.1"
  #状态管理
  provider: ^3.0.0+1
  #地区选择组件
  city_pickers: ^0.1.23
  #刷新组件
  flutter_easyrefresh: ^2.0.3
  #加载网页插件
  flutter_webview_plugin: ^0.3.7
  #图片缓存组件
  cached_network_image: ^2.0.0-rc
  #苹果风格图标
  cupertino_icons: ^0.1.2
```

```yaml
dev_dependencies:
  flutter_test:
    sdk: flutter
  #编译运行
  build_runner: ^1.6.1
  #json序列化库
  json_serializable: ^3.0.0
```

添加好配置文件后,一定要单击配置文件右上角"Packages get"按钮,以获取指定版本的插件。

23.2 目录结构

新建好项目、配置好资源后,接下来就是编码工作。项目的源码位于 lib 目录下,包含主页、分类、商品、购物车、订单、收藏、地址和用户等。按 lib 目录添加好子目录及源码文件,目录结构如下:

```
├── README.md(项目说明)
├── fonts(字体资源)
│   └── iconfont.ttf(图标字体)
├── images(图片资源)
│   ├── avatar.jpeg(头像图片)
│   ├── loading.png(启动屏图片)
│   ├── network_error.png(网络错误图片)
│   ├── no_data.png(没有数据图片)
│   └── pw.png(密码图片)
├── lib(源码)
│   ├── config(配置文件)
│   │   ├── color.dart(颜色)
│   │   ├── icon.dart(字体)
│   │   ├── index.dart(导出配置)
│   │   ├── server_url.dart(接口地址)
│   │   └── string.dart(字符串常量)
│   ├── event(事件目录)
│   │   ├── login_event.dart(登录事件)
│   │   └── refresh_event.dart(刷新事件)
│   ├── main.dart(入口文件)
│   ├── model(数据模型目录)
│   │   ├── address_model.dart(地址)
│   │   ├── cart_list_model.dart(购物车列表)
│   │   ├── category_title_model.dart(分类图标)
│   │   ├── collect_list_model.dart(收藏列表)
│   │   ├── fill_in_order_model.dart(填写订单)
│   │   ├── first_category_model.dart(一级分类)
│   │   ├── goods_detail_model.dart(商品详情)
```

```
│   │       ├── goods_model.dart(商品)
│   │       ├── home_model.dart(首页)
│   │       ├── order_detail_model.dart(订单详情)
│   │       ├── order_model.dart(订单)
│   │       ├── sub_category_model.dart(二级分类)
│   │       └── user_model.dart(用户)
│   ├── page(页面)
│   │   ├── cart(购物车)
│   │   │   └── cart_page.dart(购物车页面)
│   │   ├── category(分类)
│   │   │   ├── category_page.dart(分类页面)
│   │   │   ├── first_category.dart(一级分类)
│   │   │   └── sub_category.dart(二级分类)
│   │   ├── goods(商品)
│   │   │   ├── goods_category_page.dart(分类下的商品页面)
│   │   │   ├── goods_detail_gallery.dart(商品详情轮播图)
│   │   │   ├── goods_detail_page.dart(商品详情页面)
│   │   │   └── goods_list_page.dart(分类下的商品列表)
│   │   ├── home(主页)
│   │   │   ├── home_banner.dart(主页轮播图)
│   │   │   ├── home_category.dart(主页分类)
│   │   │   ├── home_page.dart(主页面)
│   │   │   ├── home_product.dart(主页产品)
│   │   │   └── index_page.dart(索引页面)
│   │   ├── loading(启动屏)
│   │   │   └── loading_page.dart(启动屏页面)
│   │   ├── login(登录)
│   │   │   ├── login_page.dart(登录页面)
│   │   │   └── register_page.dart(注册页面)
│   │   ├── mine(我的)
│   │   │   ├── about_us_page.dart(关于我们页面)
│   │   │   ├── address_edit_page.dart(地址编辑页面)
│   │   │   ├── address_page.dart(地址页面)
│   │   │   ├── collect_page.dart(收藏页面)
│   │   │   └── mine_page.dart(我的页面)
│   │   └── order(订单)
│   │       ├── fill_in_order_page.dart(填写订单页面)
│   │       ├── order_detail_page.dart(订单详情页面)
│   │       └── order_page.dart(订单页面)
│   ├── provider(状态管理)
│   │   └── user_info_provider.dart(用户状态)
│   ├── router(路由配置)
│   │   ├── application.dart(路由应用类)
│   │   ├── router_handlers.dart(路由 Handler)
│   │   └── routers.dart(路由路径配置)
│   ├── service(数据服务)
```

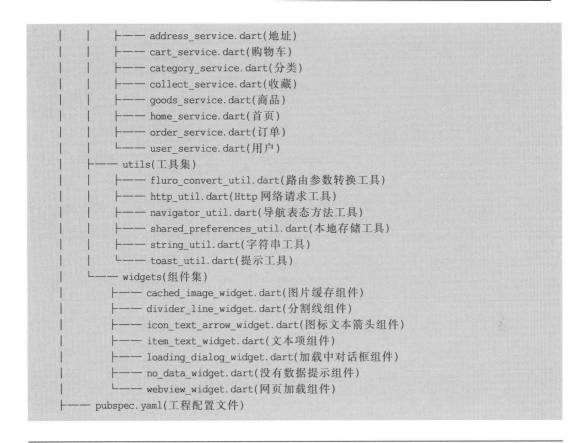

> **注意** 项目结构文件是不包含 android、ios 及 build 目录的。数据模型 model 包下的扩展文件，如 address_model.g.dart 是根据 Json 序列化工具产生的文件不必手工添加。

从项目的目录结构可以看出项目的总体框架设计。其中，page、model，以及 service 是开发中工作量最大的几块。它们的作用如下：

- page：功能页面主要用来展示页面，如分类页面。
- model：数据模型，主要用来定义数据的字段，并将后端传过来的 Json 数据转换成此对象，如分类数据。
- service：数据服务类，主要用于前后端数据交互使用。例如页面触发了一个动作，调用此服务里的请求方法，请求后端数据，数据返回后解析并转换成数据模型，最后展示在页面中。

以登录页面的登录操作为例。登录页面 LoginPage 调用数据服务 UserService 里的 login 方法，发起 Http 请求至后端，后端接口返回登录数据，数据服务 UserService 将返回的 Json 数据转换成数据模型 UserModel，数据模型 UserModel 处理完数据后将数据返回数据服务 UserService，然后数据服务 UserService 将数据返回给登录页面 LoginPage，登录页面根据数据作相应的提示或其他操作，数据流程图如图 23-2 所示。

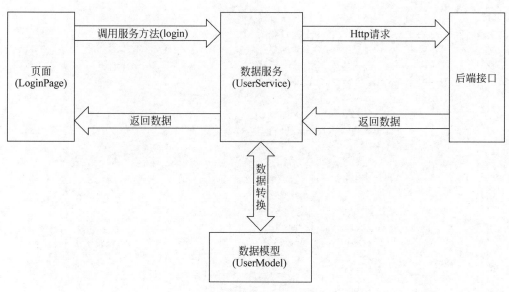

图 23-2　数据流程图

第 24 章 项目配置

项目配置为项目中抽取出来的公共可配置的选项,如图标、颜色和字符串等。它是项目里的基础部分内容,通常在开发之前需要提前准备好。项目配置文件放在 config 包下,主要包含以下几部分内容:

- color.dart(颜色);
- icon.dart(图标);
- server_url.dart(接口地址);
- string.dart(字符串常量)。

本章将详细说明项目中用到的配置文件有哪些,及它们的配置项。

24.1 颜色配置

把程序中的常用颜色抽出来统一放在一个文件里的好处是,当需要修改颜色时,可以只改一个地方,所有引用它的地方都会发生变化。打开 config/color.dart 文件,添加如下代码:

```
//config/color.dart 文件
import 'dart:ui';
import 'package:flutter/material.dart';
//主要颜色常量
class KColor{
  static const Color primaryColor = Colors.red;                      //默认主要颜色
  static const Color defaultTextColor = Colors.redAccent;            //默认文本颜色
  static const Color defaultButtonColor = Colors.redAccent;          //默认按钮颜色
  static const Color defaultSwitchColor = Colors.redAccent;          //默认切换按钮颜色
  static const Color defaultCheckBoxColor = Colors.redAccent;        //默认复选框按钮颜色
  static const Color toastBgColor = Colors.redAccent;                //Toast 提示背景颜色
  static const Color watingColor = Colors.redAccent;                 //加载数据提示颜色
  static const Color toastTextColor = Colors.white;                  //Toast 提示文本颜色
  static const Color priceColor = Colors.redAccent;                  //价格颜色
  static const Color indexTabSelectedColor = Colors.red;             //选项卡按钮选中颜色
  static const Color indexTabUnSelectedColor = Colors.grey;          //选项卡按钮未选中颜色
```

```
    static const Color bannerDefaultColor = Colors.white;              //轮播图激活小点默认颜色
    static const Color bannerActiveColor = Colors.red;                 //轮播图激活小点选中颜色
    static const Color categorySelectedColor = Colors.redAccent;       //分类选中颜色
    static const Color categoryDefaultColor = Colors.black54;          //分类默认颜色
    static const Color noDataTextColor = Colors.redAccent;             //没有数据文本颜色
    static const Color loginButtonColor = Colors.redAccent;            //登录按钮颜色
    static const Color loginIconColor = Colors.redAccent;              //登录图标颜色
    static const Color registerTextColor = Colors.redAccent;           //注册文本颜色
    static const Color registerButtonColor = Colors.redAccent;         //注册按钮颜色
    static const Color registerIconColor = Colors.redAccent;           //注册图标颜色
    static const Color collectionButtonColor = Colors.redAccent;       //收藏按钮颜色
    static const Color unCollectionButtonColor = Colors.grey;          //取消收藏按钮颜色
    static const Color addCartIconColor = Colors.redAccent;            //购物车图标颜色
    static const Color addCartButtonColor = Colors.green;              //添加至购物车按钮颜色
    static const Color buyButtonColor = Colors.red;                    //立即购买按钮颜色
    static const Color attributeTextColor = Colors.black54;            //商品属性文本颜色
    static const Color issueQuestionColor = Colors.black54;            //常见问题颜色
    static const Color issueAnswerColor = Colors.grey;                 //常见问题回答颜色
    static const Color specificationWarpColor = Colors.redAccent;      //商品规格选中颜色
}
```

24.2 图标配置

项目中会大量使用图标，Flutter 有内置的图标、图片图标，以及字体图标。常用的图标可以统一放在一个文件里。打开 config/icon.dart 文件，添加如下代码：

```
//config/icon.dart 文件
import 'package:flutter/widgets.dart';
//图标常量
class KIcon{
    static const String FONT_FAMILY = 'ShopIcon';    //字体名称
    static const IconData PASS_WORD = const IconData(0xe617, fontFamily: KIcon.FONT_FAMILY);
//密码图标
    static const IconData ADDRESS = const IconData(0xe63c, fontFamily: KIcon.FONT_FAMILY);
//地址图标
    static const IconData ORDER = const IconData(0xe634, fontFamily: KIcon.FONT_FAMILY);
//订单图标
    static const IconData COLLECTION = const IconData(0xe61e, fontFamily: KIcon.FONT_FAMILY);
//收藏图标
    static const IconData ABOUT_US = const IconData(0xe654, fontFamily: KIcon.FONT_FAMILY);
//关于我们图标
}
```

可以看到，代码中的 IconData 使用了 fontFamily 属性，这里指定字体名称为 ShopICon。字体名称需要和程序的主配置文件 pubspec.yaml 里的字体名称保持一致。

24.3 字符串配置

程序中会使用到标题、按钮文本和提示文件等,都把它们配置到 config/string.dart 文件里即可。代码如下:

```dart
//config/string.dart 文件
//字符串常量
class KString {
  static const String NO_DATA_TEXT = "暂无数据";
  static const String NEW_PRODUCT = "新品";
  static const String HOT_PRODUCT = "热卖产品";
  static const String HOME = "首页";
  static const String CATEGORY = "分类";
  static const String SHOP_CAR = "购物车";
  static const String MINE = "我的";
  static const String SERVER_EXCEPTION = "请求网络数据异常";
  static const String GOODS_DETAIL = "商品详情";
  static const String GOODS_ATTRIBUTES = "商品属性";
  static const String COMMON_PROBLEM = "常见问题";
  static const String ADD_CART = "加入购物车";
  static const String BUY = "立即购买";
  static const String LOGIN = "登录";
  static const String REGISTER = "注册";
  static const String ACCOUNT_HINT = "请输入手机号码";
  static const String PASSWORD_HINT = "请输入密码";
  static const String ACCOUNT = "账号";
  static const String PASSWORD = "密码";
  static const String ACCOUNT_RULE = "账号必须为长度为 11 的数字";
  static const String PASSWORD_RULE = "密码最少为 6 位";
  static const String NOW_REGISTER = "马上注册";
  static const String NICK_NAME = "NICK_NAME";
  static const String HEAD_URL = "HEAD_URL";
  static const String REGISTER_SUCCESS = "注册成功";
  static const String PRICE = "价格";
  static const String ALREAD_SELECTED = "已选择";
  static const String SPECIFICATIONS = "规格";
  static const String NUMBER = "数量";
  static const String ISLOGIN = "isLogin";
  static const String NICKNAME = "nickName";
  static const String AVATARURL = "avatarUrl";
  static const String LOGIN_SUCCESS = "登录成功";
  static const String SUCCESS = "success";
  static const String ADD_CART_SUCCESS = "添加成功";
  static const String CART = "购物车";
  static const String TOTAL_MONEY = "共计";
  static const String EDIT = "编辑";
```

```dart
    static const String COMPLETE = "完成";
    static const String DELETE = "删除";
    static const String BUY_NOW = "下单";
    static const String ORDER = "我的订单";
    static const String ADDRESS = "地址管理";
    static const String COLLECTION = "收藏";
    static const String ABOUT_US = "关于我们";
    static const String CLICK_LOGIN = "点击登录";
    static const String SETTLEMENT = "结算";
    static const String TIPS = "提示";
    static const String DELETE_CART_ITEM_TIPS = "是否确认删除?";
    static const String CONFIRM = "确认";
    static const String CANCEL = "取消";
    static const String DELETE_SUCCESS = "删除成功";
    static const String FILL_IN_ORDER = "填写订单";
    static const String DEFAULT = "默认";
    static const String REMARK = "备注";
    static const String GOODS_TOTAL = "商品合计";
    static const String FREIGHT = "运费";
    static const String PAY = "付款";
    static const String MY_ADDRESS = "我的收货地址";
    static const String ADD_ADDRESS = "添加新地址";
    static const String IS_DEFAULT = "默认";
    static const String ADDRESS_EDIT = "编辑";
    static const String ADDRESS_EDIT_TITLE = "编辑地址";
    static const String ADDRESS_PLEASE_INPUT_NAME = "请输入联系人姓名";
    static const String ADDRESS_PLEASE_INPUT_PHONE = "请输入联系人电话";
    static const String ADDRESS_PLEASE_SELECT_CITY = "请选择地址";
    static const String ADDRESS_PLEASE_INPUT_DETAIL = "请输入详细地址如街道、楼栋、房号等";
    static const String ADDRESS_SET_DEFAULT = "设为默认地址";
    static const String ADDRESS_DELETE = "删除收货地址";
    static const String SUBMIT = "提交";
    static const String SUBMIT_SUCCESS = "提交成功";
    static const String ADDRESS_DELETE_SUCCESS = "删除成功";
    static const String PLEASE_SELECT_ADDRESS = "请选择收货地址";
    static const String MINE_COLLECT = "我的收藏";
    static const String MINE_CANCEL_COLLECT = "是否确定取消收藏此商品?";
    static const String MINE_ABOUT_US = "关于我们";
    static const String MINE_ABOUT_US_CONTENT = "FlutterShop";
    static const String MINE_ABOUT_NAME_TITLE = "公司名称";
    static const String MINE_ABOUT_NAME = "Flutter";
    static const String MINE_ABOUT_EMAIL_TITLE = "邮箱";
    static const String MINE_ABOUT_EMAIL = "flutter@gmail.com";
    static const String MINE_ABOUT_TEL_TITLE = "联系电话";
    static const String MINE_ABOUT_TEL = "400-100-100";
    static const String MINE_ORDER = "我的订单";
    static const String HOME_TITLE = "首页";
    static const String CATEGORY_TITLE = "分类";
    static const String MINE_ORDER_TOTAL_GOODS = "共计";
```

```
  static const String MINE_ORDER_GOODS_TOTAL = "件商品,";
  static const String MINE_ORDER_PRICE = "合计¥";
  static const String MINE_ORDER_SN = "订单编号:";
  static const String MINE_ORDER_DETAIL = "订单详情";
  static const String MINE_ORDER_TIME = "订单时间";
  static const String ORDER_INFORMATION = "商品信息";
  static const String DOLLAR = "¥";
  static const String MINE_ORDER_DETAIL_TOTAL = "商品合计";
  static const String MINE_ORDER_DETAIL_PAYMENTS = "实付";
  static const String MINE_ORDER_DELETE_SUCCESS = "删除成功";
  static const String MINE_ORDER_CANCEL_SUCCESS = "取消成功";
  static const String MINE_ORDER_DELETE_TIPS = "是否确认删除此订单";
  static const String MINE_ORDER_CANCEL_TIPS = "是否确认取消此订单";
  static const String MINE_ORDER_ALREADY_CANCEL = "已取消";
  static const String PLEASE_LOGIN = "请先登录";
  static const String LOADING = "正在加载中...";
  static const String TOKEN = "X-Shop-Token";
  static const String LOGIN_OUT = "退出登录";
  static const String LOGIN_OUT_TIPS = "是否确认退出当前账号";
  static const String ORDER_TITLE = "订单";
}
```

24.4 接口地址配置

项目中用到的后端接口众多,可以把相同类的接口地址归类在一起,这样便于查阅,如地址列表、增加地址和删除地址等。打开 config/server_url.dart 文件,添加如下代码:

```
//config/server_url.dart 文件
//服务端接口地址
class ServerUrl {
  static const String BASE_URL = 'http://localhost:8080/client';      //基础地址

  static const String HOME_URL = BASE_URL + '/home/index';            //首页数据

  static const String CATEGORY_FIRST = BASE_URL + '/catalog/getfirstcategory';
                                                                      //商品分类第一级
  static const String CATEGORY_SECOND = BASE_URL + '/catalog/getsecondcategory';
                                                                      //商品分类第二级

  static const String GOODS_CATEGORY = BASE_URL + "/goods/category";  //获取商品分类数据
  static const String GOODS_LIST = BASE_URL + '/goods/list';          //获取商品数据列表
  static const String GOODS_DETAILS_URL = BASE_URL + '/goods/detail'; //商品详情

  static const String REGISTER = BASE_URL + '/auth/register';         //注册
  static const String LOGIN = BASE_URL + '/auth/login';               //登录
```

```dart
    static const String LOGIN_OUT = BASE_URL + "/auth/logout";            //退出登录

    static const String CART_ADD = BASE_URL + '/cart/add';                //加入购物车
    static const String CART_LIST = BASE_URL + '/cart/index';             //购物车数据
    static const String CART_UPDATE = BASE_URL + '/cart/update';          //更新购物车
    static const String CART_DELETE = BASE_URL + '/cart/delete';          //删除购物车数据
    static const String CART_CHECK = BASE_URL + '/cart/checked';          //购物车商品勾选
    static const String CART_BUY = BASE_URL + '/cart/checkout';           //购物车下单
    static const String FAST_BUY = BASE_URL + '/cart/fastadd';            //立即购买

    static const String ADDRESS_LIST = BASE_URL + '/address/list';        //地址列表
    static const String ADDRESS_SAVE = BASE_URL + '/address/save';        //增加地址
    static const String ADDRESS_DELETE = BASE_URL + '/address/delete';    //删除地址
    static const String ADDRESS_DETAIL = BASE_URL + '/address/detail';    //地址详情

    static const String COLLECT_LIST = BASE_URL + '/collect/list';        //收藏列表
    static const String COLLECT_ADD_DELETE = BASE_URL + '/collect/addordelete';
                                                                          //添加或取消收藏

    static const String ORDER_SUBMIT = BASE_URL + '/order/submit';        //提交订单
    static const String ORDER_LIST = BASE_URL + '/order/list';            //我的订单
    static const String ORDER_DETAIL = BASE_URL + "/order/detail";        //订单详情
    static const String ORDER_CANCEL = BASE_URL + "/order/cancel";        //取消订单
    static const String ORDER_DELETE = BASE_URL + "/order/delete";        //删除订单

}
```

地址接口主要是数据服务层在使用，当需要请求某个后台接口时使用此配置文件。

24.5 导出配置

由于配置文件较多，在编写业务模块时需要多次引用，可以在 config 包下添加一个 index.dart 文件统一导出以方便使用。

打开 config/index.dart 文件，使用关键字 export 依次导出配置文件，代码如下：

```dart
//config/index.dart 文件
//导出配置
export 'color.dart';        //导出颜色配置
export 'icon.dart';         //导出图标配置
export 'string.dart';       //导出字符串配置
export 'server_url.dart';   //导出后台接口地址配置
```

通过上面统一导出后，就可如下面方式直接导入 index.dart 文件：

```dart
import 'package:shop/config/index.dart';
```

第 25 章 工 具 集

工具集类似生活中的工具箱,我们可以在工具箱里找到各种工具以供使用。项目中的工具箱通常为程序员写代码中积累下来的常用方法集。项目的工具集放在 utils 包下,本章将详细说明各个工具的作用。

25.1 路由导航工具

商城项目中使用的路由方案为 Fluro,由于页面较多,需要来回跳转,所以使用 Fluro 这种企业路由方案。例如从商品列表页面跳转至商品详情页面就需要首先传递一个商品 Id 参数,然后再执行跳转动作。

25.1.1 路由参数处理

Fluro 在执行页面跳转或页面返回之前通常需要对参数进行一些处理才能使用。如将对象转换成字符串,或将字符串转为对象,以及中文处理等。打开 utils/fluro_convert_util.dart 文件添加如下代码:

```dart
//utils/fluro_convert_util.dart 文件
import 'dart:convert';
//Fluro 参数处理工具
class FluroConvertUtil{
  //Object 转为 String Json
  static String objectToString<T>(T t) {
    return fluroCnParamsEncode(jsonEncode(t));
  }

  //String Json 转为 Map
  static Map<String, dynamic> stringToMap(String str) {
    return json.decode(fluroCnParamsDecode(str));
  }
  //Fluro 传递中文参数前先转换,Fluro 不支持中文传递
  static String fluroCnParamsEncode(String originalCn) {
    return jsonEncode(Utf8Encoder().convert(originalCn));
```

```
    }
    //Fluro 传递后取出参数解析
    static String fluroCnParamsDecode(String encodeCn) {
      var list = List<int>();
      jsonDecode(encodeCn).forEach(list.add);
      String value = Utf8Decoder().convert(list);
      return value;
    }
}
```

上面的工具用法如下面代码所示,用户添加收货地址信息后,当返回填写订单页面时先将地址对象转换成字符串,然后再返回页面。

```
//跳转至填写订单页面
_goFillInOrder(AddressModel addressData) {
  //页面返回之前先将地址对象转换成字符串
  Navigator.of(context).pop(FluroConvertUtil.objectToString(addressData));
}
```

25.1.2 导航工具

导航工具里封装了路由导航的页面跳转方法。主要目的是为了简化处理代码,先看下面这段代码,它可以使得页面跳转至详情页,并带上商品 Id 参数:

```
//传递参数为商品 Id
Application.router.navigateTo(
    context, Routers.goodsDetail + "?goodsId= $ goodsId",
    transition: TransitionType.inFromRight);
```

上面的代码可能使用起来有些冗长,那么可以封装此段代码如下:

```
//跳转至商品详情页面
static goGoodsDetails(BuildContext context, int goodsId) {
  //传递参数为商品 Id
  Application.router.navigateTo(
      context, Routers.goodsDetail + "?goodsId= $ goodsId",
      transition: TransitionType.inFromRight);
}
```

那么使用者就可以这样调用,代码如下:

```
NavigatorUtil.goGoodsDetails(context, goods.id);
```

这样看起来就比较简洁明了,NavigatorUtil 即为导航工具类,里面提供了一系列的页

面跳转静态方法。打开 utils/navigator_util.dart 文件,添加如下代码:

```dart
//utils/navigatr_util.dart 文件
import 'package:fluro/fluro.dart';
import 'package:flutter/material.dart';
import 'package:shop/router/application.dart';
import 'package:shop/router/routers.dart';
import 'package:shop/utils/string_util.dart';
import 'package:shop/utils/fluro_convert_util.dart';
//导航工具提供静态方法,简化路由处理代码
class NavigatorUtil {
  //跳转至首页
  static goShopMainPage(BuildContext context) {
    Application.router.navigateTo(
        context,
        Routers.home,
        transition: TransitionType.inFromRight, replace: true);
  }
  //跳转至商品列表页面
  static goCategoryGoodsListPage(BuildContext context, String categoryName, int categoryId) {
    //Json 编码处理
    var categoryNameText = StringUtil.encode(categoryName);
    //传递参数为分类 Id 及分类名称
    Application.router.navigateTo(
        context,
        Routers.categoryGoodsList + "? categoryName = $categoryNameText&categoryId = $categoryId",
        transition: TransitionType.inFromRight);
  }
  //跳转至注册页面
  static goRegister(BuildContext context) {
    Application.router.navigateTo(context, Routers.register,
        transition: TransitionType.inFromRight);
  }
  //跳转至登录页面
  static goLogin(BuildContext context) {
    Application.router.navigateTo(context, Routers.login,
        transition: TransitionType.inFromRight);
  }
  //跳转至商品详情页面
  static goGoodsDetails(BuildContext context, int goodsId) {
    //传递参数为商品 Id
    Application.router.navigateTo(
        context, Routers.goodsDetail + "?goodsId = $goodsId",
        transition: TransitionType.inFromRight);
  }
  //返回注册页面
  static popRegister(BuildContext context) {
    Application.router.pop(context);
```

```dart
  }
  //跳转至填写订单页面
  static goFillInOrder(BuildContext context, int cartId) {
    //参数为购物车 Id
    Application.router.navigateTo(
        context,
        Routers.fillInOrder + "?cartId=$cartId",
        transition: TransitionType.inFromRight);
  }
  //跳转至地址页面
  static Future goAddress(BuildContext context) {
    return Application.router.navigateTo(context, Routers.address,
        transition: TransitionType.inFromRight);
  }

  //跳转至地址编辑页面
  static Future goAddressEdit(BuildContext context, int addressId) {
    //参数为地址 Id
    return Application.router.navigateTo(
        context,
        Routers.addressEdit + "?addressId=$addressId",
        transition: TransitionType.inFromRight);
  }
  //跳转至收藏页面
  static goCollect(BuildContext context) {
    Application.router.navigateTo(context, Routers.mineCollect,
        transition: TransitionType.inFromRight);
  }
  //跳转至关于我们页面
  static goAboutUs(BuildContext context) {
    Application.router.navigateTo(context, Routers.aboutUs,
        transition: TransitionType.inFromRight);
  }
  //跳转至订单页面
  static goOrder(BuildContext context) {
    Application.router.navigateTo(context, Routers.mineOrder,
        transition: TransitionType.inFromRight);
  }
  //跳转至订单详情页面
  static Future goOrderDetail(BuildContext context, int orderId, String token) {
    //参数有订单 Id 及 token
    return Application.router.navigateTo(
        context, Routers.mineOrderDetail + "?orderId=$orderId&token=$token",
        transition: TransitionType.inFromRight);
  }
  //跳转至网页页面
  static goWebView(BuildContext context, String title, String url) {
    //中文参数处理
    var titleName = FluroConvertUtil.fluroCnParamsEncode(title);
```

```
      var urlEncode = FluroConvertUtil.fluroCnParamsEncode(url);
      Application.router.navigateTo(
          context, Routers.webView + "?title=$titleName&&url=$urlEncode",
          transition: TransitionType.inFromRight);
    }
    //跳转至轮播图详情页面
    static goBrandDetail(BuildContext context, String titleName, int id) {
      var title = FluroConvertUtil.fluroCnParamsEncode(titleName);
      Application.router.navigateTo(
          context, Routers.brandDetail + "?titleName=$title&id=$id",
          transition: TransitionType.inFromRight);
    }
}
```

> **提示** 代码中封装了项目中所使用的所有页面跳转方法，这里路由还未定义，第 27 章将会详细讲到路由配置。

25.2 Http 请求工具

在商城项目中，Http 请求工具主要用的是 Dio 库，这里我们需要对其进行一些基本的封装处理。Http 请求工具主要为数据服务层提供数据请求支持，工具做了以下几部分处理：

- 连接超时设置；
- 接收超时设置；
- 工厂单例模式处理；
- 携带 token 值；
- Post 请求封装；
- Get 请求封装。

打开 utils/http_util.dart 文件，添加如下代码：

```
//utils/http_util.dart 文件
import 'package:dio/dio.dart';
import 'package:shop/config/index.dart';
import 'package:shop/utils/shared_preferences_util.dart';
//定义 dio 变量
var dio;
//Http 请求处理工具，提供了 Get 及 Post 请求封装方法
class HttpUtil {
  //获取 HttpUtil 实例
  static HttpUtil get instance => _getInstance();
  //定义 HttpUtil 实例
```

```dart
  static HttpUtil _httpUtil;
  //获取 HttpUtil 实例方法,工厂模式
  static HttpUtil _getInstance() {
    if (_httpUtil == null) {
      _httpUtil = HttpUtil();
    }
    return _httpUtil;
  }
  //构造方法
  HttpUtil() {
    //选项
    BaseOptions options = BaseOptions(
      //连接超时
      connectTimeout: 5000,
      //接收超时
      receiveTimeout: 5000,
    );
    //实例化 Dio
    dio = Dio(options);
    dio.interceptors.add(InterceptorsWrapper(onRequest: (RequestOptions options) async {
      print(" ======================= 请求数据 =================== ");
      print("url = ${options.uri.toString()}");
      print("params = ${options.data}");
      //锁住
      dio.lock();
      //获取本地 token
      await SharedPreferencesUtil.getToken().then((token) {
        //将 token 值放入请求头里
        options.headers[KString.TOKEN] = token;
      });
      //解锁
      dio.unlock();
      return options;
    }, onResponse: (Response response) {
      print(" ======================= 请求数据 =================== ");
      print("code = ${response.statusCode}");
      print("response = ${response.data}");
    }, onError: (DioError error) {
      print(" ======================= 请求错误 =================== ");
      print("message = ${error.message}");
    }));
  }

  //封装 get 请求
  Future get(String url, {Map<String, dynamic> parameters, Options options}) async {
    //返回对象
    Response response;
    //判断请求参数并发起 get 请求
    if (parameters != null && options != null) {
```

```
      response = await dio.get(url, queryParameters: parameters, options: options);
    } else if (parameters != null && options == null) {
      response = await dio.get(url, queryParameters: parameters);
    } else if (parameters == null && options != null) {
      response = await dio.get(url, options: options);
    } else {
      response = await dio.get(url);
    }
    //返回数据
    return response.data;
  }

  //封装 post 请求
  Future post(String url, {Map<String, dynamic> parameters, Options options}) async {
    //返回对象
    Response response;
    //判断请求参数并发起 post 请求
    if (parameters != null && options != null) {
      response = await dio.post(url, data: parameters, options: options);
    } else if (parameters != null && options == null) {
      response = await dio.post(url, data: parameters);
    } else if (parameters == null && options != null) {
      response = await dio.post(url, options: options);
    } else {
      response = await dio.post(url);
    }
    //返回数据
    return response.data;
  }
}
```

这里采用单例模式的原因是只需要创建一个 Dio 实例即可，不必每发起一次请求就要创建一个 Dio 对象，同时设置请求及返回时长。同时我们可以看到 token 值被放置在 Http 请求的头里，这样后端根据请求的头便可取出 token 值。

25.3 本地存储工具

本地存储是一种持久化的操作，如记录登录用户的名称、头像，以及 token 等信息，当用户下次打开应用时可以从本地提取这些数据。

打开 utils/shared_preferences_util.dart 文件，添加如下代码：

```
//utils/shared_preferences_util.dart 文件
import 'package:shared_preferences/shared_preferences.dart';
import 'package:shop/config/index.dart';
//本地存储工具
```

```
class SharedPreferencesUtil {
  //token 字符串
  static String token = "";
  //获取 token 值
  static Future getToken() async {
    if (token == null || token.isEmpty) {
      //从本地取出 token 值
      SharedPreferences sharedPreferences = await SharedPreferences.getInstance();
      token = sharedPreferences.getString(KString.TOKEN) ?? null;
    }
    return token;
  }
  //获取头像 URL
  static Future getImageHead() async {
    SharedPreferences sharedPreferences = await SharedPreferences.getInstance();
    return sharedPreferences.get(KString.HEAD_URL);
  }
  //获取昵称
  static Future getUserName() async {
    SharedPreferences sharedPreferences = await SharedPreferences.getInstance();
    return sharedPreferences.get(KString.NICK_NAME);
  }
}
```

上面的方法中多处用到 getToken，一般用于数据请求时获取本地 token 值的处理。

> **提示** 这里的本地存储工具里的方法提供的是一些数据取的操作，你可以根据需要添加存的操作方法。

25.4 字符串处理工具

字符串处理是编程必备技能，如字符的截取、获取长度、大小写转换和排序等。商城项目里我们主要做了 Json 的编解码处理。打开 utils/string_util.dart 文件，添加如下代码：

```
//utils/string_util.dart 文件
import 'dart:convert';
//字符串处理工具
class StringUtil {
  //字符串 json 编码
  static String encode(String originalCn) {
    return jsonEncode(Utf8Encoder().convert(originalCn));
  }
  //字符串 json 解码
  static String decode(String encodeCn) {
```

```
    var list = List<int>();
    jsonDecode(encodeCn).forEach(list.add);
    String value = Utf8Decoder().convert(list);
    return value;
  }
}
```

25.5 Toast 提示工具

Toast 提示非常实用,如登录成功后提示、注册成功后提示或添加商品至购物车成功后提示。居中飘一个文本,过两秒后消失。打开 utils/toast_util.dart 文件,添加如下代码:

```
//utils/toast_util.dart 文件
import 'package:fluttertoast/fluttertoast.dart';
import 'package:flutter_screenutil/flutter_screenutil.dart';
import 'package:shop/config/index.dart';
//Toast 提示组件显示工具
class ToastUtil {
  static showToast(String message) {
    Fluttertoast.showToast(
        //提示消息
        msg: message,
        toastLength: Toast.LENGTH_SHORT,
        //居中
        gravity: ToastGravity.CENTER,
        timeInSecForIos: 1,
        //背景色
        backgroundColor: KColor.toastBgColor,
        //文本颜色
        textColor: KColor.toastTextColor,
        fontSize: ScreenUtil.instance.setSp(28.0));
  }
}
```

这里之所以没有把提示工具列为组件包的原因是它不需要继承,也不需要添加至界面,而只需要在使用时调用 showToast 方法即可。

第 26 章 组 件 封 装

本章将封装一些商城项目中使用到的组件。封装的好处是简化使用,统一风格。另外当引用的第三方组件升级时,只需要升级封装过的组件,而不必每个页面都调整一下代码,这样便于维护。

本章将对这些组件的封装方法做详细阐述,涉及的组件有以下几个:
- 缓存图片组件;
- 数量加减组件;
- 分割线组件;
- 图文组件;
- 文本组件;
- 加载数据组件;
- 没有数据提示组件;
- 网页加载组件。

26.1 缓存图片组件

缓存图片组件是对 CachedNetworkImage 组件的一个基本的封装,使用时只需要传入宽度、高度,以及图片地址即可。使用缓存图片组件加载图片可以使得应用不必每次重复加载图片数据。打开 widgets/cached_image_widget.dart 文件,添加如下代码:

```
//widgets/cached_image_widget.dart 文件
import 'package:flutter/material.dart';
import 'package:cached_network_image/cached_network_image.dart';
import 'package:flutter_screenutil/flutter_screenutil.dart';
import 'package:shop/config/string.dart';
//缓存图片组件
class CachedImageWidget extends StatelessWidget {
  //宽度
  double width;
```

```dart
//高度
double height;
//图片地址
String url;
//构造方法
CachedImageWidget(this.width, this.height, this.url);

@override
Widget build(BuildContext context) {
  return Container(
    width: this.width,
    height: this.height,
    alignment: Alignment.center,
    //使用 CachedNetworkImage 组件
    child: CachedNetworkImage(
      //图片地址
      imageUrl: this.url,
      //填充方式
      fit: BoxFit.cover,
      width: this.width,
      height: this.height,
      //等待提示
      placeholder: (BuildContext context, String url) {
        return Container(
          width: this.width,
          color: Colors.grey[350],
          height: this.height,
          alignment: Alignment.center,
          //加载中
          child: Text(
            KString.LOADING,
            style: TextStyle(
                fontSize: ScreenUtil.instance.setSp(26.0),
                color: Colors.white),
          ),
        );
      },
    ),
  );
}
```

商城项目中首页的 Banner 和商品图片等地方使用了缓存图片组件，效果如图 26-1 中箭头指向的图片所示。

图 26-1　缓存图片效果

26.2　数量加减组件

数量加减组件属于一个特定的组件,仅用于购物车页面的商品项中。可以使用其完成商品的增加、减少,以及计数的功能。具体实现步骤如下:

步骤 1:定义商品数量,改变回调方法,用于通知数量改变。代码如下:

```
typedef OnNumberChange(int number);
```

步骤 2:定义监听数量改变事件,此方法的作用是监听外部发出的购物车数量事件,组件同时修改其商品数量。监听代码如下:

```
_listener() {
  cartNumberEventBus.on<CartNumberEvent>().listen((CartNumberEvent cartNumberEvent) {
    //设置当前商品数量
    setState(() {
      goodsNumber = cartNumberEvent.number;
    });
  });
}
```

> **提示** 这里并未实现 CartNumberEvent 购物数量改变事件,可以先将其注释起来,在购物车模块实现后再解除注释即可。

步骤3:添加增加及减少商品数量的方法。大致代码如下:

```
//减少处理
_reduce() {
  //判断当前商品数量是否大于1
  ...
  //调用回调方法
  onNumberChange(goodsNumber);
}

//增加处理
_add() {
  ...
  //调用回调方法
  onNumberChange(goodsNumber);
}
```

当增加或减少商品数量后,需要调用回调方法来通知购物车页面进行下一步的处理。

步骤4:添加界面处理代码,采用水平方式布局:左侧为减号,中间为商品数量,右侧为加号。打开 widgets/cart_number_widget.dart 文件,添加如下完整代码:

```
//widgets/cart_number_widget.dart 文件
import 'package:flutter/material.dart';
import 'package:flutter_screenutil/flutter_screenutil.dart';
import 'package:shop/event/cart_number_event.dart';

//定义数量改变回调方法
typedef OnNumberChange(int number);

//购物车数量加减组件
class CartNumberWidget extends StatefulWidget {
  //数量改变回调方法
  OnNumberChange onNumberChange;
  //计数
  var _number;
  //构造方法,传入初始值及回调方法
  CartNumberWidget(this._number, this.onNumberChange);

  @override
  _CartNumberWidgetState createState() => _CartNumberWidgetState();
}

class _CartNumberWidgetState extends State<CartNumberWidget> {
  //商品数量
```

```dart
var goodsNumber;
//回调方法
OnNumberChange onNumberChange;

@override
void initState() {
  super.initState();
  goodsNumber = widget._number;
  onNumberChange = widget.onNumberChange;
}

//监听数量改变事件
_listener() {
  cartNumberEventBus.on<CartNumberEvent>().listen((CartNumberEvent cartNumberEvent) {
    //设置当前商品数量
    setState(() {
      goodsNumber = cartNumberEvent.number;
    });
  });
}

@override
Widget build(BuildContext context) {
  //调用事件监听
  _listener();
  return Container(
    //设置宽、高
    width: ScreenUtil.instance.setWidth(150),
    height: ScreenUtil.instance.setWidth(50),
    //水平布局
    child: Row(
      children: <Widget>[
        InkWell(
          //减少
          onTap: () => _reduce(),
          child: Container(
            width: ScreenUtil.instance.setWidth(50),
            height: double.infinity,
            alignment: Alignment.center,
            //添加边框样式
            decoration: ShapeDecoration(
                shape: Border(
                    left: BorderSide(color: Colors.grey, width: 1.0),
                    top: BorderSide(color: Colors.grey, width: 1.0),
                    right: BorderSide(color: Colors.grey, width: 1.0),
                    bottom: BorderSide(color: Colors.grey, width: 1.0))),
            //减少符号
            child: Text(
              "-",
              style: TextStyle(
                  color: Colors.black54,
```

```
                    fontSize: ScreenUtil.instance.setSp(26.0)),
              ),
            )),
        //中间数量容器
        Container(
          alignment: Alignment.center,
          height: double.infinity,
          width: ScreenUtil.instance.setWidth(50),
          //添加边框样式
          decoration: ShapeDecoration(
              shape: Border(
                  top: BorderSide(color: Colors.grey, width: 1.0),
                  bottom: BorderSide(color: Colors.grey, width: 1.0))),
          //商品数量
          child: Text(
            "${goodsNumber}",
            style: TextStyle(
                color: Colors.black54,
                fontSize: ScreenUtil.instance.setSp(26.0)),
          ),
        ),
        InkWell(
          //增加
          onTap: () => _add(),
          child: Container(
            alignment: Alignment.center,
            width: ScreenUtil.instance.setWidth(50),
            height: double.infinity,
            //添加边框样式
            decoration: ShapeDecoration(
                shape: Border(
                    left: BorderSide(color: Colors.grey, width: 1.0),
                    top: BorderSide(color: Colors.grey, width: 1.0),
                    right: BorderSide(color: Colors.grey, width: 1.0),
                    bottom: BorderSide(color: Colors.grey, width: 1.0))),
            //增加符号
            child: Text(
              "+",
              style: TextStyle(
                  color: Colors.black54,
                  fontSize: ScreenUtil.instance.setSp(26.0)),
            ),
          )),
      ],
    ),
  );
}

//减少处理
_reduce() {
  //判断当前商品数量是否大于1
```

```
      if (goodsNumber > 1) {
        setState(() {
          goodsNumber = goodsNumber - 1;
        });
      }
      print("${widget._number}");
      //调用回调方法
      onNumberChange(goodsNumber);
    }

    //增加处理
    _add() {
      setState(() {
        goodsNumber = goodsNumber + 1;
      });
      print("${goodsNumber}");
      //调用回调方法
      onNumberChange(goodsNumber);
    }
  }
```

界面部分针对组件做了一个边框处理。此组件改造后可以作为Flutter项目中的计数器使用，组件使用的效果如图26-2所示。

图26-2 数量加减组件

26.3 分割线组件

分割线主要是对 Divider 组件的一种封装，一般用于列表项下面的线条处理，打开 widgets/divider_line_widget.dart 文件，添加如下代码：

```dart
//widgets/divider_line_widget.dart 文件
import 'package:flutter/material.dart';
import 'package:flutter_screenutil/flutter_screenutil.dart';
//分割线组件
class DividerLineWidget extends StatelessWidget {
  @override
  Widget build(BuildContext context) {
    return Divider(
      //颜色
      color: Colors.grey[350],
      //高度
      height: ScreenUtil.instance.setHeight(1.0),
    );
  }
}
```

使用效果如图 26-3 所示。

图 26-3　分割线

26.4 图文组件

图文组件主要用于有图片加文本的地方，如我的页面里的列表项都是统一的结构；左侧为图标，紧挨着是文本，右侧是一个向右的箭头。

布局采用水平方式布局，组件需要传递一个回调方法，此方法的作用是当单击整个组件时,可以进行下一步的处理,如在我的页面里单击我的订单列表项,可以进入我的订单列表页面。打开 widgets/icon_text_arrow_widget.dart 文件,加入如下代码:

```dart
//widgets/icon_text_arrow_widget.dart 文件
import 'package:flutter/material.dart';
import 'package:flutter_screenutil/flutter_screenutil.dart';

//带有文本的箭头图标组件
class IconTextArrowWidget extends StatelessWidget {
  //图标数据
  final IconData iconData;
  //标题
  final title;
  //单击回调方法
  final VoidCallback callback;
  //图标颜色
  final Color color;

  //构造方法
  IconTextArrowWidget(
    this.iconData,
    this.title,
    this.color,
    this.callback,
  );

  @override
  Widget build(BuildContext context) {
    return Container(
        height: ScreenUtil.getInstance().setHeight(100.0),
        width: double.infinity,
        //单击整个组件回调此方法
        child: InkWell(
          onTap:callback,
          //水平布局
          child: Row(
            //次轴居中
            crossAxisAlignment: CrossAxisAlignment.center,
            children: <Widget>[
              Container(
                margin: EdgeInsets.only(left: 10.0),
                //左侧图标
                child: Icon(
                  iconData,
                  size: ScreenUtil.getInstance().setWidth(40.0),
                  color: color,
                ),
              ),
```

```
        Padding(
          padding: EdgeInsets.only(
              left: ScreenUtil.getInstance().setWidth(20.0)),
        ),
        //标题
        Text(
          title,
          style: TextStyle(
              fontSize: ScreenUtil.getInstance().setSp(26.0),
              color: Colors.black54),
        ),
        //右侧箭头
        Expanded(
          child: Container(
            alignment: Alignment.centerRight,
            margin: EdgeInsets.only(
                right: ScreenUtil.getInstance().setWidth(30.0)),
            child: Icon(
              Icons.arrow_forward_ios,
              color: Colors.grey,
              size: ScreenUtil.getInstance().setWidth(30),
            ),
          ),
        )
      ],
    ),
  ));
  }
}
```

组件在项目中应用的效果如图 26-4 所示。

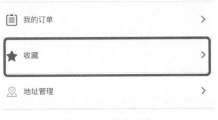

图 26-4　图文组件

26.5　文本组件

文本组件是简化版的图文组件,用于列表项的展示,左侧及右侧为文本内容,单击组件可以触发回调方法。打开 widgets/item_text_widget.dart 文件,添加如下代码:

```dart
//widgets/item_text_widget.dart 文件
import 'package:flutter/material.dart';
import 'package:flutter_screenutil/flutter_screenutil.dart';

//文本组件
class ItemTextWidget extends StatelessWidget {
  //左侧文本
  var leftText;
  //右侧文本
  var rightText;
  //回调方法
  VoidCallback callback;
  //构造方法,回调方法为可选参数
  ItemTextWidget(this.leftText, this.rightText, {this.callback});

  @override
  Widget build(BuildContext context) {
    return InkWell(
      //单击回调
      onTap: () {
        callback();
      },
      child: Container(
        //左右内边距
        padding: EdgeInsets.only(
            left: ScreenUtil.instance.setWidth(20.0),
            right: ScreenUtil.instance.setWidth(20.0)),
        height: ScreenUtil.instance.setHeight(80.0),
        //水平布局
        child: Row(
          children: <Widget>[
            //左侧文本
            Text(
              leftText,
              style: TextStyle(
                  color: Colors.black,
                  fontSize: ScreenUtil.instance.setSp(26.0)),
            ),
            Expanded(
                child: Container(
              alignment: Alignment.centerRight,
              //右侧文本
              child: Text(
                rightText,
                style: TextStyle(
                    color: Colors.grey,
                    fontSize: ScreenUtil.instance.setSp(26.0)),
              ),
            ))
```

```
          ],
        ),
      ),
    );
  }
}
```

在"关于我们"页面里使用了此组件,效果如图 26-5 所示。

图 26-5 文本组件

26.6 加载数据组件

当网络传输不好或者后台接口返回数据较慢时,需要一个加载数据的组件来提示用户。项目中封装了 SpinKitFoldingCube 这个组件来实现加载数据组件。

打开 widgets/loading_dialog_widget.dart 文件,添加如下代码:

```
//widgets/loading_dialog_widget.dart 文件
import 'package:flutter/material.dart';
import 'package:flutter_spinkit/flutter_spinkit.dart';
import 'package:flutter_screenutil/flutter_screenutil.dart';
import 'package:shop/config/index.dart';
//加载数据组件
class LoadingDialogWidget extends StatefulWidget {
  @override
  _LoadingDialogWidgetState createState() => _LoadingDialogWidgetState();
}

class _LoadingDialogWidgetState extends State<LoadingDialogWidget> {
  @override
  Widget build(BuildContext context) {
    return Container(
      child: Center(
        //加载数据组件
        child: SpinKitFoldingCube(
          //大小
          size: ScreenUtil.instance.setWidth(60.0),
          //颜色
```

```
            color: KColor.watingColor,
          ),
        ),
      );
    }
}
```

组件的使用方式如下面代码所示，可以用三元表达式来判断是否有数据：

```
data == null
    //当没有数据时显示正在加载中
    ? LoadingDialogWidget()
    //当有数据时显示内容
    : Container(
        //...
    )
```

组件运行的效果如图 26-6 中间部分所示，几个小点反复地显示和隐藏。

图 26-6　加载数据组件

26.7 没有数据提示组件

当打开的页面没有数据时,可以使用没有数据提示组件来提示用户。这里需要自定义一个组件,使用图片加文本的显示方式。打开 widgets/no_data_widget.dart 文件,添加如下代码:

```dart
//widgets/no_data_widget.dart 文件
import 'package:flutter/material.dart';
import 'package:flutter_screenutil/flutter_screenutil.dart';
import 'package:shop/config/index.dart';
//没有数据提示组件
class NoDataWidget extends StatelessWidget {
  @override
  Widget build(BuildContext context) {
    return Container(
      width: double.maxFinite,
      height: double.maxFinite,
      alignment: Alignment.center,
      //居中显示
      child: Center(
        //垂直布局
        child: Column(
          //垂直居中
          mainAxisAlignment: MainAxisAlignment.center,
          //水平居中
          crossAxisAlignment: CrossAxisAlignment.center,
          children: <Widget>[
            //图片提示
            Image.asset(
              "images/no_data.png",
              height: ScreenUtil.instance.setWidth(120.0),
              width: ScreenUtil.instance.setWidth(120.0),
            ),
            Padding(
              padding: EdgeInsets.only(
                  top: ScreenUtil.instance.setHeight(20.0)),
            ),
            //文本提示
            Text(
              KString.NO_DATA_TEXT,
              style: TextStyle(
                  fontSize: ScreenUtil.instance.setSp(28.0),
                  color: KColor.defaultTextColor),
            )
          ],
        ),
```

```
          ),
        );
      }
    }
```

如包包分类下的皮包,当没有商品时可以使用此组件提示,效果如图 26-7 所示。

图 26-7 没有数据提示组件

26.8 网页加载组件

网页加载组件可以用来加载某个网站或指定网页的 URL,如首页的轮播图通常是商城的广告图片,单击它可以跳转至某个广告页,这个广告页就可以是一个网页,此时就需要用到网页加载组件。

网页加载组件封装了 WebviewScaffold 组件,打开 widgets/webview_widget.dart 文件,添加如下代码:

```
//widgets/webview_widget.dart 文件
import 'package:flutter/material.dart';
```

```dart
import 'package:flutter_webview_plugin/flutter_webview_plugin.dart';

//网页加载组件
class WebViewWidget extends StatelessWidget {
  //路径
  var url;
  //标题
  var title;
  //构造方法
  WebViewWidget(this.url, this.title);

  @override
  Widget build(BuildContext context) {
    return Scaffold(
      appBar: AppBar(
        //显示网页标题
        title: Text(title),
        centerTitle: true,
      ),
      //使用 WebviewScaffold 加载网页
      body: WebviewScaffold(
        url: url,
        withZoom: false,
        withLocalStorage: true,
        withJavascript: true,
      ),
    );
  }
}
```

第 27 章 入口与路由配置

当项目的基础工作准备完成后,就可以从入口程序开始编写应用逻辑代码。本章将处理以下几部分内容:
- 入口程序实现;
- 路由配置;
- 启动页面。

27.1 入口程序

所有的应用都有一个入口程序,通常是 main 函数引导进入应用程序。入口程序主要做以下几方面的处理:
- 自定义主题;
- Fluro 路由初始化;
- Provider 状态管理配置。

入口程序实现步骤如下:

步骤 1:通过自定义将主题应用的导航栏、标题栏菜单等定义为红色风格。代码如下:

```
primaryColor: Colors.red
```

步骤 2:在 main.dart 文件里我们并不做具体的路由设置,只是将路由做一些初始化处理,详细的设置在 router 包里。初始化代码如下:

```
//定义路由
final router = Router();
//配置路由
Routers.configureRoutes(router);
//指定路由至 Application 对象,便于调用
Application.router = router;
```

初始化完成后,需要将 MaterialApp 的生成路由的回调函数指定为 Fluro 库里的

Router 对象的生成器 Application.router.generator。代码如下：

```
//生成路由的回调函数,当导航命名路由的时候,会使用这个来生成界面
onGenerateRoute: Application.router.generator,
```

步骤 3：状态管理顶层配置，用于 Provider 管理多个共享数据类，每当我们实现一个状态管理类则需要将如下模块的代码添加至中括号里。

```
return MultiProvider(
  providers: [
    //各个模块状态管理类
    ...
  ],
```

步骤 4：编写用户信息状态管理部分，打开 provider/user_info.dart 文件，添加如下代码：

```
//provider/user_info.dart 文件
import 'package:flutter/material.dart';
import 'package:shop/model/user_model.dart';

//用户信息状态管理
class UserInfoProvider with ChangeNotifier {
  //用户数据模型
  UserModel userModel;

  //更新用户信息
  updateInfo(UserModel userModel) {
    this.userModel = userModel;
    //通知刷新数据
    notifyListeners();
  }
}
```

这里 UserModel 数据模型还并未实现，代码里添加了一个更新用户信息的方法，当此方法被调用后会使用 notifyListeners 方法来通知页面进行刷新处理。

步骤 5：打开 main.dart 文件，按照上面几个步骤编写如下完整代码：

```
//main.dart 文件
import 'package:flutter/material.dart';
import 'package:fluro/fluro.dart';
import 'package:shop/router/routers.dart';
import 'package:shop/router/application.dart';
import 'package:shop/provider/user_info.dart';
import 'package:provider/provider.dart';
```

```dart
//入口程序
void main() {
  //启动应用
  runApp(ShopApp());
}

//根组件
class ShopApp extends StatelessWidget {
  ShopApp() {
    //定义路由
    final router = Router();
    //配置路由
    Routers.configureRoutes(router);
    //指定路由至 Application 对象,便于调用
    Application.router = router;
  }

  @override
  Widget build(BuildContext context) {
    //管理多个共享数据类
    return MultiProvider(
      providers: [
        //用户信息
        ChangeNotifierProvider(builder: (_) => UserInfoProvider()),
      ],
      child: MaterialApp(
        debugShowCheckedModeBanner: false,
        //生成路由的回调函数,当导航命名路由的时候,会使用这个来生成界面
        onGenerateRoute: Application.router.generator,
        //定义主题
        theme: ThemeData(
          primaryColor: Colors.red,
        ),
      ),
    );
  }
}
```

> **提示** 以上代码中的路由处理并未实现,会在后面一步步添加进来。其中 MultiProvider 一定要放在顶层组件外层,否则将无法共享数据。

27.2 路由配置

此项目里路由采用 Fluro 企业级路由方案,主要完成以下几部分功能:

❑ 应用模块访问路由 Router 对象;

- 路由路径配置；
- 路由 Handler 定义；
- 路由 Handler 实现；
- 路由参数传递；
- 返回路径对应页面。

具体处理步骤如下。

步骤 1：打开 router/application.dart 文件，添加静态路由 Router 对象以便应用的所有模块可以通过以下方式进行访问。

```
Application.router.navigateTo(context, "/");
```

application.dart 完整代码如下：

```dart
//router/application.dart 文件
import 'package:fluro/fluro.dart';

//此类主要用于存放 Router 对象
class Application {
  //Router 对象
  static Router router;
}
```

步骤 2：打开 router/routers.dart 文件，定义路由路径及跳转页面 Handler。完整代码如下：

```dart
//router/routers.dart
import 'package:fluro/fluro.dart';
import 'package:flutter/material.dart';
import 'package:shop/router/router_handlers.dart';

class Routers {
  //根路径,打开启动屏页面
  static String root = "/";
  //主页路径
  static String home = "/home";
  //分类商品页面路径
  static String categoryGoodsList = "/categoryGoodsList";
  //商品详情页面路径
  static String goodsDetail = "/goodsDetail";
  //登录页面路径
  static String login = "/login";
  //注册页面路径
  static String register = "/register";
  //填写订单页面路径
```

```dart
static String fillInOrder = "/fillInOrder";
//地址页面路径
static String address = "/myAddress";
//地址编辑页面路径
static String addressEdit = "/addressEdit";
//我的收藏页面路径
static String mineCollect = "/mineCollect";
//关于我们页面路径
static String aboutUs = "/aboutUs";
//我的订单页面路径
static String mineOrder = "/mineOrder";
//我的订单详情页面路径
static String mineOrderDetail = "/mineOrderDetail";
//网页加载路径
static String webView = "/webView";
//轮播图路径
static String brandDetail = "/brandDetail";

//配置路由
static void configureRoutes(Router router) {

  //找不到路径对应 Handler
  router.notFoundHandler = Handler(handlerFunc:
      (BuildContext context, Map<String, List<String>> parameters) {
    print("没有找到");
  });

  //定义路径对应的 Handler
  router.define(root, handler: loadingHandler);
  router.define(home, handler: homeHandler);
  router.define(categoryGoodsList, handler: categoryGoodsListHandler);
  router.define(login, handler: loginHandler);
  router.define(register, handler: registerHandler);
  router.define(goodsDetail, handler: goodsDetailsHandler);
  router.define(fillInOrder, handler: fillInOrderHandler);
  router.define(address, handler: addressHandler);
  router.define(addressEdit, handler: addressEditHandler);
  router.define(mineCollect, handler: collectHandler);
  router.define(aboutUs, handler: aboutHandler);
  router.define(mineOrder, handler: orderHandler);
  router.define(mineOrderDetail, handler: orderDetailHandler);
  router.define(webView, handler: webViewHandler);
  }
}
```

提示 以上代码定义了一个特殊的 Handler，当程序路径指定错误时，会提示"没有找到"。

步骤 3：打开 router/router_handler.dart 文件，处理各个路由的 Handler。主要处理如下三部分：

- 路由参数获取；
- 路由参数传递；
- 返回路由页面。

完整代码如下：

```dart
//router/router_handlers.dart 文件
import 'package:fluro/fluro.dart';
import 'package:flutter/material.dart';
import 'package:shop/page/home/index_page.dart';
import 'package:shop/page/loading/loading_page.dart';
import 'package:shop/page/goods/goods_category_page.dart';
import 'package:shop/page/login/register_page.dart';
import 'package:shop/page/login/login_page.dart';
import 'package:shop/utils/string_util.dart';
import 'package:shop/page/goods/goods_detail_page.dart';
import 'package:shop/page/order/fill_in_order_page.dart';
import 'package:shop/page/mine/address_page.dart';
import 'package:shop/page/mine/address_edit_page.dart';
import 'package:shop/page/mine/collect_page.dart';
import 'package:shop/page/mine/about_us_page.dart';
import 'package:shop/page/order/order_page.dart';
import 'package:shop/page/order/order_detail_page.dart';
import 'package:shop/widgets/webview_widget.dart';
import 'package:shop/utils/fluro_convert_util.dart';

//首页
var homeHandler = Handler(handlerFunc: (BuildContext context, Map<String, List<String>> parameters) {
  return IndexPage();
});

//加载页
var loadingHandler = Handler(handlerFunc: (BuildContext context, Map<String, List<String>> parameters) {
  return LoadingPage();
});

//商品列表
var categoryGoodsListHandler = Handler(handlerFunc: (BuildContext context, Map<String, List<Object>> parameters) {
  //分类名称
  var categoryName = StringUtil.decode(parameters["categoryName"].first).toString();
  print("categoryName" + categoryName);
  print("categoryId" + parameters["categoryId"].first);
  //分类 Id
```

```dart
    var categoryId = int.parse(parameters["categoryId"].first);
    return GoodsCategoryPage(categoryName: categoryName, categoryId: categoryId);
});

//注册
var registerHandler = Handler(handlerFunc: (BuildContext context, Map<String, List<String>> parameters) {
    return RegisterPage();
});

//登录
var loginHandler = Handler(handlerFunc: (BuildContext context, Map<String, List<String>> parameters) {
    return LoginPage();
});

//商品详情
var goodsDetailsHandler = Handler(handlerFunc: (BuildContext context, Map<String, List<Object>> parameters) {
    //商品 Id
    var goodsId = int.parse(parameters["goodsId"].first);
    return GoodsDetailPage(goodsId: goodsId);
});

//填写订单
var fillInOrderHandler = Handler(handlerFunc: (BuildContext context, Map<String, List<String>> parameters) {
    //购物车 Id
    var cartId = int.parse(parameters["cartId"].first);
    return FillInOrderPage(cartId);
});

//地址
var addressHandler = Handler(handlerFunc: (BuildContext context, Map<String, List<String>> parameters) {
    return AddressPage();
});

//地址修改
var addressEditHandler = Handler(handlerFunc: (BuildContext context, Map<String, List<String>> parameters) {
    //地址 Id
    var addressId = int.parse(parameters["addressId"].first);
    return AddressEditPage(addressId);
});

//收藏
var collectHandler = Handler(handlerFunc: (BuildContext context, Map<String, List<String>> parameters) {
```

```
  return CollectPage();
});

//关于我们
var aboutHandler = Handler(handlerFunc: (BuildContext context, Map<String, List<String>>
parameters) {
  return AboutUsPage();
});

//订单
var orderHandler = Handler(handlerFunc: (BuildContext context, Map<String, List<String>>
parameters) {
  return OrderPage();
});

//订单详情
var orderDetailHandler = Handler(handlerFunc: (BuildContext context, Map<String, List
<String>> parameters) {
  //订单Id
  int orderId = int.parse(parameters["orderId"].first);
  //token值
  String token = parameters["token"].first;
  return OrderDetailPage(orderId, token);
});

//网页加载
var webViewHandler = Handler(handlerFunc: (BuildContext context, Map<String, List<String>>
parameters) {
  //标题
  var title = FluroConvertUtil.fluroCnParamsDecode(parameters["title"].first);
  //路径
  var url = FluroConvertUtil.fluroCnParamsDecode(parameters["url"].first);
  return WebViewWidget(url, title);
});
```

> **提示** 以上步骤中定义的路由页面如果还没有实现,可以先注释起来,这并不影响程序的运行。

27.3 加载页面

其实加载页面和普通的页面并没有什么两样,唯一的区别是加载页面伴随着应用程序的加载而完成。例如:以商城App为例,应用程序需要在此阶段完成连接socket、下载资源文件,以及请求广告信息等。由于这个过程是需要时间处理的,所以这个页面需要停留一定的时间,通常设置成几秒即可。代码如下:

```
Future.delayed(Duration(seconds: 3),(){
  //TODO
}
```

上面的代码使用 Future.delayed 函数,使得加载页面停留 3 秒,3 秒过后页面马上要跳转到应用程序页面,否则应用将永远停留在加载页面。应用跳转采用 Navigator 导航器实现,代码如下:

```
NavigatorUtil.goShopMainPage(context);
```

加载页面通常为广告页面及产品介绍页面。例如用户首次使用 App 时要引导用户如何使用主要功能。在这里我们用一张漂亮的背景图片作为加载页面的全部展示内容。这里我们需要打开 page/loading/loading_page.dart 文件,添加下面的所有代码:

```
//page/loading/loading_page.dart 文件
import 'package:flutter/material.dart';
import 'package:shop/utils/navigator_util.dart';

//加载页面
class LoadingPage extends StatefulWidget {
  @override
  _LoadingPageState createState() => _LoadingPageState();
}

class _LoadingPageState extends State<LoadingPage> {
  @override
  void initState() {
    super.initState();
    //延迟 3 秒执行
    Future.delayed(Duration(seconds: 3), () {
      //跳转至应用首页
      NavigatorUtil.goShopMainPage(context);
    });
  }

  @override
  Widget build(BuildContext context) {
    return Scaffold(
      body: Container(
        color: Colors.white,
        //添加加载图片,即广告图
        child: Image.asset(
          "images/loading.png",
          width: double.infinity,
          height: double.infinity,
          //指定填充模式
          fit: BoxFit.fill,
```

```
        ),
      ),
    );
  }
}
```

> **注意** 加载页面停顿要放在 initState 函数里处理,原因是必须等待页面渲染完成,否则加载的画面内容就看不到了。

这里读者可能会有个疑问,为什么应用启动时首先打开的是加载页面,这主要是因为我们在路由的根路径配置的是加载页面,代码如下:

```
//根路径
static String root = "/";
//根路径对应的 Handler
router.define(root, handler: loadingHandler);
//加载页 Handler
var loadingHandler = Handler(handlerFunc: (BuildContext context, Map<String, List<String>> parameters) {
  //返回加载页面
  return LoadingPage();
});
```

加载页面的效果如图 27-1 所示。

图 27-1 加载页面

第 28 章 首 页

首页是商城应用的第一个功能页面,用来展示商城主要的信息,由广告 Banner、分类、最新商品和最热商品等组成。

从本章开始一步步带领大家实现商城的功能模块。

28.1 索引页面

在编写首页之前先实现一个索引页面。索引页面采用当前主流的底部导航栏式的布局,即把这个主要页面通过选项卡 Tab 的方式进行切换并且可以打开。例如,微信、支付宝、淘宝和京东等 App 均采用这种布局方式。

索引页所对应的页面目前均未实现,你可以先建一个空页面,保证不出错即可,页面如下所示:

- 首页;
- 分类;
- 购物车;
- 我的。

打开 page/home/index_page.dart 文件,添加如下代码:

```
//page/home/index_page.dart 文件
import 'package:flutter/material.dart';
import 'package:shop/page/home/home_page.dart';
import 'package:shop/config/index.dart';
import 'package:shop/page/category/category_page.dart';
import 'package:shop/page/cart/cart_page.dart';
import 'package:shop/page/mine/mine_page.dart';

//索引页面用于切换几个主要页面
class IndexPage extends StatefulWidget {
  @override
  _IndexPageState createState() => _IndexPageState();
}
```

```dart
class _IndexPageState extends State<IndexPage> {
  //当前选项卡索引
  int _selectedIndex = 0;
  //主要页面列表
  List<Widget> _list = List();

  @override
  void initState() {
    super.initState();
    //添加主要页面至列表里,使用..级联操作
    _list
      ..add(HomePage())
      ..add(CategoryPage())
      ..add(CartPage())
      ..add(MinePage());
  }

  //选项卡按钮单击回调
  void _onItemTapped(int index) {
    setState(() {
      //设置当前索引状态值
      _selectedIndex = index;
    });
  }

  @override
  Widget build(BuildContext context) {
    return Scaffold(
      //页面根据索引进行切换
      body: IndexedStack(
        //当前索引
        index: _selectedIndex,
        //绑定页面列表
        children: _list,
      ),
      //页面底部选项卡
      bottomNavigationBar: BottomNavigationBar(
        type: BottomNavigationBarType.fixed,
        items: const <BottomNavigationBarItem>[
          //首页
          BottomNavigationBarItem(
            icon: Icon(Icons.home),
            title: Text(KString.HOME),
          ),
          //分类
          BottomNavigationBarItem(
            icon: Icon(Icons.category),
            title: Text(KString.CATEGORY),
          ),
          //购物车
          BottomNavigationBarItem(
```

```
          icon: Icon(Icons.shopping_cart),
          title: Text(KString.SHOP_CAR),
        ),
        //我的
        BottomNavigationBarItem(
          icon: Icon(Icons.person),
          title: Text(KString.MINE),
        ),
      ],
      //绑定当前索引
      currentIndex: _selectedIndex,
      //选中项的颜色
      selectedItemColor: KColor.indexTabSelectedColor,
      //未选中项的颜色
      unselectedItemColor: KColor.indexTabUnSelectedColor,
      //单击回调方法
      onTap: _onItemTapped,
    ),
  );
}
```

上面的代码通过索引来切换不同的页面，这里还不用做任何数据方面的处理，效果如图 28-1 所示。

图 28-1　索引页面

28.2 首页数据模型

在编写首页界面之前首先准备好数据模型。有了数据模型,我们在编写界面时才可以把字段值赋值进去。数据模型包含以下几个列表数据:

- 新品列表;
- 分类列表;
- 轮播图列表;
- 热卖商品列表。

首页数据模型还包括几个列表数据对应的模型,具体实现步骤如下:
步骤1:打开首页数据应用接口地址,URL如下:

```
http://localhost:8080/client/home/index
```

通过返回的Json数据查看其具备哪些字段并将其整理出来,如下面数据:

```
{
    errno: 0,
    //数据
    data: {
        //新品数据列表
        newGoodsList: [
            {
                id: 1181007,
                name: "羊皮衣真皮外套",
                brief: "复古廓形机车进口绵羊皮衣真皮外套女 E142",
                picUrl: "http://localhost:8080/client/storage/fetch/ln3n7j2ucxqn38m8elbd.jpg",
                isNew: true,
                isHot: true,
                counterPrice: 588,
                retailPrice: 388
            },
            //...
        ],
        //分类数据列表
        channel: [
            {
                id: 1036008,
                name: "服装",
                iconUrl: "http://localhost:8080/client/storage/fetch/yky3aox7jme0u2bjfbnz.png"
            },
            //...
        ],
        //轮播数据列表
```

```
            banner: [
                {
                    id: 5,
                    name: "首页轮播图 1",
                    link: "",
                    url: "http://localhost:8080/client/storage/fetch/u4kbfavifvb7cyfiysut.jpeg",
                    position: 1,
                    content: "首页轮播图 1",
                    enabled: true,
                    addTime: "2019 - 10 - 31 11:56:14",
                    updateTime: "2019 - 10 - 31 11:56:14",
                    deleted: false
                },
                //...
            ],
            //热卖商品列表
            hotGoodsList: [
                {
                    id: 1181007,
                    name: "羊皮衣真皮外套",
                    brief: "复古廓形机车进口绵羊皮衣真皮外套女 E142",
                    picUrl: "http://localhost:8080/client/storage/fetch/ln3n7j2ucxqn38m8elbd.jpg",
                    isNew: true,
                    isHot: true,
                    counterPrice: 588,
                    retailPrice: 388
                },
                //...
            ]
        },
        //消息
        errmsg: "成功"
}
```

提示 后端接口返回的 Json 数据是你编写数据模型的依据，字段一定要相同并且不能写错。后面的数据如分类、商品详情和订单等数据均采用这种方式分析即可。

步骤 2：编写数据模型，根据接口返回的 Json 数据添加字段，同时添加每个模型对应的 fromJson 和 toJson 方法。打开 model/home_model.dart 文件，添加如下代码：

```
//model/home_model.dart 文件
import 'package:json_annotation/json_annotation.dart';

part 'home_model.g.dart';

//首页数据模型
```

```dart
@JsonSerializable()
class HomeModel extends Object {

  //新品列表
  @JsonKey(name: 'newGoodsList')
  List<Goods> newGoodsList;

  //分类列表
  @JsonKey(name: 'channel')
  List<Channel> channel;

  //轮播图列表
  @JsonKey(name: 'banner')
  List<BannerModel> banner;

  //热卖商品列表
  @JsonKey(name: 'hotGoodsList')
  List<Goods> hotGoodsList;

  //构造方法
  HomeModel(this.newGoodsList,this.channel,this.banner,this.hotGoodsList,);

  factory HomeModel.fromJson(Map<String, dynamic> srcJson) => _$HomeModelFromJson(srcJson);

  Map<String, dynamic> toJson() => _$HomeModelToJson(this);

}

//商品数据模型
@JsonSerializable()
class Goods extends Object {

  //商品 id
  @JsonKey(name: 'id')
  int id;

  //商品名称
  @JsonKey(name: 'name')
  String name;

  //商品简介
  @JsonKey(name: 'brief')
  String brief;

  //商品图片
  @JsonKey(name: 'picUrl')
  String picUrl;
```

```dart
  //是否为新品
  @JsonKey(name: 'isNew')
  bool isNew;

  //是否为热卖商品
  @JsonKey(name: 'isHot')
  bool isHot;

  //专柜价格
  @JsonKey(name: 'counterPrice')
  double counterPrice;

  //零售价格
  @JsonKey(name: 'retailPrice')
  double retailPrice;

  Goods(this.id,this.name,this.brief,this.picUrl,this.isNew,this.isHot,this.counterPrice,this.retailPrice,);

  factory Goods.fromJson(Map<String, dynamic> srcJson) => _$GoodsFromJson(srcJson);

  Map<String, dynamic> toJson() => _$GoodsToJson(this);

}

//分类数据
@JsonSerializable()
class Channel extends Object {
  //分类 id
  @JsonKey(name: 'id')
  int id;
  //分类名称
  @JsonKey(name: 'name')
  String name;
  //分类图标地址
  @JsonKey(name: 'iconUrl')
  String iconUrl;

  Channel(this.id,this.name,this.iconUrl,);

  factory Channel.fromJson(Map<String, dynamic> srcJson) => _$ChannelFromJson(srcJson);

  Map<String, dynamic> toJson() => _$ChannelToJson(this);

}

//轮播图数据模型
@JsonSerializable()
```

```dart
class BannerModel extends Object {

  //轮播图 id
  @JsonKey(name: 'id')
  int id;

  //名称
  @JsonKey(name: 'name')
  String name;

  //轮播图链接
  @JsonKey(name: 'link')
  String link;

  //轮播图地址
  @JsonKey(name: 'url')
  String url;

  //位置
  @JsonKey(name: 'position')
  int position;

  //内容
  @JsonKey(name: 'content')
  String content;

  //是否启用
  @JsonKey(name: 'enabled')
  bool enabled;

  //添加时间
  @JsonKey(name: 'addTime')
  String addTime;

  //更新时间
  @JsonKey(name: 'updateTime')
  String updateTime;

  //是否删除
  @JsonKey(name: 'deleted')
  bool deleted;

  BannerModel(this.id,this.name,this.link,this.url,this.position,this.content,this.enabled,this.addTime,this.updateTime,this.deleted,);

  factory BannerModel.fromJson(Map<String,dynamic> srcJson) => _$BannerModelFromJson(srcJson);

  Map<String, dynamic> toJson() => _$BannerModelToJson(this);

}
```

步骤3：每个模型对应的 fromJson 和 toJson 并未实现，这个需要使用命令来为其生成扩展代码。进入 shop-flutter 项目根目录，使用如下命令生成即可：

```
flutter packages pub run build_runner build -- delete-conflicting-outputs
```

生成后会多出一个 home_model.g.dart 文件，此文件是 home_model.dart 文件的组成部分。具体生成的步骤和遇到的问题参考 20.4 节"元数据应用"。

提示　定义数据模型，并使用命令生成扩展文件的做法是一种软件开发工程化的体现。因为数据接口较多，并且可能会变更，不可能每次都手动添加处理代码。后面数据模型生成的命令统称为"build_runner"工具。

28.3　首页布局拆分

商城首页展示的内容较多，界面相对复杂，所以需要对整体布局做一个拆分处理。这样按照拆分的情况首先实现各个模块，然后再将其串联起来，加上数据处理即可完成整个页面的渲染。

首先看一下首页的整体布局拆分图，如图 28-2 中三个矩形框所示。

图 28-2　首页布局拆分

从拆分图上可以看出首页整体采用垂直布局，分为如下几大部分：
- 轮播广告图：采用左右循环切换图片的方式展示；
- 首页分类：展示商城一级分类信息，只显示前 10 大分类；
- 最新商品：展示商品后台勾选为最新商品的数据，采用两列多行的形式；
- 最热商品：展示商品后台勾选为最热商品的数据，采用两列多行的形式。

其中，最新商品和最热商品展示的形式一致。由于首页内容较多，页面可能会很长，所以需要在最外层包裹一个滚动组件。在顶部标题的最右侧可以实现搜索功能。

28.4 轮播图实现

首页轮播图主要用来轮播商城广告图片，这里需要封装一个轮播图组件。使用 Swiper 组件给其指定图片数量、图片 URL 和间隔时长，这样就可以实现按照固定时间左右切换广告图的效果。打开 home/home_banner.dart 文件，添加如下代码：

```dart
//home/home_banner.dart 文件
import 'package:flutter_swiper/flutter_swiper.dart';
import 'package:flutter/material.dart';
import 'package:flutter_screenutil/flutter_screenutil.dart';
import 'package:shop/config/index.dart';
import 'package:shop/model/home_model.dart';
import 'package:shop/utils/navigator_util.dart';
import 'package:shop/widgets/cached_image_widget.dart';

//首页轮播图组件
class HomeBannerWidget extends StatelessWidget {
  //轮播图数据
  List<BannerModel> bannerData = List();
  //数量
  int size;
  //视图高度
  double viewHeight;
  //构造方法,传入轮播图数据、大小,以及视图高度
  HomeBannerWidget(this.bannerData, this.size, this.viewHeight);

  @override
  Widget build(BuildContext context) {
    return Container(
      height: viewHeight,
      width: double.infinity,
      child: bannerData == null || bannerData.length == 0
          //没有数据提示
          ? Container(
              height: ScreenUtil.instance.setHeight(400.0),
              color: Colors.grey,
```

```
            //居中显示
            alignment: Alignment.center,
            //没有数据文本提示
            child: Text(KString.NO_DATA_TEXT),
          )
        //轮播组件
        : Swiper(
            //单击跳转
            onTap: (index) {
              NavigatorUtil.goWebView(context, bannerData[index].name, bannerData[index].link);
            },
            //轮播图数量
            itemCount: bannerData.length,
            //指定滚动方向
            scrollDirection: Axis.horizontal,
            //滚动方向,设置为 Axis.vertical 则垂直滚动
            loop: true,
            //无限轮播模式开关
            index: 0,
            //初始的时候下标位置
            autoplay: false,
            itemBuilder: (BuildContext buildContext, int index) {
              //使用缓存图片组件显示轮播图
              return CachedImageWidget(
                  double.infinity, double.infinity, bannerData[index].url);
            },
            //滚动时长
            duration: 10000,
            //指定分页组件
            pagination: SwiperPagination(
                //底部居中展示
                alignment: Alignment.bottomCenter,
                //构建小圆点
                builder: DotSwiperPaginationBuilder(
                  size: 8.0,
                  //未单击颜色
                  color: KColor.bannerDefaultColor,
                  //单击颜色
                  activeColor: KColor.bannerActiveColor)),
          ),
    );
  }
}
```

通过上述代码可以看出,轮播图组件可以提取一个公共组件供使用,例如商品详情的主图显示也可以使用它。这里需要使用缓存图片组件 CachedImageWidget,因为广告图一般较大,避免重复下载,最终运行效果如图 28-3 所示。单击小点可以切换到对应的图片,也

可以左右滑动图片进行切换。

图 28-3　轮播图效果

28.5　首页分类实现

首页分类的作用是使得用户可以快速打开分类相关商品的页面,挑选想购买的商品。此组件采用网格布局,取一级分类前 10 项数据,两行五列进行排列分类项。单击单元格可以跳转至分类商品列表页面。

打开 page/home/home_category.dart 文件,添加如下代码:

```dart
//page/home/home_category.dart 文件
import 'package:flutter/material.dart';
import 'package:flutter_screenutil/flutter_screenutil.dart';
import 'package:shop/model/home_model.dart';
import 'package:shop/utils/navigator_util.dart';
import 'package:shop/widgets/cached_image_widget.dart';
//首页分类组件
class HomeCategoryWidget extends StatelessWidget {
  //首页分类列表数据
  List<Channel> categoryList;
  //传入分类列表
  HomeCategoryWidget(this.categoryList);
  //跳转到指定分类的商品列表页面
  _goCategoryView(BuildContext context, Channel channel) {
    //传入分类名称及分类 Id
    NavigatorUtil.goCategoryGoodsListPage(context, channel.name, channel.id);
  }

  @override
  Widget build(BuildContext context) {
    return Container(
      //网格布局,2 行 5 列
```

```dart
        child: GridView.builder(
          physics: NeverScrollableScrollPhysics(),
          shrinkWrap: true,
          //分类项个数
          itemCount: categoryList.length,
          //项构建器
          itemBuilder: (BuildContext context, int index) {
            //返回单元格项
            return _getGridViewItem(context, categoryList[index]);
          },
          gridDelegate: SliverGridDelegateWithFixedCrossAxisCount(
            //单个子 Widget 的水平最大宽度
            crossAxisCount: 5,
            //水平单个子 Widget 之间间距
            mainAxisSpacing: ScreenUtil.instance.setWidth(20.0),
            //垂直单个子 Widget 之间间距
            crossAxisSpacing: ScreenUtil.instance.setWidth(20.0),
          ),
        ));
  }

  //网格单元格,展示一个分类信息
  Widget _getGridViewItem(BuildContext context, Channel channel) {
    return Center(
        //单击跳转到分类商品页面
        child: InkWell(
          onTap: () => _goCategoryView(context, channel),
          child: Column(
            children: <Widget>[
              //分类图标
              CachedImageWidget(
                  ScreenUtil.instance.setWidth(60.0),
                  ScreenUtil.instance.setWidth(60.0),
                  channel.iconUrl
              ),
              Padding(
                padding: EdgeInsets.only(top: ScreenUtil.instance.setWidth(10.0)),
              ),
              //分类名称
              Text(
                channel.name,
                style: TextStyle(
                    fontSize: ScreenUtil.instance.setSp(26.0),
                    color: Colors.black87),
              )
            ],
          ),
        ),
    );
  }
}
```

代码中，categoryList 分类列表数据是由父组件首页传递过来的，分类效果如图 28-4 所示。

图 28-4　首页分类

28.6　首页产品实现

首页的产品包括新品和热卖商品，这两种商品的布局及数据模型是一样的，所以可以封装成一个组件。布局采用两列多行的形式，通过传入的商品数据列表 productList 渲染数据。当单击某一个单元格时可以跳转至商品详情页面。打开 home/home_product.dart 文件，添加如下代码：

```
//home/home_product.dart 文件
import 'package:flutter/material.dart';
import 'package:flutter_screenutil/flutter_screenutil.dart';
import 'package:shop/model/home_model.dart';
import 'package:shop/utils/navigator_util.dart';
import 'package:shop/widgets/cached_image_widget.dart';
import 'package:shop/config/index.dart';

//首页产品组件,用于显示最新产品及热卖产品
class HomeProductWidget extends StatelessWidget {
  //产品数据列表
  List<Goods> productList;
  //构造方法
  HomeProductWidget(this.productList);

  @override
  Widget build(BuildContext context) {
    return Container(
      padding: EdgeInsets.all(5.0),
      //网格组件
      child: GridView.builder(
        shrinkWrap: true,
        //产品个数
        itemCount: productList.length,
        physics: NeverScrollableScrollPhysics(),
```

```
      gridDelegate: SliverGridDelegateWithFixedCrossAxisCount(
        //列的数量
        crossAxisCount: 2,
        //单元格宽高比系数
        childAspectRatio: 0.90
      ),
      //根据索引构建单元格
      itemBuilder: (BuildContext context, int index) {
        //返回单元格,传入商品数据
        return _getGridItemWidget(context, productList[index]);
      }),
  );
}

//跳转至产品详情页面
_goGoodsDetail(BuildContext context, Goods goods) {
  NavigatorUtil.goGoodsDetails(context, goods.id);
}

//返回单元格,传入商品数据
Widget _getGridItemWidget(BuildContext context, Goods productModel) {
  return Container(
    //单击跳转至商品详情页面
    child: InkWell(
      onTap: () => _goGoodsDetail(context, productModel),
      child: Card(
        elevation: 2.0,
        margin: EdgeInsets.all(6.0),
        //垂直布局
        child: Column(
          //垂直方向顶部对齐
          mainAxisAlignment: MainAxisAlignment.start,
          children: <Widget>[
            Container(
              margin: EdgeInsets.all(5.0),
              //商品图片
              child: CachedImageWidget(
                ScreenUtil.instance.setHeight(200.0),
                ScreenUtil.instance.setHeight(200.0),
                //图片路径
                productModel.picUrl
              ),
            ),
            Padding(
              padding: EdgeInsets.only(top: 4.0),
            ),
            //商品名称
            Container(
              padding: EdgeInsets.only(left: 4.0, top: 4.0),
              alignment: Alignment.centerLeft,
              child: Text(
```

```
              productModel.name,
              maxLines: 1,
              overflow: TextOverflow.ellipsis,
              style: TextStyle(color: Colors.black54, fontSize: 14.0),
            ),
          ),
          Padding(
            padding: EdgeInsets.only(top: 4.0),
          ),
          //商品零售价
          Container(
            padding: EdgeInsets.only(left: 4.0, top: 4.0),
            alignment: Alignment.center,
            child: Text(
              "￥${productModel.retailPrice}",
              style: TextStyle(color: KColor.priceColor, fontSize: 12.0),
            ),
          )
        ],
      ),
    ),
  );
 }
}
```

以热卖商品为例，展示效果如图 28-5 所示。

图 28-5　首页产品布局

28.7 首页数据服务

首页的拆分组件实现后,还不能马上组装首页,需要先把首页的数据服务写好供首页调用。

数据服务需要定义几个回调方法,成功了则返回数据,失败了则返回错误信息,尤其是服务端返回的错误信息一定要提示到位,这样才能在开发或维护过程中更快速地解决问题。定义方法如下:

```
//定义成功返回列表数据
typedef OnSuccessList<T>(List<T> list);
//定义成功返回数据
typedef OnSuccess<T>(T t);
//定义返回失败消息
typedef OnFail(String message);
```

这里关于成功的回调方法有两个,可以返回列表数据和非列表数据。失败回调方法一个即可,以便返回错误信息。参数采用范型 T,这样可以指定不同的数据模型。

接下来编写请求服务端数据的方法,以及返回数据后如何处理。打开 service/home_service.dart 文件,添加如下完整代码:

```
//service/home_service.dart 文件
import 'package:shop/utils/http_util.dart';
import 'package:shop/config/index.dart';
import 'package:shop/config/string.dart';
import 'package:shop/model/home_model.dart';

//定义成功返回列表数据
typedef OnSuccessList<T>(List<T> list);
//定义成功返回数据
typedef OnSuccess<T>(T t);
//定义返回失败消息
typedef OnFail(String message);

//首页数据服务
class HomeService {

  //查询首页数据,异步处理
  Future queryHomeData(OnSuccess onSuccess, OnFail onFail) async {
    try {
      //首页数据接口调用
      var response = await HttpUtil.instance.get(ServerUrl.HOME_URL);
      if (response['errno'] == 0) {
        //Json 数据转换成 HomeModel
```

```
            HomeModel homeModel = HomeModel.fromJson(response['data']);
            //调用成功回调方法,返回数据
            onSuccess(homeModel);
        } else {
            //调用失败回调方法,返回错误信息
            onFail(response['errmsg']);
        }
    } catch (e) {
        print(e);
        //调用失败回调方法,返回服务器异常信息
        onFail(KString.SERVER_EXCEPTION);
    }
  }
}
```

从以上代码可以看到,这里 queryHomeData 方法需要传入成功与失败回调方法。方法加了 async 表示方法里有异步处理。因为需要等待服务器数据返回,所以在 HttpUtil. instance.get 方法加入了 await 关键字。当服务端返回 Json 数据后,再将其转换成数据模型 HomeModel,然后调用成功回调方法后返回最终的数据,否则返回错误信息。

> **提示** 首页数据服务是一个标准的写法,如回调方法定义、参数传递、后端请求处理,以及返回数据转换等。商城项目中其他服务均采用这种标准写法。

28.8 组装首页

当首页数据模型、拆分组件及数据服务全部准备完毕后,就可以组装首页,并展示首页数据了。具体步骤如下:

步骤 1:打开 home/home_page.dart 文件,新建 HomePage 页面,定义首页数据服务和首页数据模型,代码如下:

```
//首页数据接口服务
HomeService _homeService = HomeService();
//首页数据模型
HomeModel _homeModel;
```

步骤 2:在页面初始化完成后,调用首页数据服务方法以便开始查询数据,大致处理如下:

```
_queryHomeData() {
  _homeService.queryHomeData((success) {
```

```
      //设置状态..
    });
  }, (error) {
    //错误揭示...
  });
}
```

步骤3：在首页的 build 方法添加屏幕适配工具 ScreenUtil 的初始化处理，它的作用是初始一个固定的宽和高，这样在不同的机型屏幕下可以根据一个系数进行缩放处理。代码如下：

```
ScreenUtil.instance = ScreenUtil(width: 750, height: 1334)..init(context);
```

步骤4：添加 Column 组件，将各个拆分组件组装起来，大致结构如下：

```
//轮播图组件
HomeBannerWidget(_homeModel.banner, _homeModel.banner.length,...),
//...
//首页分类组件
HomeCategoryWidget(_homeModel.channel),
//...
//最新产品组件
HomeProductWidget(_homeModel.newGoodsList),
//...
//最热产品组件
HomeProductWidget(_homeModel.hotGoodsList),
```

步骤5：当这些组件组装完成后，还需要在外围套一层滚动组件 SingleChildScrollView，这样首页就可以上下滚动页面了。另外还需要再套一层刷新组件 EasyRefresh，用户可以上下拖动以便刷新首页数据。具体细节参看下面完整代码：

```
//home/home_page.dart 文件
import 'package:flutter/material.dart';
import 'package:flutter_screenutil/flutter_screenutil.dart';
import 'package:flutter_easyrefresh/bezier_bounce_footer.dart';
import "package:flutter_easyrefresh/easy_refresh.dart";
import 'package:flutter_easyrefresh/bezier_circle_header.dart';
import 'package:shop/service/home_service.dart';
import 'package:shop/page/home/home_banner.dart';
import 'package:shop/page/home/home_category.dart';
import 'package:shop/config/index.dart';
import 'package:shop/page/home/home_product.dart';
import 'package:shop/model/home_model.dart';
import 'package:shop/utils/toast_util.dart';
import 'package:shop/widgets/loading_dialog_widget.dart';
```

```dart
import 'package:shop/utils/shared_preferences_util.dart';
//首页
class HomePage extends StatefulWidget {
  @override
  _HomePageState createState() => _HomePageState();
}

class _HomePageState extends State<HomePage> {
  //首页数据接口服务
  HomeService _homeService = HomeService();
  //首页数据模型
  HomeModel _homeModel;
  //刷新组件控制器
  EasyRefreshController _controller = EasyRefreshController();

  @override
  void initState() {
    super.initState();
    SharedPreferencesUtil.getToken().then((token){
    });
    //查询首页数据
    _queryHomeData();
  }
  //查询首页数据
  _queryHomeData() {
    _homeService.queryHomeData((success) {
      setState(() {
        _homeModel = success;
      });
      _controller.finishRefresh();
    }, (error) {
      //弹出错误信息
      ToastUtil.showToast(error);
      //结束刷新
      _controller.finishRefresh();
    });
  }

  @override
  Widget build(BuildContext context) {
    //初始化屏幕适配工具
    ScreenUtil.instance = ScreenUtil(width: 750, height: 1334)..init(context);
    return Scaffold(
      //页面顶部组件
      appBar: AppBar(
        //首页标题
        title: Text(KString.HOME_TITLE),
        centerTitle: true,
        actions: <Widget>[
```

```dart
            IconButton(
                icon: Icon(
                  Icons.search,
                  color: Colors.white,
                ),
                onPressed: (){},
            )
          ],
        ),
        body: contentWidget());
}

Widget contentWidget() {
    return _homeModel == null
        //当没有数据时显示正在加载中
        ? LoadingDialogWidget()
        : Container(
            //数据刷新组件
            child: EasyRefresh(
              //控制器
              controller: _controller,
              //刷新组件头部
              header: BezierCircleHeader(backgroundColor: Colors.redAccent),
              //刷新组件底部
              footer: BezierBounceFooter(backgroundColor: Colors.redAccent),
              enableControlFinishRefresh: true,
              enableControlFinishLoad: false,
              //滚动组件,这样首页可以上下滚动
              child: SingleChildScrollView(
                  child: Column(
                    children: <Widget>[
                      //轮播图组件
                      HomeBannerWidget(_homeModel.banner, _homeModel.banner.length,
                          ScreenUtil.instance.setHeight(360.0)),
                      Padding(
                        padding: EdgeInsets.only(top: 10.0),
                      ),
                      //首页分类组件
                      HomeCategoryWidget(_homeModel.channel),
                      Container(
                        height: 40.0,
                        alignment: Alignment.center,
                        child: Text(KString.NEW_PRODUCT),
                      ),
                      //最新商品组件
                      HomeProductWidget(_homeModel.newGoodsList),
                      Container(
                        height: 40.0,
                        alignment: Alignment.center,
```

```
                  child: Text(KString.HOT_PRODUCT),
                ),
                //最热商品组件
                HomeProductWidget(_homeModel.hotGoodsList),
              ],
          )),
          //刷新回调方法
          onRefresh: () async {
            //重新查询首页数据
            _queryHomeData();
            //调用完成刷新方法
            _controller.finishRefresh();
          },
        ),
      );
  }
}
```

运行项目后首页的显示效果如图 28-6 所示。

图 28-6　首页展示

第 29 章 分 类

分类是商城项目中重要的模块之一,通常有一级、二级,甚至三级分类。由于商城中商品较多,用户可以通过分类快速找到自己想要的产品。

本章将详细介绍分类的布局、数据模型,以及数据服务等内容。分类下的商品列表展示将放在商品模块进行讲解。

29.1 分类数据模型

在编写分类界面之前首先准备好数据模型。有了数据模型,我们在编写界面时才可以把字段值赋值进去。数据模型包含以下几部分:

- 一级分类数据模型;
- 一级分类数据列表模型;
- 二级分类模型;
- 二级分类数据列表模型。

商城项目里主要做了两级分类,具体实现步骤如下:

步骤 1:首先获取一级分类数据,URL 如下:

```
http://localhost:8080/client/catalog/getfirstcategory
```

通过返回的 Json 数据查看其具备哪些字段并将其整理出来,如下所示:

```
{
  errno: 0,
  data: [
    {
      id: 1036008,
      name: "服装",
      keywords: "服装",
      desc: "服装",
      pid: 0,
      iconUrl: "http://localhost:8080/client/storage/fetch/yky3aox7jme0u2bjfbnz.png",
```

```
        picUrl: "http://localhost:8080/client/storage/fetch/wk46d6qbxbh3vhes421y.png",
        level: "L1",
        sortOrder: 50,
        addTime: "2019 - 10 - 31 12:00:13",
        updateTime: "2019 - 10 - 31 17:08:39",
        deleted: false
    },
    //...
  ],
  errmsg: "成功"
}
```

具体每个字段的含义可以参考数据库里的分类表,然后再查询二级分类数据,URL 如下所示:

```
http://localhost:8080/client/catalog/getsecondcategory?id = 1036008
```

二级分类的参数多了一个 id,即父 id 的值,返回的 Json 数据如下:

```
{
    errno: 0,
    data: [
        {
            id: 1036010,
            name: "女装",
            keywords: "女装",
            desc: "女装",
            pid: 1036008,
            iconUrl: "http://localhost:8080/client/storage/fetch/51uq8a5ib3ow36sbxuhj.png",
            picUrl: "http://localhost:8080/client/storage/fetch/hxzr4v4j41rair56pe6k.png",
            level: "L2",
            sortOrder: 50,
            addTime: "2019 - 10 - 31 12:01:44",
            updateTime: "2019 - 10 - 31 16:46:12",
            deleted: false
        },
        //...
    ],
    errmsg: "成功"
}
```

父 id 是服装分类对应的 id,上面查询的结果即为服务下的所有分类。

步骤 2:编写数据模型,根据接口返回的 Json 数据添加字段,同时添加每个模型对应的 fromJson 和 toJson 方法。打开 model/first_category_model.dart 文件,添加一级分类数据模型,代码如下:

```dart
//model/first_category_model.dart 文件
import 'package:json_annotation/json_annotation.dart';

//扩展文件
part 'first_category_model.g.dart';

//一级分类数据列表模型
class FirstListCategoryModel {
  //一级分类数据列表
  List<FirstCategoryModel> firstLevelCategory;
  //构造方法
  FirstListCategoryModel(this.firstLevelCategory);

  factory FirstListCategoryModel.fromJson(List<dynamic> parseJson) {
    List<FirstCategoryModel> firstLevelCategorys;
    firstLevelCategorys =
        parseJson.map((i) => FirstCategoryModel.fromJson(i)).toList();
    return FirstListCategoryModel(firstLevelCategorys);
  }
}

//一级分类数据模型
@JsonSerializable()
class FirstCategoryModel extends Object {
  //id
  @JsonKey(name: 'id')
  int id;
  //名称
  @JsonKey(name: 'name')
  String name;
  //关键字
  @JsonKey(name: 'keywords')
  String keywords;
  //描述
  @JsonKey(name: 'desc')
  String desc;
  //父分类 id
  @JsonKey(name: 'pid')
  int pid;
  //分类图标
  @JsonKey(name: 'iconUrl')
  String iconUrl;
  //分类图片
  @JsonKey(name: 'picUrl')
  String picUrl;
  //分类等级
  @JsonKey(name: 'level')
  String level;
  //排序规则
```

```dart
  @JsonKey(name: 'sortOrder')
  int sortOrder;
  //添加时间
  @JsonKey(name: 'addTime')
  String addTime;
  //更新时间
  @JsonKey(name: 'updateTime')
  String updateTime;
  //是否删除
  @JsonKey(name: 'deleted')
  bool deleted;
  //构造方法
  FirstCategoryModel(
    this.id,
    this.name,
    this.keywords,
    this.desc,
    this.pid,
    this.iconUrl,
    this.picUrl,
    this.level,
    this.sortOrder,
    this.addTime,
    this.updateTime,
    this.deleted,
  );

  factory FirstCategoryModel.fromJson(Map<String, dynamic> srcJson) =>
      _$FirstCategoryModelFromJson(srcJson);

  Map<String, dynamic> toJson() => _$FirstCategoryModelToJson(this);
}
```

打开 model/sub_category_model.dart 文件，添加二级分类数据模型，代码如下：

```dart
//model/sub_category_model.dart 文件
import 'package:json_annotation/json_annotation.dart';

//二级分类扩展文件
part 'sub_category_model.g.dart';

//二级分类数据列表模型
class SubCategoryListModel {
  //二级分类数据列表
  List<SubCategoryModel> subCategoryModels;
  //构造方法
  SubCategoryListModel(this.subCategoryModels);
```

```dart
  factory SubCategoryListModel.fromJson(List<dynamic> parseJson) {
    List<SubCategoryModel> productModels;
    productModels =
        parseJson.map((i) => SubCategoryModel.fromJson(i)).toList();
    return SubCategoryListModel(productModels);
  }
}

//二级分类数据模型
@JsonSerializable()
class SubCategoryModel extends Object {
  //id
  @JsonKey(name: 'id')
  int id;
  //名称
  @JsonKey(name: 'name')
  String name;
  //关键字
  @JsonKey(name: 'keywords')
  String keywords;
  //描述
  @JsonKey(name: 'desc')
  String desc;
  //父类 id
  @JsonKey(name: 'pid')
  int pid;
  //分类图标
  @JsonKey(name: 'iconUrl')
  String iconUrl;
  //二级分类图片
  @JsonKey(name: 'picUrl')
  String picUrl;
  //分类等级
  @JsonKey(name: 'level')
  String level;
  //排序规则
  @JsonKey(name: 'sortOrder')
  int sortOrder;
  //添加时间
  @JsonKey(name: 'addTime')
  String addTime;
  //更新时间
  @JsonKey(name: 'updateTime')
  String updateTime;
  //是否删除
  @JsonKey(name: 'deleted')
  bool deleted;
  //构造方法
  SubCategoryModel(
```

```
    this.id,
    this.name,
    this.keywords,
    this.desc,
    this.pid,
    this.iconUrl,
    this.picUrl,
    this.level,
    this.sortOrder,
    this.addTime,
    this.updateTime,
    this.deleted,
  );

  factory SubCategoryModel.fromJson(Map<String, dynamic> srcJson) =>
      _$SubCategoryModelFromJson(srcJson);

  Map<String, dynamic> toJson() => _$SubCategoryModelToJson(this);
}
```

步骤3：每个模型对应的fromJson和toJson并未实现，需要使用命令为其生成扩展代码。进入shop-flutter项目根目录，再次使用如下命令生成即可：

```
flutter packages pub run build_runner build --delete-conflicting-outputs
```

生成后会多出first_category_model.g.dart及sub_category_model.g.dart两个文件。

29.2　分类数据服务

数据模型生成后，再编写分类数据服务并提供如下两个方法：
- ❏ 获取一级分类数据；
- ❏ 获取二级分类数据。

数据服务需要定义几个回调方法，可以参考首页数据服务的定义，定义的内容一样。接下来编写请求服务端数据的方法，以及返回数据后如何处理。打开service/category_service.dart文件，添加如下完整代码：

```
//service/category_service.dart 文件
import 'package:shop/utils/http_util.dart';
import 'package:shop/config/index.dart';
import 'package:shop/model/first_category_model.dart';
import 'package:shop/model/sub_category_model.dart';
//定义成功返回列表数据
typedef OnSuccessList<T>(List<T> list);
```

```dart
//定义成功返回数据
typedef OnSuccess<T>(T t);
//定义返回失败消息
typedef OnFail(String message);

//分类数据服务
class CategoryService {

  //获取一级分类数据,传入回调方法,不需要传入参数
  Future getFirstCategoryData(OnSuccessList onSuccessList, {OnFail onFail}) async {
    try {
      var responseList = [];
      //请求一级分类数据
      var response = await HttpUtil.instance.get(ServerUrl.CATEGORY_FIRST);
      if (response['errno'] == 0) {
        responseList = response['data'];
        //将返回的Json数据转换为FirstLevelListCategory模型
        FirstListCategoryModel firstLevelListCategory = FirstListCategoryModel.fromJson(responseList);
        //调用成功回调方法,返回一级分类数据
        onSuccessList(firstLevelListCategory.firstLevelCategory);
      } else {
        //返回服务端错误信息
        onFail(response['errmsg']);
      }
    } catch (e) {
      print(e);
      //返回服务端错误信息
      onFail(KString.SERVER_EXCEPTION);
    }
  }
  //获取二级分类数据,传入分类Id,查询此分类下的所有分类
  Future getSubCategoryData(
      //请求参数,参数为分类Id
      Map<String, dynamic> parameters,
      OnSuccessList onSuccessList,
      {OnFail onFail}) async {
    try {
      var responseList = [];
      //查询二级分类数据
      var response = await HttpUtil.instance.get(ServerUrl.CATEGORY_SECOND, parameters: parameters);
      if (response['errno'] == 0) {
        responseList = response['data'];
        //将返回的Json数据转换为SubCategoryListModel模型
        SubCategoryListModel subCategoryListModel = SubCategoryListModel.fromJson(responseList);
        //调用成功回调方法,返回二级分类数据
        onSuccessList(subCategoryListModel.subCategoryModels);
```

```
        } else {
          //返回服务端错误信息
          onFail(response['errmsg']);
        }
      } catch (e) {
        print(e);
        //返回服务端错误信息
        onFail(KString.SERVER_EXCEPTION);
      }
    }
}
```

从上面代码可以看到，分类数据服务和首页数据服务用法大同小异，唯一不同的是获取二级分类列表数据方法需要传入参数 parameters，此参数为 Map 类型，用法如下：

```
var params = {"id": id};
```

29.3 一级分类组件实现

一级分类组件为分类页面的左侧部分，采用垂直布局的方式展示。当单击某个分类时，底部加一条横线表示其处于选中状态，同时发出事件并通知二级分类组件刷新数据。具体实现步骤如下：

步骤 1：打开 page/category/first_category.dart 文件，创建一个 FirstCategoryWidget 组件，然后初始化分类数据服务及分类数据列表，代码如下：

```
//分类数据服务
CategoryService categoryService = CategoryService();
//一级分类数据列表
List<FirstCategoryModel> firstCategoryList = List();
```

步骤 2：调用分类数据服务的 getFirstCategoryData 方法获取一级分类数据。成功返回数据后首先派发分类事件，通知二级分类刷新数据，然后再设置一级分类数据列表。大致逻辑如下：

```
categoryService.getFirstCategoryData((list) {
  //派发事件,二级分类监听此事件
  eventBus.fire(CategoryEvent(分类参数...));
  //设置一级分类数据列表
});
```

步骤 3：打开 event/category_event.dart 文件，添加分类事件：主要传递分类 id、分类名称，以及分类图片数据，代码如下：

```
//event/category_event.dart 文件
import 'package:event_bus/event_bus.dart';

EventBus eventBus = EventBus();
//分类事件,用于单击一级分类后触发获取二级分类数据
class CategoryEvent {
  //分类 id
  int id;
  //分类名称
  String categoryName;
  //分类图片
  String categoryImage;
  //构造方法
  CategoryEvent(this.id, this.categoryName, this.categoryImage);
}
```

步骤 4:添加一级分类项单击事件处理方法,首先更改当前选择分类索引,然后触发事件,通过二级分类页面刷新数据。大致代码如下:

```
_itemClick(int index) {
    setState(() {
     //设置当前选择分类索引
     _selectIndex = index;
    });
    //触发事件,二级分类监听此事件用于获取二级分类数据
    eventBus.fire(CategoryEvent(分类参数...));
}
```

步骤 5:使用 ListView 组件来组装一级分类列表数据,完整代码如下:

```
//page/category/first_category.dart 文件
import 'package:flutter/material.dart';
import 'package:shop/model/first_category_model.dart';
import 'package:shop/event/category_event.dart';
import 'package:shop/service/category_service.dart';
import 'package:shop/config/index.dart';

//一级分类组件
class FirstCategoryWidget extends StatefulWidget {
  @override
  _FirstCategoryWidgetState createState() => _FirstCategoryWidgetState();
}

class _FirstCategoryWidgetState extends State<FirstCategoryWidget> {
  //分类数据服务
  CategoryService categoryService = CategoryService();
  //一级分类数据列表
```

```
List<FirstCategoryModel> firstCategoryList = List();
//当前分类选择索引
int _selectIndex = 0;

@override
void initState() {
  super.initState();
  //获取一级分类数据
  categoryService.getFirstCategoryData((list) {
    //派发事件,二级分类监听此事件
    eventBus.fire(CategoryEvent(list[0].id, list[0].name, list[0].picUrl));
    setState(() {
      //设置一级分类列表数据
      firstCategoryList = list;
    });
  });
}

@override
Widget build(BuildContext context) {
  return Container(
      color: Colors.white,
      //一级分类列表
      child: ListView.builder(
          //一级分类长度
          itemCount: firstCategoryList.length,
          //一级分类项构造器
          itemBuilder: (BuildContext context, int index) {
            return _getFirstLevelItemWidget(firstCategoryList[index], index);
          }));
}

//单击一类分类项
_itemClick(int index) {
  setState(() {
    _selectIndex = index;
  });
  //触发事件,二级分类监听此事件用于获取二级分类数据
  eventBus.fire(CategoryEvent(
      firstCategoryList[index].id,
      firstCategoryList[index].name,
      firstCategoryList[index].picUrl));
}

//返回一级分类项组件
Widget _getFirstLevelItemWidget(FirstCategoryModel firstLevelCategory, int index) {
  return GestureDetector(
    //单击一级分类项
    onTap: () => _itemClick(index),
    child: Container(
        width: 100.0,
```

```
                    height: 50.0,
                    alignment: Alignment.center,
                    //垂直布局
                    child: Column(
                      children: <Widget>[
                        Container(
                          height: 48,
                          alignment: Alignment.center,
                          //分类名称
                          child: Text(firstLevelCategory.name,
                              //判断索引是否为当前选择的索引
                              style: index == _selectIndex
                                  ? TextStyle(
                                      fontSize: 14.0, color: KColor.categorySelectedColor)
                                  : TextStyle(fontSize: 14.0, color: KColor.categoryDefaultColor)),
                        ),
                        //分类分割线
                        index == _selectIndex
                            ? Divider(
                                height: 2.0,
                                color: KColor.categorySelectedColor,
                              )
                            : Divider(
                                color: Colors.white,
                                height: 1.0,
                              )
                      ],
                    )),
                );
              }
            }
```

上面的代码根据当前选择索引 _selectIndex 来变换分类字体颜色及分割线颜色，一级分类的效果如图 29-1 所示。

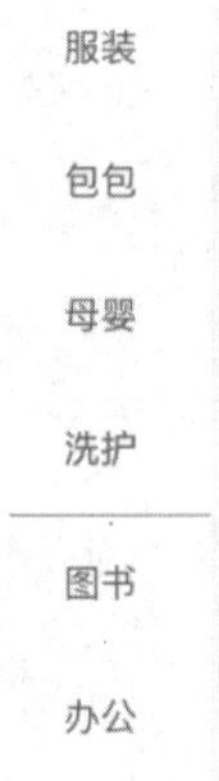

服装

包包

母婴

洗护

图书

办公

图 29-1　一级分类效果

29.4 二级分类组件实现

一级分类组件实现后,紧接着就需要实现二级分类组件。两个组件是有联动的,当用户单击一级分类后需要刷新二级分类数据,这里是通过 CategoryEvent 事件来完成通知并传递参数的。具体实现步骤如下:

步骤1:打开 page/category/sub_category.dart 文件,创建一个 SubCategoryWidget 组件,然后初始化分类数据服务及二级分类数据列表,代码如下:

```
//分类数据服务
CategoryService categoryService = CategoryService();
//二级分类数据列表
List < SubCategoryModel > subCategoryModels = List();
```

步骤2:添加分类事件监听处理,代码如下:

```
_listener() {
eventBus.on<CategoryEvent>().listen((CategoryEvent event) => _updateView(event));
}
```

步骤3:根据监听到的一级分类数据,获取二级分类数据,进而刷新二级分类页面。大致处理过程如下:

```
//更新二级分类视图
_updateView(CategoryEvent categoryEvent) {
    //...
    //根据一级分类 id 获取二级分类数据
    _getSubCategory(categoryEvent.id);
}

//根据 id 获取二级分类数据
_getSubCategory(int id) {
    //...
    //获取二级分类数据
    categoryService.getSubCategoryData(params, (subCategoryModelList) {
     //...
    }, onFail: (value) {});
}
```

步骤4:编写二级分类界面,整体采用垂直布局,上面为一级分类图片,下面为二级分类列表。二级分类采用网格布局。完整代码如下:

```dart
//page/category/sub_category.dart 文件
import 'package:flutter/material.dart';
import 'package:flutter_screenutil/flutter_screenutil.dart';
import 'package:shop/model/sub_category_model.dart';
import 'package:shop/event/category_event.dart';
import 'package:shop/service/category_service.dart';
import 'package:shop/utils/navigator_util.dart';

//二级分类组件
class SubCategoryWidget extends StatefulWidget {
  @override
  _SubCategoryWidgetState createState() => _SubCategoryWidgetState();
}

class _SubCategoryWidgetState extends State<SubCategoryWidget> {
  //分类数据服务
  CategoryService categoryService = CategoryService();
  //二级分类数据列表
  List<SubCategoryModel> subCategoryModels = List();

  //一级分类名称,图片,以及分类 Id
  var categoryName, categoryImage, categoryId;
  bool flag = true;

  @override
  void initState() {
    super.initState();
  }

  //监听事件,更新视图
  _listener() {
    eventBus.on<CategoryEvent>().listen((CategoryEvent event) => _updateView(event));
  }

  //更新二级分类视图
  _updateView(CategoryEvent categoryEvent) {
    if (flag) {
      flag = false;
      print("_updateView");
      setState(() {
        //一级分类名称
        categoryName = categoryEvent.categoryName;
        //一级分类图片
        categoryImage = categoryEvent.categoryImage;
        //一级分类 id
        categoryId = categoryEvent.id;
      });
      //根据一级分类 id 获取二级分类数据
      _getSubCategory(categoryEvent.id);
    }
  }
```

```
//根据 id 获取二级分类数据
_getSubCategory(int id) {
  var params = {"id": id};
  print(params);
  print("_getSubCategory");
  //获取二级分类数据
  categoryService.getSubCategoryData(params, (subCategoryModelList) {
    flag = true;
    setState(() {
      //设置二级分类数据
      subCategoryModels = subCategoryModelList;
    });
  }, onFail: (value) {});
}

@override
Widget build(BuildContext context) {
  //监听事件
  _listener();

  return Container(
    //垂直布局
    child: Column(
      children: <Widget>[
        //一级分类图片容器
        Container(
          padding: EdgeInsets.all(ScreenUtil.instance.setWidth(20.0)),
          height: ScreenUtil.instance.setHeight(200.0),
          //一级分类图片
          child: categoryImage != null ? Image.network(
            categoryImage,
            fit: BoxFit.fill,
          ) : Container(),
        ),
        Padding(
          padding: EdgeInsets.only(top: 4.0),
        ),
        Center(
          //一级分类名称
          child: Text(
            categoryName ?? "",
            style: TextStyle(fontSize: 14.0, color: Colors.black54),
          ),
        ),
        //网格布局
        GridView.builder(
          physics: NeverScrollableScrollPhysics(),
          //二级分类数量
          itemCount: subCategoryModels.length,
          shrinkWrap: true,
          gridDelegate: SliverGridDelegateWithFixedCrossAxisCount(
```

```dart
              //3 列
              crossAxisCount: 3,
              mainAxisSpacing: 20.0,
              childAspectRatio: 0.85,
              //垂直单个子 Widget 之间间距
              crossAxisSpacing: 20.0),
          //分类项构造器
          itemBuilder: (BuildContext context, int index) {
            return getItemWidget(subCategoryModels[index]);
          }),
      Padding(
        padding: EdgeInsets.only(top: 10.0),
      ),
    ],
  ),
  );
}

//单击二级分类项,导航至分类商品列表页面
_itemClick(int id) {
  NavigatorUtil.goCategoryGoodsListPage(context,categoryName,id);
}

//二级分类项组件
Widget getItemWidget(SubCategoryModel categoryName) {
  return GestureDetector(
    child: Container(
      alignment: Alignment.center,
      child: Card(
        //垂直布局,上面为分类图标,下面为分类名称
        child: Column(
          children: <Widget>[
            //二级分类图标
            Image.network(
              categoryName.picUrl ?? "",
              fit: BoxFit.fill,
              height: 60,
            ),
            Padding(
              padding: EdgeInsets.only(top: 5.0),
            ),
            //二级分类名称
            Text(
              categoryName.name,
              style: TextStyle(fontSize: 14.0, color: Colors.black54),
            ),
          ],
        ),
      )),
    //单击处理
    onTap: () => _itemClick(categoryName.id),
```

```
      );
    }
}
```

当单击二级分类图标则页面跳转至分类商品列表页面,二级分类组件的显示效果如图 29-2 所示。

图 29-2　二级分类组件

29.5　组装分类页面

当一级分类组件和二级分类组件完成后,可以组装分类页面了。分类页面里不需要再做数据处理,只需要做一个水平布局即可,左侧为一级分类,右侧为二级分类。

打开 page/category/category_page.dart 文件,添加如下完整代码:

```
//page/category/category_page.dart 文件
import 'package:flutter/material.dart';
import 'package:shop/page/category/first_category.dart';
import 'package:shop/page/category/sub_category.dart';
import 'package:shop/config/string.dart';
//分类页面
class CategoryPage extends StatefulWidget {
  @override
  _CategoryPageState createState() => _CategoryPageState();
}

class _CategoryPageState extends State<CategoryPage> {

  @override
  Widget build(BuildContext context) {
```

```
        return Scaffold(
            appBar: AppBar(
              //分类标题
              title: Text(KString.CATEGORY_TITLE),
              centerTitle: true,
            ),
            body: Container(
              //水平布局
              child: Row(
                children: <Widget>[
                  //左侧为一级分类
                  Expanded(
                    flex: 2,
                    child: FirstCategoryWidget(),
                  ),
                  //右侧为二级分类
                  Expanded(
                    flex: 8,
                    child: SubCategoryWidget(),
                  ),
                ],
              ),
            ),
        );
    }
}
```

分类页面运行后的显示效果如图 29-3 所示。

图 29-3　分类页面

第 30 章 登录注册

商城的注册与登录是必备的功能。首页与分类页面不需要登录也可以查看,它们是给用户检索商品数据使用的。当用户将商品添加至购买车或者要下单时,就需要登录才可以继续操作,另外还有"我的地址""我的收藏"等模块也需要用户登录。注册账号并登录后就可以使用这些功能了。

本章将详细阐述用户数据模型、用户数据服务,以及注册登录页面的实现。

30.1 用户数据模型

注册登录模块使用的是用户相关的数据,如账户、密码、头像和昵称等。登录、登出,以及注册的接口如下:

```
//登录接口
http://localhost:8080/client/auth/login
//登出接口
http://localhost:8080/client/auth/logout
//注册接口
http://localhost:8080/client/auth/register
```

根据这些接口整理出用户数据模型,打开 model/user_model.dart 文件,添加代码如下:

```
//model/user_model.dart 文件
import 'package:json_annotation/json_annotation.dart';

//数据模型扩展文件
part 'user_model.g.dart';

//用户数据模型
@JsonSerializable()
class UserModel extends Object {
```

```dart
    //用户基本信息
    @JsonKey(name: 'userInfo')
    UserInfo userInfo;

    //服务端返回的 token 值
    @JsonKey(name: 'token')
    String token;

    //构造方法
    UserModel(this.userInfo,this.token,);

    factory UserModel.fromJson(Map<String, dynamic> srcJson) => _$UserModelFromJson(srcJson);

    Map<String, dynamic> toJson() => _$UserModelToJson(this);

}

//用户信息数据模型
@JsonSerializable()
class UserInfo extends Object {

    //昵称
    @JsonKey(name: 'nickName')
    String nickName;

    //头像地址
    @JsonKey(name: 'avatarUrl')
    String avatarUrl;

    //构造方法
    UserInfo(this.nickName,this.avatarUrl,);

    factory UserInfo.fromJson(Map<String, dynamic> srcJson) => _$UserInfoFromJson(srcJson);

    Map<String, dynamic> toJson() => _$UserInfoToJson(this);

}
```

接下来打开命令行工具，进入 shop-flutter 项目根目录使用 build_runner 工具生成 user_model.dart 的扩展文件 user_model.g.dart 文件。

30.2 用户数据服务

用户数据服务里的登录和注册模块提供访问后端接口的方法，主要有以下三种方法：
❑ 登录方法；

- 登出方法；
- 注册方法。

打开 service/user_service.dart 文件，添加这三个方法，代码如下：

```dart
//service/user_service.dart 文件
import 'package:shop/config/index.dart';
import 'package:shop/utils/http_util.dart';
import 'package:shop/model/user_model.dart';
//定义成功返回列表数据
typedef OnSuccessList<T>(List<T> list);
//定义成功返回数据
typedef OnSuccess<T>(T t);
//定义返回失败消息
typedef OnFail(String message);

//用户数据服务
class UserService {

  //注册请求方法,需要传递账号、密码等参数
  Future register(Map<String, dynamic> parameters, OnSuccess onSuccess, OnFail onFail) async {
    try {
      //post 请求注册地址
      var response = await HttpUtil.instance.post(ServerUrl.REGISTER, parameters: parameters);
      if (response['errno'] == 0) {
        //成功返回
        onSuccess("");
      } else {
        //注册失败
        onFail(response['errmsg']);
      }
    } catch (e) {
      print(e);
      //服务器接口异常
      onFail(KString.SERVER_EXCEPTION);
    }
  }

  //登录请求方法,需要传递账号、密码等参数
  Future login(Map<String, dynamic> parameters, OnSuccess onSuccess, OnFail onFail) async {
    try {
      var response = await HttpUtil.instance.post(ServerUrl.LOGIN, parameters: parameters);
      if (response['errno'] == 0) {
        //返回 token 及用户基本信息
        UserModel userModel = UserModel.fromJson(response['data']);
        //登录成功
        onSuccess(userModel);
```

```
        } else {
          //登录失败,返回失败信息
          onFail(response['errmsg']);
        }
      } catch (e) {
        print(e);
        onFail(KString.SERVER_EXCEPTION);
      }
    }

    //登出请求方法
    Future loginOut(OnSuccess onSuccess, OnFail onFail) async {
      try {
        var response = await HttpUtil.instance.post(ServerUrl.LOGIN_OUT);
        if (response['errno'] == 0) {
          //登出成功
          onSuccess(KString.SUCCESS);
        } else {
          //登出失败
          onFail(response['errmsg']);
        }
      } catch (e) {
        print(e);
        onFail(KString.SERVER_EXCEPTION);
      }
    }
```

其中,登录的作用主要是为了获取 token 值,这样有了这个令牌再去做与用户相关的操作,这样用户进行添加购物车和结算等操作时,就有了一个通行的"身份",同时在登录及注册失败时会返回服务端错误提示信息,如用户名不存在或密码不正确等。

30.3 登录页面实现

登录页面可以从"我的页面"单击"登录"按钮跳转过来,也可以当某个操作需要登录时跳转至登录页面,如添加至购物车当用户登录完成后返回至登录之前的页面继续操作。

前面已经实现了用户数据模型及用户数据服务,接下来就可以编写登录页面了。具体步骤如下:

步骤1:打开 page/login/login_page.dart 文件,添加 LoginPage 组件,然后初始化用户数据模型及用户数据服务,代码如下:

```
//用户数据服务
UserService userService = UserService();
//用户数据模型
UserModel userModel;
```

步骤2：添加验证方法，验证账号和密码是否为空，以及它们的长度，方法如下：

```
//验证账号
String _validatorAccount(String value) {
    //值不能为空并且长度要大于等于11
}
//验证密码
String _validatorPassWord(String value) {
    //值不能为空并且长度要大于等于6
}
```

步骤3：编写登录方法，登录方法的处理流程是：提供登录参数、账号及密码→调用用户数据服务的登录方法→服务端返回数据→保存 token 等信息至本地→触发登录事件，通知"购物车"或"我的页面"此用户已经登录成功。实现此方法的结构如下：

```
_login() {
    //登录之前执行验证
    if (registerFormKey.currentState.validate()) {
      //...
      //登录参数
      Map<String, dynamic> map = Map();
      //...
      //调用用户服务，执行登录方法
      userService.login(map, (success) {
        //返回登录数据，赋值给用户数据模型
        userModel = success;
        _saveUserInfo();
        //...
        //触发登录事件，通知"购物车"或"我的页面"此用户已经登录成功
        loginEventBus.fire(LoginEvent(登录事件参数...));
      }, (onFail) {
        //...
      });
    } else {
      //...
    }
}
```

步骤4：打开 event/login_event.dart 文件，实现登录事件代码如下：

```
//event/login_event.dart 文件
import 'package:event_bus/event_bus.dart';

EventBus loginEventBus = EventBus();
//登录事件
class LoginEvent {
  //是否登录
```

```
    bool isLogin;
    //昵称
    String nickName;
    //头像 url
    String url;
    //构造方法
    LoginEvent(this.isLogin, {this.nickName, this.url});
}
```

步骤5：保存用户信息及 token 至本地，这样用户下次打开商城应用时可以直接提取用户基本信息数据。

步骤6：编写登录界面，登录界面采用垂直布局的方式：上面是账户输入框，下面是密码输入框。参考如下完成布局代码：

```
//page/login/login_page.dart
import 'package:flutter/material.dart';
import 'package:flutter_screenutil/flutter_screenutil.dart';
import 'package:shared_preferences/shared_preferences.dart';
import 'package:fluttertoast/fluttertoast.dart';
import 'package:shop/config/index.dart';
import 'package:shop/service/user_service.dart';
import 'package:shop/model/user_model.dart';
import 'package:shop/utils/navigator_util.dart';
import 'package:shop/event/login_event.dart';
import 'package:shop/utils/shared_preferences_util.dart';

//登录页面
class LoginPage extends StatefulWidget {
  @override
  _LoginPageState createState() => _LoginPageState();
}

class _LoginPageState extends State<LoginPage> {
  //账号文本控制器
  TextEditingController _accountTextControl = TextEditingController();
  //密码文本控制器
  TextEditingController _passwordTextControl = TextEditingController();
  //用户数据服务
  UserService userService = UserService();
  //用户数据模型
  UserModel userModel;
  //是否自动验证
  bool _autovalidator = false;
  final registerFormKey = GlobalKey<FormState>();

  @override
  Widget build(BuildContext context) {
```

```
        return Scaffold(
          body: SafeArea(
            child: Container(
              alignment: Alignment.centerLeft,
              //登录框居中
              child: Center(
                child: SingleChildScrollView(
                  //登录框容器
                  child: Container(
                    margin: EdgeInsets.fromLTRB(ScreenUtil().setWidth(30.0), 0, ScreenUtil().setWidth(30.0), 0),
                    height: ScreenUtil.instance.setHeight(800.0),
                    //外框样式
                    decoration: BoxDecoration(
                      color: Colors.white,
                      borderRadius: BorderRadius.circular(10.0)),
                    //Form 表单
                    child: Form(
                      key: registerFormKey,
                      //垂直布局,上面为账号,下面为密码
                      child: Column(
                        children: <Widget>[
                          Padding(
                            padding: EdgeInsets.only(top: ScreenUtil.instance.setHeight(60.0))),
                          Container(
                            margin: EdgeInsets.all(ScreenUtil.instance.setWidth(30.0)),
                            //账号输入框
                            child: TextFormField(
                              //单行
                              maxLines: 1,
                              maxLength: 11,
                              //自动验证
                              autovalidate: _autovalidator,
                              //弹出键盘为数字
                              keyboardType: TextInputType.phone,
                              //验证回调方法
                              validator: _validatorAccount,
                              //边框样式
                              decoration: InputDecoration(
                                icon: Icon(
                                  Icons.person,
                                  color: KColor.loginIconColor,
                                  size: ScreenUtil.instance.setWidth(60.0),
                                ),
                                //提示输入账号
                                hintText: KString.ACCOUNT_HINT,
                                //提示文本样式
                                hintStyle: TextStyle(
```

```dart
              color: Colors.grey,
              fontSize: ScreenUtil.instance.setSp(28.0)),
          //账号文本样式
          labelStyle: TextStyle(
              color: Colors.black54,
              fontSize: ScreenUtil.instance.setSp(28.0)),
          //账号文本
          labelText: KString.ACCOUNT,
        ),
        //控制器
        controller: _accountTextControl,
      ),
    ),
    Container(
      margin: EdgeInsets.all(ScreenUtil.instance.setWidth(30.0)),
      //密码输出框
      child: TextFormField(
        //单行
        maxLines: 1,
        maxLength: 12,
        //密码显示
        obscureText: true,
        //自动验证
        autovalidate: _autovalidator,
        //验证回调方法
        validator: _validatorPassWord,
        //边框样式
        decoration: InputDecoration(
          icon: Icon(
            Icons.lock,
            color: KColor.loginIconColor,
            size: ScreenUtil.instance.setWidth(60.0),
          ),
          //密码提示文本
          hintText: KString.PASSWORD_HINT,
          //密码提示文本样式
          hintStyle: TextStyle(
              color: Colors.grey,
              fontSize: ScreenUtil.instance.setSp(28.0)),
          //密码标签样式
          labelStyle: TextStyle(
              color: Colors.black54,
              fontSize: ScreenUtil.instance.setSp(28.0)),
          //密码标签
          labelText: KString.PASSWORD,
        ),
        //控制器
        controller: _passwordTextControl,
      ),
```

```dart
            ),
            Container(
                margin: EdgeInsets.all(ScreenUtil.instance.setWidth(30.0)),
                child: SizedBox(
                  height: ScreenUtil.instance.setHeight(80.0),
                  width: ScreenUtil.instance.setWidth(600.0),
                  //登录按钮
                  child: RaisedButton(
                    //登录方法
                    onPressed: _login,
                    color: KColor.loginButtonColor,
                    child: Text(
                      KString.LOGIN,
                      style: TextStyle(
                        color: Colors.white,
                        fontSize: ScreenUtil.instance.setSp(28.0)),
                    ),
                  ),
                )),
            Container(
              margin: EdgeInsets.all(ScreenUtil.instance.setWidth(20.0)),
              alignment: Alignment.centerRight,
              child: InkWell(
                //单击跳转至注册页面
                onTap: () => _register(),
                child: Text(
                  //马上注册提示文本
                  KString.NOW_REGISTER,
                  style: TextStyle(
                    color: KColor.registerTextColor,
                    fontSize: ScreenUtil.instance.setSp(24.0)),
                ),
              ),
            )
          ],
        ),
      ),
    ))),
  ),
 );
}

//跳转至注册页面
_register() {
  NavigatorUtil.goRegister(context);
}
//验证账号
String _validatorAccount(String value) {
  //值不能为空并且长度要大于等于 11
  if (value == null || value.length < 11) {
```

```dart
        return KString.ACCOUNT_RULE;
      }
      return null;
    }
    //验证密码
    String _validatorPassWord(String value) {
      //值不能为空并且长度要大于等于6
      if (value == null || value.length < 6) {
        return KString.PASSWORD_HINT;
      }
      return null;
    }
    //登录
    _login() {
      //登录之前执行验证
      if (registerFormKey.currentState.validate()) {
        registerFormKey.currentState.save();
        //登录参数
        Map<String, dynamic> map = Map();
        //设置密码参数
        map.putIfAbsent("username", () => _accountTextControl.text.toString());
        //设置账号参数
        map.putIfAbsent("password", () => _passwordTextControl.text.toString());
        //调用用户服务执行登录方法
        userService.login(map, (success) {
          print(success);
          //返回登录数据,赋值给用户数据模型
          userModel = success;
          _saveUserInfo();
          //登录成功提示
          _showToast(KString.LOGIN_SUCESS);
          //触发登录事件,通知"购物车"或"我的页面"此用户已经登录成功
          loginEventBus.fire(LoginEvent(true, url: userModel.userInfo.avatarUrl, nickName: userModel.userInfo.nickName));
          Navigator.pop(context);
        }, (onFail) {
          print(onFail);
          //弹出错误信息
          _showToast(onFail);
        });
      } else {
        setState(() {
          _autovalidator = true;
        });
      }
    }

    //提示消息
    _showToast(message) {
      Fluttertoast.showToast(
        msg: message,
```

```
          toastLength: Toast.LENGTH_SHORT,
          gravity: ToastGravity.CENTER,
          timeInSecForIos: 1,
          backgroundColor: KColor.toastBgColor,
          textColor: KColor.toastTextColor,
          fontSize: ScreenUtil.instance.setSp(28.0));
    }

    //保存用户信息至本地
    _saveUserInfo() async {
      //获取本地存储对象
      SharedPreferences sharedPreferences = await SharedPreferences.getInstance();
      //将用户的 token 值存在本地
      SharedPreferencesUtil.token = userModel.token;
      await sharedPreferences.setString(KString.TOKEN, userModel.token);
      //存储头像
      await sharedPreferences.setString(KString.HEAD_URL, userModel.userInfo.avatarUrl);
      //存储昵称
      await sharedPreferences.setString(KString.NICK_NAME, userModel.userInfo.nickName);
    }
}
```

登录页面布局采用了 Form 表单,在登录容器的右下角还提供了"马上注册"按钮,单击此按钮可以跳转至注册页面,页面显示效果如图 30-1 所示。

图 30-1　登录页面

30.4 注册页面

注册页面使用的数据模型和数据服务与登录页面是一样的,界面布局方式也基本一致,另外注册之前的输入框验证也使用的是同样的方法,所以这里不做重点介绍。

这里重点看一看注册方法,注册的流程是:验证输入框参数→调用用户服务的注册方法→服务端返回数据→注册成功提示。注册方法的结构如下:

```
_register() {
    //提交注册前先验证
    if (registerFormKey.currentState.validate()) {
        //注册参数
        Map<String, dynamic> map = Map();
        //用户名
        //密码
        //电话号码
        //验证码
        //调用用户服务的注册方法
        userService.register(map, (success) {
            //注册成功后提示
        }, (onFail) {
            //...
        });
    } else {
        //...
    }
}
```

从上面代码可以看到,注册方法比登录方法多了手机号码及验证码两个参数。

提示 注册的实际使用流程:输入手机号,填写收到的验证码,提交后完成注册。这里模拟这个过程并把验证码设置为一个固定值。

打开 page/login/register_page.dart 文件,添加注册页面完整代码如下:

```
//page/login/register_page.dart
import 'package:flutter/material.dart';
import 'package:flutter_screenutil/flutter_screenutil.dart';
import 'package:fluttertoast/fluttertoast.dart';
import 'package:shop/config/index.dart';
import 'package:shop/config/icon.dart';
import 'package:shop/service/user_service.dart';
import 'package:shop/utils/navigator_util.dart';
//注册页面
```

```dart
class RegisterPage extends StatefulWidget {
  @override
  _RegisterPageState createState() => _RegisterPageState();
}

class _RegisterPageState extends State<RegisterPage> {
  //账号文本控制器
  TextEditingController _accountTextControl = TextEditingController();
  //密码文本控制器
  TextEditingController _passwordTextControl = TextEditingController();
  //用户数据服务
  UserService userService = UserService();
  //是否自动验证
  bool _autovalidator = false;
  final registerFormKey = GlobalKey<FormState>();

  @override
  Widget build(BuildContext context) {
    return Scaffold(
      body: SafeArea(
        child: Container(
          //注册框居中
          child: Center(
            child: Container(
              margin: EdgeInsets.fromLTRB(ScreenUtil().setWidth(30.0), 0, ScreenUtil().setWidth(30.0), 0),
              height: ScreenUtil.instance.setHeight(700.0),
              //边框样式
              decoration: BoxDecoration(
                  color: Colors.white, borderRadius: BorderRadius.circular(10.0)),
              //注册表单
              child: Form(
                key: registerFormKey,
                //垂直布局
                child: Column(
                  children: <Widget>[
                    Padding(
                      padding: EdgeInsets.only(top: ScreenUtil.instance.setHeight(60.0))),
                    Container(
                      margin: EdgeInsets.all(ScreenUtil.instance.setWidth(30.0)),
                      //账号输入表单
                      child: TextFormField(
                        //单行显示
                        maxLines: 1,
                        maxLength: 11,
                        //自动验证
                        autovalidate: _autovalidator,
                        //数字键盘
                        keyboardType: TextInputType.phone,
```

```dart
                    //账号验证方法
                    validator: _validatorAccount,
                    //边框样式
                    decoration: InputDecoration(
                      icon: Icon(
                        Icons.account_circle,
                        color: KColor.registerIconColor,
                        size: ScreenUtil.instance.setWidth(60.0),
                      ),
                      //提示文本
                      hintText: KString.ACCOUNT_HINT,
                      //提示文本样式
                      hintStyle: TextStyle(
                          color: Colors.grey,
                          fontSize: ScreenUtil.instance.setSp(28.0)),
                      //账号标签文本样式
                      labelStyle: TextStyle(
                          color: Colors.black54,
                          fontSize: ScreenUtil.instance.setSp(28.0)),
                      //账号标签文本
                      labelText: KString.ACCOUNT,
                    ),
                    //控制器
                    controller: _accountTextControl,
                  ),
                ),
                Container(
                  margin: EdgeInsets.all(ScreenUtil.instance.setWidth(30.0)),
                  //密码输入表单
                  child: TextFormField(
                    //单行
                    maxLines: 1,
                    maxLength: 12,
                    //自动验证
                    autovalidate: _autovalidator,
                    //验证方法
                    validator: _validatorPassWord,
                    //边框样式
                    decoration: InputDecoration(
                      icon: Icon(
                        KIcon.PASS_WORD,
                        color: KColor.registerIconColor,
                        size: ScreenUtil.instance.setWidth(60.0),
                      ),
                      //请输入密码提示
                      hintText: KString.PASSWORD_HINT,
                      //请输入密码提示样式
                      hintStyle: TextStyle(
                          color: Colors.grey,
```

```
                    fontSize: ScreenUtil.instance.setSp(28.0)),
                //密码标签文本样式
                labelStyle: TextStyle(
                    color: Colors.black54,
                    fontSize: ScreenUtil.instance.setSp(28.0)),
                //密码标签文本
                labelText: KString.PASSWORD,
              ),
              //控制器
              controller: _passwordTextControl,
            ),
          ),
          Container(
            margin: EdgeInsets.all(ScreenUtil.instance.setWidth(30.0)),
            child: SizedBox(
              height: ScreenUtil.instance.setHeight(80.0),
              width: ScreenUtil.instance.setWidth(600.0),
              //注册按钮
              child: RaisedButton(
                onPressed: _register,
                color: KColor.registerButtonColor,
                child: Text(
                  //注册文本
                  KString.REGISTER,
                  style: TextStyle(
                      color: Colors.white,
                      fontSize: ScreenUtil.instance.setSp(28.0)),
                ),
              ),
            )),
        ],
      ),
    ),
   ),
  ),
 )),
);
}
//验证账号
String _validatorAccount(String value) {
  //值不能为空并且长度要大于等于11
  if (value == null || value.length < 11) {
    return KString.ACCOUNT_RULE;
  }
  return null;
}
//验证密码
String _validatorPassWord(String value) {
  //值不能为空并且长度要大于等于6
```

```dart
      if (value == null || value.length < 6) {
        return KString.PASSWORD_HINT;
      }
      return null;
    }

    //提示注册消息
    _showToast(String message) {
      Fluttertoast.showToast(
          msg: message,
          toastLength: Toast.LENGTH_SHORT,
          gravity: ToastGravity.CENTER,
          timeInSecForIos: 1,
          backgroundColor: KColor.toastBgColor,
          textColor: KColor.toastTextColor,
          fontSize: ScreenUtil.instance.setSp(28.0));
    }

    //注册
    _register() {
      //提交注册前先验证
      if (registerFormKey.currentState.validate()) {
        registerFormKey.currentState.save();
        //注册参数
        Map<String, dynamic> map = Map();
        //用户名
        map.putIfAbsent("username", () => _accountTextControl.text.toString());
        //密码
        map.putIfAbsent("password", () => _passwordTextControl.text.toString());
        //电话号码
        map.putIfAbsent("mobile", () => _accountTextControl.text.toString());
        //验证码
        map.putIfAbsent("code", () => "0000");
        //测试验证码,为了模拟此功能,此处设置为固定值8888
        //调用用户服务的注册方法
        userService.register(map, (success) {
          print(success);
          //注册成功后提示
          _showToast(KString.REGISTER_SUCCESS);
          //注册成功,返回上级页面
          NavigatorUtil.popRegister(context);
        }, (onFail) {
          print(onFail);
          _showToast(onFail);
        });
      } else {
        setState(() {
          _autovalidator = true;
        });
```

```
        }
      }
}
```

单击登录页面的"马上注册"按钮,跳转至注册页面,显示效果如图 30-2 所示。

图 30-2　注册页面

第 31 章 商　品

商品是商城应用的核心模块之一，包括商品列表和详情页面。当用户挑选好商品后，就可以添加至购物车或者立即下单。

本章关于商品的内容如下：
- 商品分类页面；
- 商品列表；
- 商品详情页；
- 商品数据模型；
- 商品分类标题数据模型；
- 商品数据服务。

31.1　商品列表数据模型

商品列表数据模型 GoodListModel 主要是用来给商品列表提供渲染数据的，展示了某个分类下的所有商品信息，其访问接口如下所示：

```
//获取商品数据列表
http://localhost:8080/client/goods/list
```

打开 model/goods_model.dart 文件，添加如下代码：

```
//model/goods_model.dart 文件
import 'package:json_annotation/json_annotation.dart';
//商品数据模型扩展
part 'goods_model.g.dart';

//商品列表数据模型
class GoodsListModel {
  //商品列表
  List<GoodsModel> goodsModels;
  //构造方法
  GoodsListModel(this.goodsModels);
```

```dart
  factory GoodsListModel.fromJson(List<dynamic> parseJson) {
    List<GoodsModel> goodsModels;
    goodsModels = parseJson.map((i) => GoodsModel.fromJson(i)).toList();
    return GoodsListModel(goodsModels);
  }
}

//商品数据模型
@JsonSerializable()
class GoodsModel extends Object {
  //id
  @JsonKey(name: 'id')
  int id;
  //名称
  @JsonKey(name: 'name')
  String name;
  //简介
  @JsonKey(name: 'brief')
  String brief;
  //图片地址
  @JsonKey(name: 'picUrl')
  String picUrl;
  //是否为新品
  @JsonKey(name: 'isNew')
  bool isNew;
  //是否热卖
  @JsonKey(name: 'isHot')
  bool isHot;
  //专柜价格
  @JsonKey(name: 'counterPrice')
  double counterPrice;
  //零售价格
  @JsonKey(name: 'retailPrice')
  double retailPrice;
  //构造方法
  GoodsModel(
    this.id,
    this.name,
    this.brief,
    this.picUrl,
    this.isNew,
    this.isHot,
    this.counterPrice,
    this.retailPrice,
  );

  factory GoodsModel.fromJson(Map<String, dynamic> srcJson) =>
      _$GoodsModelFromJson(srcJson);

  Map<String, dynamic> toJson() => _$GoodsModelToJson(this);
}
```

接下来打开命令行工具，进入 shop-flutter 项目根目录并使用 build_runner 工具生成 good_model.dart 的扩展文件 good_model.g.dart 文件。

31.2 商品分类标题数据模型

当用户对首页分类栏或者分类页的二级分类部分单击时，可以跳转至商品分类页面。商品分类页首先提取商品分类数据，这里需要创建其数据模型。根据分类所处的位置可以分为以下几种情况：

- 当前分类：用户当前选择的分类；
- 同级分类：同一个分类层级的分类；
- 父分类：当前分类的上一级分类。

以服装分类为例，如果用户选择的是女装，那么当前分类即为女装，同级分类为女装、男装、童装，以及内衣，它的父分类则为服装。

商品分类使用的数据接口如下：

```
//获取商品分类数据
http://localhost:8080/client/goods/category
```

打开 model/category_title_model.dart 文件，添加如下数据模型代码：

```dart
//model/category_title_model.dart 文件
import 'package:json_annotation/json_annotation.dart';
//扩展文件
part 'category_title_model.g.dart';
//分类标题数据模型
@JsonSerializable()
class CategoryTitleModel extends Object {
  //当前分类
  @JsonKey(name: 'currentCategory')
  CategoryModel currentCategory;
  //同级分类
  @JsonKey(name: 'brotherCategory')
  List<CategoryModel> brotherCategory;
  //父分类
  @JsonKey(name: 'parentCategory')
  CategoryModel parentCategory;

  CategoryTitleModel(this.currentCategory,this.brotherCategory,this.parentCategory,);

  factory CategoryTitleModel.fromJson(Map<String, dynamic> srcJson) =>
      _$CategoryTitleModelFromJson(srcJson);
```

```dart
  Map<String, dynamic> toJson() => _$CategoryTitleModelToJson(this);
}

//当前分类
@JsonSerializable()
class CategoryModel extends Object {
  //分类 id
  @JsonKey(name: 'id')
  int id;
  //名称
  @JsonKey(name: 'name')
  String name;
  //关键字
  @JsonKey(name: 'keywords')
  String keywords;
  //描述
  @JsonKey(name: 'desc')
  String desc;
  //父 id
  @JsonKey(name: 'pid')
  int pid;
  //图标地址
  @JsonKey(name: 'iconUrl')
  String iconUrl;
  //图片地址
  @JsonKey(name: 'picUrl')
  String picUrl;
  //等级
  @JsonKey(name: 'level')
  String level;
  //排序规则
  @JsonKey(name: 'sortOrder')
  int sortOrder;
  //添加时间
  @JsonKey(name: 'addTime')
  String addTime;
  //更新时间
  @JsonKey(name: 'updateTime')
  String updateTime;
  //是否删除
  @JsonKey(name: 'deleted')
  bool deleted;

  CategoryModel(this.id,this.name,this.keywords,this.desc,this.pid,this.iconUrl,this.picUrl,this.level,this.sortOrder,this.addTime,this.updateTime,this.deleted,);

  factory CategoryModel.fromJson(Map<String, dynamic> srcJson) => _$CategoryModelFromJson(srcJson);
```

```
    Map<String, dynamic> toJson() => _$CategoryModelToJson(this);
}
```

接下来打开命令行工具，进入 shop-flutter 项目根目录并使用 build_runner 工具生成 category_title_model.dart 的扩展文件 category_title_model.g.dart 文件。

31.3 商品数据服务

商品数据服务是给商品分类、商品列表，以及商品详情提供数据接口的类，它提供以下三种方法：

- 获取商品分类数据；
- 获取商品列表数据；
- 获取商品详情数据。

数据服务的实现可以参考前面几章的描述，打开 service/goods_service.dart 文件，添加如下代码：

```
//service/goods_service.dart 文件
import 'package:shop/utils/http_util.dart';
import 'package:shop/config/server_url.dart';
import 'package:shop/model/goods_model.dart';
import 'package:shop/model/goods_detail_model.dart';
import 'package:shop/config/index.dart';
import 'package:shop/model/category_title_model.dart';
//定义成功返回列表数据
typedef OnSuccessList<T>(List<T> list);
//定义成功返回数据
typedef OnSuccess<T>(T t);
//定义返回失败消息
typedef OnFail(String message);

//商品数据服务
class GoodsService {
  //获取商品分类数据,参数为分类 id
  Future getGoodsCategory(Map<String, dynamic> parameters, OnSuccess onSuccess, OnFail onFail) async {
    try {
      var response = await HttpUtil.instance.get(ServerUrl.GOODS_CATEGORY, parameters: parameters);
      if (response['errno'] == 0) {
        //返回 Json 数据并转换成 CategoryTitleModel 数据模型
        CategoryTitleModel categoryTitleModel = CategoryTitleModel.fromJson(response["data"]);
        //成功回调
```

```
        onSuccess(categoryTitleModel);
      } else {
        //失败返回错误消息
        onFail(response['errmsg']);
      }
    } catch (e) {
      print(e);
      onFail(KString.SERVER_EXCEPTION);
    }
  }
  //获取商品列表数据,参数为分类 id
  Future getGoodsList(Map < String, dynamic > parameters, OnSuccessList onSuccessList, {OnFail onFail}) async {
    try {
      var responseList = [];
      var response = await HttpUtil.instance.get(ServerUrl.GOODS_LIST, parameters: parameters);
      if (response['errno'] == 0) {
        //获取返回数据
        responseList = response['data']['list'];
        //根据返回 Json 数据转换成 GoodsListModel 数据模型
        GoodsListModel goodsListModels = GoodsListModel.fromJson(responseList);
        //成功回调
        onSuccessList(goodsListModels.goodsModels);
      } else {
        //失败返回错误消息
        onFail(response['errmsg']);
      }
    } catch (e) {
      print(e);
      onFail(KString.SERVER_EXCEPTION);
    }
  }
  //获取商品详情数据,参数为商品 id
  Future getGoodsDetailData(Map < String, dynamic > parameters, OnSuccess onSuccess, {OnFail onFail}) async {
    try {
      var response = await HttpUtil.instance.get(ServerUrl.GOODS_DETAILS_URL, parameters: parameters);
      if (response['errno'] == 0) {
        //根据返回 Json 数据转换成 GoodsDetailModel 数据模型
        GoodsDetailModel goodsDetailModel = GoodsDetailModel.fromJson(response['data']);
        //成功回调
        onSuccess(goodsDetailModel);
      } else {
        //失败返回错误消息
        onFail(response['errmsg']);
      }
    } catch (e) {
```

```
        print(e);
        onFail(KString.SERVER_EXCEPTION);
      }
    }
  }
```

> **提示** 这里的 GoodsDetailModel 尚未实现,当实现商品详情功能时再添加,可以先将获取商品详情方法注释起来。

31.4 商品分类页面实现

商品分类页面包括商品分类标题及商品列表两部分。通常一级分类下面会有多个二级分类,采用选项卡的方式展示,单击某个分类时展示此分类下的商品列表数据。

商品分类页面实现的步骤如下:

步骤1:首先打开 page/goods/goods_category_page.dart 文件,新建 GoodsCategoryPage 页面。初始化商品数据模型、分类标题数据模型,以及同级分类数据列表,代码如下:

```
//数据服务
GoodsService _goodsService = GoodsService();
//分类标题数据模型
CategoryTitleModel _categoryTitleModel;
//同级(二级)分类数据列表
List<CategoryModel> brotherCategory = List();
```

步骤2:在页面的 initState 方法里调用商品服务的 getGoodsCategory 方法,获取商品分类数据,实现代码大致如下:

```
categoryFuture = _goodsService.getGoodsCategory({"id": widget.categoryId},
(categoryTitles) {
  //分类标题
  _categoryTitleModel = categoryTitles;
  //同级分类标题
  brotherCategory = _categoryTitleModel.brotherCategory;
  //当前选项卡索引
  //...
}
```

步骤3:根据分类数据,添加分类选项卡及选项卡对应的视图,主要实现以下几个方法:

```
//获取 TabBar
List<Widget> getTabBars() {
```

```
//...
}

//获取TabBarView
List<Widget> getTabBarViews() {
  //...
  //添加分类商品列表组件
  tabBarView.add(GoodsListPage(brotherCategory[i].id));
  //...
  return tabBarView;
}

//获取Tab
Widget getTabBarWidget(CategoryModel category) {
  //...
}
```

其中选项卡对应的视图即为商品列表组件。

步骤4：组装选项卡及选项卡对应的视图，GoodsCategoryPage 页面完整代码如下：

```
//page/goods/goods_category_page.dart
import 'package:flutter/material.dart';
import 'package:shop/page/goods/goods_list_page.dart';
import 'package:shop/service/goods_service.dart';
import 'package:shop/model/category_title_model.dart';
//商品分类页面
class GoodsCategoryPage extends StatefulWidget {
  //一级分类名称
  String categoryName;
  //一级分类 Id
  int categoryId;

  //构造方法,传入一级分类名称和分类 Id
  GoodsCategoryPage({Key key, @required this.categoryName, @required this.categoryId})
      : super(key: key);

  @override
  _GoodsCategoryPageState createState() => _GoodsCategoryPageState();
}

class _GoodsCategoryPageState extends State<GoodsCategoryPage> with TickerProviderStateMixin {
  ScrollController _scrollController;
  //二级分类选项卡
  TabController _tabController;
  //数据服务
  GoodsService _goodsService = GoodsService();
  //分类标题数据模型
```

```dart
CategoryTitleModel _categoryTitleModel;
//同级(二级)分类数据列表
List<CategoryModel> brotherCategory = List();
var categoryFuture;
//当前选中的二级分类索引
var currentIndex = 0;

@override
void initState() {
  super.initState();
  //根据分类 id 获取二级分类标题
  categoryFuture = _goodsService.getGoodsCategory({"id": widget.categoryId}, (categoryTitles) {
    //分类标题
    _categoryTitleModel = categoryTitles;
    //同级分类标题
    brotherCategory = _categoryTitleModel.brotherCategory;
    //当前选项卡索引
    currentIndex = getCurrentIndex();
  }, (error) {});
}

//获取当前分类选项卡索引
getCurrentIndex() {
  //循环同级分类
  for (int i = 0; i < brotherCategory.length; i++) {
    //判断,如果分类 id 相等,表示查找到当前分类索引
    if (brotherCategory[i].id == _categoryTitleModel.currentCategory.id) {
      return i;
    }
  }
}

@override
void dispose() {
  _scrollController.dispose();
  _tabController.dispose();
  super.dispose();
}

@override
Widget build(BuildContext context) {
  return Container(
    //异步构建组件
    child: FutureBuilder(
      future: categoryFuture,
      //构建器
      builder: (BuildContext context, AsyncSnapshot asyncSnapshot) {
        //滚动控制
```

```
            _scrollController = ScrollController();
            //选项卡控制器
            _tabController = TabController(
                //初始化索引
                initialIndex: currentIndex,
                //选项个数
                length: brotherCategory.length,
                vsync: this
            );
            return Scaffold(
                appBar: AppBar(
                    //一级分类名称
                    title: Text(widget.categoryName),
                    centerTitle: true,
                    //选项卡,用于放二级分类名称
                    bottom: TabBar(
                        isScrollable: true,
                        controller: _tabController,
                        //指示器颜色
                        indicatorColor: Colors.white,
                        //指定选项卡集
                        tabs: getTabBars()),
                ),
                //二级分类选项卡对应的视图
                body: TabBarView(
                    children: getTabBarViews(),
                    controller: _tabController,
                ),
            );
        }),
    );
}

//获取 TabBar
List<Widget> getTabBars() {
    List<Widget> tabBar = List();
    for (var category in brotherCategory) {
        tabBar.add(getTabBarWidget(category));
    }
    return tabBar;
}

//获取 TabBarView
List<Widget> getTabBarViews() {
    List<Widget> tabBarView = List();
    for (var i = 0; i < brotherCategory.length; i++) {
        //添加分类商品列表组件
        tabBarView.add(GoodsListPage(brotherCategory[i].id));
    }
```

```
      return tabBarView;
    }

    //获取 Tab
    Widget getTabBarWidget(CategoryModel category) {
      return Tab(
        //分类名称
        text: category.name,
      );
    }
```

步骤 5：实现商品列表组件，打开 page/goods/goods_list_page.dart 文件，添加 GoodListPage 组件，此组件在获取商品列表数据后，使用两列多行的形式展示商品信息。首先定义商品数据服务及商品数据模型，代码如下：

```
//商品数据服务
GoodsService goodsService = GoodsService();
//商品数据模型
List<GoodsModel> goodsModels = List();
```

步骤 6：编写根据分类 Id 查询商品列表数据方法，代码如下：

```
_getGoodsData(int categoryId) {
  //参数为分类 Id、页码和查询数量
  goodsService.getGoodsList ({ "categoryId": categoryId, "page": 1, "limit": 100 },
  (goodsModelList) {
    //...
  });
}
```

从上面代码可以看到指定了页码及查询的数量，因为某个分类下的商品数据可能很多，一次只查询其中一部分即可。

步骤 7：使用网格方式展示商品列表数据，采用两列多行形式显示，商品列表组件完整代码如下：

```
//page/goods/goods_list_page.dart 文件
import 'package:flutter/material.dart';
import 'package:flutter_screenutil/flutter_screenutil.dart';
import 'package:shop/service/goods_service.dart';
import 'package:shop/model/goods_model.dart';
import 'package:shop/config/index.dart';
import 'package:shop/utils/navigator_util.dart';
import 'package:shop/widgets/cached_image_widget.dart';
//商品列表组件
```

```dart
class GoodsListPage extends StatefulWidget {
  //二级分类 Id
  int categoryId;
  //构造方法,传入二级分类 Id
  GoodsListPage(this.categoryId);

  @override
  _GoodsListPageState createState() => _GoodsListPageState();
}

class _GoodsListPageState extends State<GoodsListPage> {
  //商品数据服务
  GoodsService goodsService = GoodsService();
  //商品数据模型
  List<GoodsModel> goodsModels = List();
  //二级分类 Id,用于查询二级分类的商品列表数据
  var categoryId;

  //根据二级分 Id 获取商品数据
  _getGoodsData(int categoryId) {
    goodsService.getGoodsList({"categoryId": categoryId, "page": 1, "limit": 100},
        (goodsModelList) {
      //判断当前页面是否存在
      if (mounted) {
        setState(() {
          goodsModels = goodsModelList;
        });
      }
    });
  }

  @override
  void initState() {
    super.initState();
    if (goodsModels == null || goodsModels.length == 0) {
      categoryId = widget.categoryId;
      //获取商品列表数据
      _getGoodsData(categoryId);
    }
  }

  @override
  void dispose() {
    super.dispose();
  }

  @override
  Widget build(BuildContext context) {
    return Scaffold(
```

```dart
        body: Container(
          child: Center(
            child: goodsModels != null && goodsModels.length != 0
                ? GridView.builder(
                    itemCount: goodsModels == null ? 0 : goodsModels.length,
                    gridDelegate: SliverGridDelegateWithFixedCrossAxisCount(
                      //两列
                      crossAxisCount: 2,
                      //间距
                      mainAxisSpacing: 6.0,
                      //间距
                      crossAxisSpacing: 6.0,
                      childAspectRatio: 1.0
                    ),
                    //构建商品列表项
                    itemBuilder: (BuildContext context, int index) {
                      //返回商品项
                      return getGoodsItemWidget(goodsModels[index]);
                    })
                : Center(
                    //垂直布局
                    child: Column(
                      mainAxisAlignment: MainAxisAlignment.center,
                      children: <Widget>[
                        //没有数据图片提示
                        Image.asset(
                          "images/no_data.png",
                          height: 80,
                          width: 80,
                        ),
                        Padding(
                          padding: EdgeInsets.only(top: 10.0),
                        ),
                        //没有数据文本提示
                        Text(
                          KString.NO_DATA_TEXT,
                          style: TextStyle(
                              fontSize: 16.0, color: KColor.noDataTextColor),
                        )
                      ],
                    ),
                  ),
          ),
        ));
  }

  //获取商品项组件
  Widget getGoodsItemWidget(GoodsModel goodsModel) {
    return GestureDetector(
```

```dart
            child: Container(
              alignment: Alignment.center,
              child: SizedBox(
                  width: 320,
                  height: 460,
                  //卡片显示
                  child: Card(
                      //垂直布局
                      child: Column(
                        children: <Widget>[
                          //商品图片
                          CachedImageWidget(
                            double.infinity,
                            ScreenUtil.getInstance().setHeight(200.0),
                            goodsModel.picUrl,
                          ),
                          Padding(
                            padding: EdgeInsets.only(top: 5.0),
                          ),
                          //商品名称
                          Text(
                            goodsModel.name,
                            maxLines: 1,
                            overflow: TextOverflow.ellipsis,
                            style: TextStyle(fontSize: 14.0, color: Colors.black54),
                          ),
                          Padding(
                            padding: EdgeInsets.only(top: 5.0),
                          ),
                          //商品价格
                          Text(
                            "¥ ${goodsModel.retailPrice}",
                            maxLines: 1,
                            overflow: TextOverflow.ellipsis,
                            style: TextStyle(
                                fontSize: 14.0, color: KColor.priceColor),
                          ),
                        ],
                      ),
                   )),
            ),
            //单击处理
            onTap: () => _itemClick(goodsModel.id),
        );
    }

    //单击跳转至商品详情页面
    _itemClick(int id) {
        NavigatorUtil.goGoodsDetails(context, id);
    }
}
```

当单击其中某一项商品时可以跳转至商品详情页面,商品分类页显示的效果如图 31-1 所示。

图 31-1　商品分类页面

31.5　商品详情需求分析

商品详情是用来展示商品的详细信息的页面,当用户浏览了商品后,可能会将商品添加至购物车或立即购买。商品详情由以下几部分组成:

- 商品主图:商品主要展示图片,轮播显示;
- 商品基本信息:包括价格、标题和描述;
- 商品属性:如衣服的面料、图案和款式等;
- 商品详细信息:如衣服各个角度拍摄的图片;
- 商品规格:如衣服的大码、均码和小码等;
- 常见问题:商品购买的常见问题描述;
- 操作按钮:加入购物车和立即购买。

了解了商品详情的基本功能后,看一下商品详情的界面拆分情况,如图 31-2 所示。从上至下依次排列,由于信息量较大整体页面需要上下滚动。

当用户单击商品并加入购物车或立即购买时会从底部弹出一个对话框,让用户选择规格及填写数量,然后单击加入购物车或立即购物,排列方式如图 31-3 所示。

图 31-2　商品详情界面拆分

图 31-3　商品详情弹出框

31.6　商品详情数据模型

通过商品详情的需求分析,可以分析出商品详情具有以下几个数据模型:
- 商品详情数据模型;
- 常见问题数据模型;
- 商品属性数据模型;
- 商品规格数据模型;
- 商品信息数据模型。

商品详情的访问接口如下:

```
http://localhost:8080/client/goods/detail
```

访问这个接口需要添加商品 Id 参数。打开 model/goods_detail_model.dart 文件,添加如下代码：

```dart
//model/goods_detail_model.dart 文件
import 'package:json_annotation/json_annotation.dart';
//扩展文件
part 'goods_detail_model.g.dart';
//商品详情数据模型
@JsonSerializable()
class GoodsDetailModel extends Object {
  //常见问题
  @JsonKey(name: 'issue')
  List<IssueModel> issue;
  //商品属性
  @JsonKey(name: 'attribute')
  List<AttributeModel> attribute;
  //商品规格
  @JsonKey(name: 'productList')
  List<ProductModel> productList;
  //商品信息
  @JsonKey(name: 'info')
  InfoModel info;
  //构造方法
  GoodsDetailModel(this.issue,this.attribute,this.productList,this.info,);

  factory GoodsDetailModel.fromJson(Map<String, dynamic> srcJson) => _$GoodsDetailModelFromJson(srcJson);

  Map<String, dynamic> toJson() => _$GoodsDetailModelToJson(this);

}
//常见问题
@JsonSerializable()
class IssueModel extends Object {
  //id
  @JsonKey(name: 'id')
  int id;
  //问
  @JsonKey(name: 'question')
  String question;
  //答
  @JsonKey(name: 'answer')
  String answer;
  //添加时间
  @JsonKey(name: 'addTime')
  String addTime;
  //更新时间
  @JsonKey(name: 'updateTime')
  String updateTime;
  //是否删除
```

```dart
    @JsonKey(name: 'deleted')
    bool deleted;
    //构造方法

    IssueModel(this.id, this.question, this.answer, this.addTime, this.updateTime, this.deleted,);

    factory IssueModel.fromJson(Map<String, dynamic> srcJson) => _$IssueModelFromJson(srcJson);

    Map<String, dynamic> toJson() => _$IssueModelToJson(this);

}
//商品属性
@JsonSerializable()
class AttributeModel extends Object {
    //id
    @JsonKey(name: 'id')
    int id;
    //商品Id
    @JsonKey(name: 'goodsId')
    int goodsId;
    //属性
    @JsonKey(name: 'attribute')
    String attribute;
    //值
    @JsonKey(name: 'value')
    String value;
    //添加时间
    @JsonKey(name: 'addTime')
    String addTime;
    //更新时间
    @JsonKey(name: 'updateTime')
    String updateTime;
    //是否删除
    @JsonKey(name: 'deleted')
    bool deleted;
    //构造方法

    AttributeModel(this.id, this.goodsId, this.attribute, this.value, this.addTime, this.updateTime, this.deleted,);

    factory AttributeModel.fromJson(Map<String, dynamic> srcJson) => _$AttributeModelFromJson(srcJson);

    Map<String, dynamic> toJson() => _$AttributeModelToJson(this);

}
//商品规格数据模型
@JsonSerializable()
class ProductModel extends Object {
```

```dart
  //id
  @JsonKey(name: 'id')
  int id;
  //商品Id
  @JsonKey(name: 'goodsId')
  int goodsId;
  //规格值
  @JsonKey(name: 'specifications')
  List<String> specifications;
  //价格
  @JsonKey(name: 'price')
  double price;
  //数量
  @JsonKey(name: 'number')
  int number;
  //图片路径
  @JsonKey(name: 'url')
  String url;
  //添加时间
  @JsonKey(name: 'addTime')
  String addTime;
  //更新时间
  @JsonKey(name: 'updateTime')
  String updateTime;
  //是否删除
  @JsonKey(name: 'deleted')
  bool deleted;
  //

  ProductModel(this.id, this.goodsId, this.specifications, this.price, this.number, this.url, this.addTime, this.updateTime, this.deleted,);

  factory ProductModel.fromJson(Map<String, dynamic> srcJson) => _$ProductModelFromJson(srcJson);

  Map<String, dynamic> toJson() => _$ProductModelToJson(this);

}
//商品信息
@JsonSerializable()
class InfoModel extends Object {
  //id
  @JsonKey(name: 'id')
  int id;
  //商品编号
  @JsonKey(name: 'goodsSn')
  String goodsSn;
  //名称
  @JsonKey(name: 'name')
  String name;
```

```dart
//分类 Id
@JsonKey(name: 'categoryId')
int categoryId;
//图片
@JsonKey(name: 'gallery')
List<String> gallery;
//关键字
@JsonKey(name: 'keywords')
String keywords;
//简介
@JsonKey(name: 'brief')
String brief;
//是否在售
@JsonKey(name: 'isOnSale')
bool isOnSale;
//排序规则
@JsonKey(name: 'sortOrder')
int sortOrder;
//图片地址
@JsonKey(name: 'picUrl')
String picUrl;
//分享地址
@JsonKey(name: 'shareUrl')
String shareUrl;
//是否为新品
@JsonKey(name: 'isNew')
bool isNew;
//是否为热卖商品
@JsonKey(name: 'isHot')
bool isHot;
//单位
@JsonKey(name: 'unit')
String unit;
//专柜价格
@JsonKey(name: 'counterPrice')
double counterPrice;
//零售价格
@JsonKey(name: 'retailPrice')
double retailPrice;
//添加时间
@JsonKey(name: 'addTime')
String addTime;
//更新时间
@JsonKey(name: 'updateTime')
String updateTime;
//是否删除
@JsonKey(name: 'deleted')
bool deleted;
//详情
```

```
@JsonKey(name: 'detail')
String detail;
//构造方法

InfoModel(this.id, this.goodsSn, this.name, this.categoryId, this.gallery, this.keywords,
this.brief, this.isOnSale, this.sortOrder, this.picUrl, this.shareUrl, this.isNew, this.isHot,
this.unit, this.counterPrice, this.retailPrice, this.addTime, this.updateTime, this.deleted,
this.detail,);

  factory InfoModel.fromJson(Map<String, dynamic> srcJson) => _$InfoModelFromJson(srcJson);

  Map<String, dynamic> toJson() => _$InfoModelToJson(this);

}
```

31.7 商品详情轮播图

在商品详情的最上面会有几张产品或广告图片来回切换,这里可以使用 Swiper 组件。具体的用法和首页的轮播图是一样的,只是传递过来的是商品详情的 galleryData 值。

参照首页轮播图实现步骤,打开 page/goods/goods_detail_gallery.dart 文件,添加如下代码:

```
//page/goods/goods_detail_gallery.dart 文件
import 'package:flutter_swiper/flutter_swiper.dart';
import 'package:flutter/material.dart';
import 'package:flutter_screenutil/flutter_screenutil.dart';
import 'package:shop/config/index.dart';
import 'package:shop/widgets/cached_image_widget.dart';
//商品详情主图轮播组件
class GoodsDetailGalleryWidget extends StatelessWidget {
  //轮播图数据
  List<String> galleryData = List();
  //数量
  int size;
  //视图高度
  double viewHeight;
  //构造方法,传入轮播图数据,数量及视图高度
  GoodsDetailGalleryWidget(this.galleryData, this.size, this.viewHeight);

  @override
  Widget build(BuildContext context) {
    return Container(
      height: viewHeight,
      child: galleryData == null || galleryData.length == 0
          ? Container(
```

```dart
            height: ScreenUtil.instance.setHeight(200.0),
            color: Colors.grey,
            alignment: Alignment.center,
            //没有数据文本提示
            child: Text(KString.NO_DATA_TEXT),
          )
        : Swiper(
            //轮播图数量
            itemCount: galleryData.length,
            //滚动方向
            scrollDirection: Axis.horizontal,
            //滚动方向,设置为 Axis.vertical 则为垂直滚动
            loop: true,
            //无限轮播模式开关
            index: 0,
            //初始的时候下标位置
            autoplay: false,
            itemBuilder: (BuildContext buildContext, int index) {
              //缓存图片
              return CachedImageWidget(
                  double.infinity, double.infinity, galleryData[index]);
            },
            //动画时长
            duration: 10000,
            //分页
            pagination: SwiperPagination(
              //小圆点放置底部中间
              alignment: Alignment.bottomCenter,
              //小圆点构建器
              builder: DotSwiperPaginationBuilder(
                size: 8.0,
                //圆点默认颜色
                color: KColor.bannerDefaultColor,
                //圆点单击颜色
                activeColor: KColor.bannerActiveColor)),
          ),
  );
 }
}
```

31.8 商品详情页面实现

商品详情数据模型及数据服务已经实现,接下来根据商品详情的需求分析把这些部分串联起来。具体步骤如下:

步骤 1:打开 page/goods/goods_detail_page.dart 文件,添加 GoodsDetailPage 组件。定义商品数据服务、购物车数据服务、商品收藏数据服务,以及商品详情数据模型,代码

如下：

```
//商品数据服务
GoodsService _goodsService = GoodsService();
//购物车数据服务
CartService _cartService = CartService();
//商品收藏数据服务
CollectService _collectService = CollectService();
//商品详情数据模型
GoodsDetailModel _goodsDetail;
```

这里的购物车数据服务和商品收藏数据服务还没有实现，但是商品详情页面需要调用其方法，以便将商品添加至购物车或收藏商品，可以先注释起来。

步骤2：页面初始化完成后，需要立即调用商品服务以便获取商品详情数据。此方法需要传递商品Id，代码如下：

```
var params = {"id": goodsId};
//获取商品详情数据
_goodsDetailFuture = _goodsService.getGoodsDetailData(params, (goodsDetail) {
  //...
});
```

步骤3：从商品数据服务返回的对象_goodsDetailFuture是一个异步对象，可以将其赋值给异步构建组件FutureBuilder的feature属性。异步构建组件会返回几个状态，用这些状态可以判断数据是否在加载或加载完成，如下面分支语句代码：

```
switch (asyncSnapshot.connectionState) {
  //无状态
  case ConnectionState.none:
  //等待加载数据
  case ConnectionState.waiting:
  //展示商品详情
  default:
}
```

步骤4：添加商品详情页使用到的方法。方法如下：

```
添加至购物车：_addCart
立即购买：_buy
收藏商品：_collection
收藏和取消收藏商品：_addOrDeleteCollect
```

这四个方法需要依赖购物车服务与收藏商品数据服务，可以先把它们注释起来。另外它们都需要判断用户是否登录，以添加至购物车为例，首先获取本地token值，如果有值则调用方法，如果没有则跳转到登录页面，处理结构代码如下：

```
_addCart() {
    //获取 token 值
    SharedPreferencesUtil.getToken().then((value) {
      if (value != null) {
        //参数
        parameters = {
          //...
        };
        //调用购物车数据服务的 addCart 方法
        _cartService.addCart(parameters, (value) {
          //...
        }, );
      } else {
        //如果没有 token 值则跳转至登录框
        NavigatorUtil.goLogin(context);
      }
    });
}
```

步骤 5：编写商品详情页面，根据商品详情页面拆分及数据模型编写，完整的代码如下：

```
//page/goods/goods_detail_page.dart 文件
import 'package:flutter/material.dart';
import 'package:dio/dio.dart';
import 'package:flutter_html/flutter_html.dart';
import 'package:flutter_spinkit/flutter_spinkit.dart';
import 'package:flutter_screenutil/flutter_screenutil.dart';
import 'package:shop/service/goods_service.dart';
import 'package:shop/service/cart_service.dart';
import 'package:shop/model/goods_detail_model.dart';
import 'package:shop/config/index.dart';
import 'package:shop/page/goods/goods_detail_gallery.dart';
import 'package:shop/utils/navigator_util.dart';
import 'package:shop/utils/shared_preferences_util.dart';
import 'package:shop/utils/toast_util.dart';
import 'package:shop/widgets/cart_number_widget.dart';
import 'package:shop/event/refresh_event.dart';
import 'package:shop/service/collect_service.dart';
import 'package:shop/widgets/cached_image_widget.dart';
//商品详情页面
class GoodsDetailPage extends StatefulWidget {
  //商品 Id
  int goodsId;
  //构造方法，商品 Id 为必传参数
  GoodsDetailPage({Key key, @required this.goodsId}) : super(key: key);
```

```dart
  @override
  _GoodsDetailPageState createState() => _GoodsDetailPageState();
}

class _GoodsDetailPageState extends State<GoodsDetailPage> {
  //商品Id
  int goodsId;
  //商品数据服务
  GoodsService _goodsService = GoodsService();
  //购物车数据服务
  CartService _cartService = CartService();
  //商品收藏数据服务
  CollectService _collectService = CollectService();
  //商品详情数据模型
  GoodsDetailModel _goodsDetail;
  //参数对象
  var parameters;
  //规格索引
  int _specificationIndex = 0;
  //商品数量
  int _number = 1;
  //请求商品详情接口返回的Future对象
  var _goodsDetailFuture;
  //token
  var token;
  //是否收藏
  var _isCollection = false;

  @override
  void initState() {
    super.initState();
    //通过widget获取商品Id
    goodsId = widget.goodsId;
    //组装参数
    var params = {"id": goodsId};
    //获取商品详情数据
    _goodsDetailFuture = _goodsService.getGoodsDetailData(params, (goodsDetail) {
      _goodsDetail = goodsDetail;
    });
  }

  @override
  Widget build(BuildContext context) {
    return Container(
      child: Scaffold(
        appBar: AppBar(
          //页面标题
          title: Text(KString.GOODS_DETAIL),
          centerTitle: true,
```

```
      ),
      //异步构建组件
      body: FutureBuilder(
        //future 对象
        future: _goodsDetailFuture,
        //构建器
        builder: (BuildContext context, AsyncSnapshot asyncSnapshot) {
          //连接状态
          switch (asyncSnapshot.connectionState) {
            case ConnectionState.none:
            case ConnectionState.waiting:
              //等待加载数据
              return Container(
                child: Center(
                  //显示等待组件
                  child: SpinKitFoldingCube(
                    size: 40.0,
                    color: KColor.watingColor,
                  ),
                ),
              );
            default:
              //服务器出错提示
              if (asyncSnapshot.hasError)
                return Container(
                  child: Center(
                    child: Text(
                      KString.SERVER_EXCEPTION,
                      style: TextStyle(fontSize: 16.0),
                    ),
                  ),
                );
              else
                //商品详情组件
                return _detailWidget();
          }
        },
      ),
      //页面底部组件
      bottomNavigationBar: BottomAppBar(
        child: Container(
          height: 50.0,
          child: Row(
            children: <Widget>[
              //是否收藏组件
              Expanded(
                flex: 1,
                child: Container(
                  color: Colors.white,
```

```dart
          child: InkWell(
            onTap: () => _collection(),
            child: Icon(
              Icons.star_border,
              color: _isCollection
                  ? KColor.collectionButtonColor
                  : KColor.unCollectionButtonColor,
              size: 30.0,
            ),
          ),
        )),
        //购物车图标
        Expanded(
            flex: 1,
            child:Icon(
              Icons.add_shopping_cart,
              color: KColor.addCartIconColor,
              size: 30.0,
            ),
        ),
        //添加至购物车按钮
        Expanded(
            flex: 2,
            child: Container(
              color: KColor.addCartButtonColor,
              child: InkWell(
                //打开底部弹出框
                onTap: () => openBottomSheet(
                    context, _goodsDetail.productList[0], 1),
                child: Center(
                  child: Text(
                    KString.ADD_CART,
                    style: TextStyle(
                        color: Colors.white, fontSize: 14.0),
                  ),
                )),
            )),
        //立即购买按钮
        Expanded(
          flex: 2,
          child: Container(
              color: KColor.buyButtonColor,
              child: InkWell(
                //打开底部弹出框
                onTap: () => openBottomSheet(
                    context, _goodsDetail.productList[0], 2),
                child: Center(
                  child: Text(
                    KString.BUY,
```

```dart
                    style: TextStyle(
                        color: Colors.white, fontSize: 14.0),
                  ),
                ))),
          ),
        ],
      ),
    ),
   ),
  ),
 );
}

//打开底部弹出框,showType 为 1 则表示添加至购物车,为 2 则表示立即购买
openBottomSheet(BuildContext context, ProductModel productList, int showType) {
  showModalBottomSheet(
      context: context,
      builder: (BuildContext context) {
        //安全区域
        return SafeArea(
          child: SizedBox(
            width: double.infinity,
            height: ScreenUtil.instance.setHeight(630.0),
            child: Container(
              //垂直布局
              child: Column(
                //此轴即水平方向靠左对齐
                crossAxisAlignment: CrossAxisAlignment.start,
                children: <Widget>[
                  Container(
                    margin: EdgeInsets.all(ScreenUtil.instance.setWidth(20.0)),
                    //水平布局
                    child: Row(
                      crossAxisAlignment: CrossAxisAlignment.center,
                      children: <Widget>[
                        //商品图片
                        CachedImageWidget(
                            ScreenUtil.instance.setWidth(120.0),
                            ScreenUtil.instance.setWidth(120.0),
                            productList.url),
                        //垂直布局,价格及选择规格
                        Column(
                          crossAxisAlignment: CrossAxisAlignment.start,
                          children: <Widget>[
                            //价格
                            Text(
                              KString.PRICE + ": " + "＄{productList.price}",
                              style: TextStyle(
                                  color: Colors.black54,
```

```
                              fontSize: ScreenUtil.instance.setSp(24.0)),
                        ),
                        Padding(
                          padding: EdgeInsets.only(
                              top: ScreenUtil.instance.setHeight(10.0)),
                        ),
                        //选择规格
                        Text(KString.ALREAD_SELECTED +
                            ": " + _goodsDetail.productList[0].specifications[_specificationIndex])
                      ],
                    ),
                  ),
                  Expanded(
                    child: Container(
                      alignment: Alignment.centerRight,
                      //删除按钮
                      child: IconButton(
                        icon: Icon(Icons.delete),
                        //单击返回
                        onPressed: () {
                          Navigator.pop(context);
                        },
                      ),
                    )),
                ],
              ),
            ),
            Container(
              margin: EdgeInsets.all(ScreenUtil.instance.setWidth(10.0)),
              //商品规格提示
              child: Text(
                KString.SPECIFICATIONS,
                style: TextStyle(
                    color: Colors.black54,
                    fontSize: ScreenUtil.instance.setSp(30.0)),
              ),
            ),
            //商品规格
            Wrap(
                children: _specificationsWidget(productList.specifications)),
            Padding(
              padding: EdgeInsets.only(
                  top: ScreenUtil.instance.setHeight(10.0)),
            ),
            Container(
              margin: EdgeInsets.all(ScreenUtil.instance.setWidth(10.0)),
              //商品数量
              child: Text(
                KString.NUMBER,
```

```dart
              style: TextStyle(
                  color: Colors.black54,
                  fontSize: ScreenUtil.instance.setSp(30.0)),
            ),
          ),
          Container(
              margin:
              EdgeInsets.all(ScreenUtil.instance.setWidth(10.0)),
              height: ScreenUtil.instance.setHeight(80),
              alignment: Alignment.centerLeft,
              //商品数量加减组件
              child: CartNumberWidget(1, (number) {
                setState(() {
                  //设置当前商品数据
                  _number = number;
                });
              })),
          Expanded(
              child: Stack(
                alignment: Alignment.bottomLeft,
                children: <Widget>[
                  SizedBox(
                    height: ScreenUtil.instance.setHeight(100.0),
                    width: double.infinity,
                    child: InkWell(
                      //单击添加至购物车或立即购买
                      onTap: () => showType == 1 ? _addCart() : _buy(),
                      child: Container(
                        alignment: Alignment.center,
                        color: KColor.defaultButtonColor,
                        //添加至购物车或立即购买文本
                        child: Text(
                          showType == 1 ? KString.ADD_CART : KString.BUY,
                          style: TextStyle(
                              color: Colors.white,
                              fontSize: ScreenUtil.instance.setSp(30.0)),
                        ),
                      )),
                  ),
                ],
              ))
        ],
      ),
    ),
  );
});
}
```

```dart
//商品规格
List<Widget> _specificationsWidget(List<String> specifications) {
  List<Widget> specificationsWidget = List();
  //循环迭代出所有规格
  for (int i = 0; i < specifications.length; i++) {
    specificationsWidget.add(Container(
      padding: EdgeInsets.all(ScreenUtil.instance.setWidth(10.0)),
      child: InkWell(
        child: Chip(
          //规格名称
          label: Text(
            specifications[i],
            //选中与未选中使用不同颜色区分
            style: TextStyle(
              color: i == _specificationIndex
                  ? Colors.white
                  : Colors.black54,
              fontSize: ScreenUtil.instance.setSp(24.0)),
          ),
          //选中与未选中使用不同颜色区分
          backgroundColor: i == _specificationIndex
              ? KColor.specificationWarpColor
              : Colors.grey,
        ),
      )));
  }
  return specificationsWidget;
}

//添加至购物车
_addCart() {
  //获取 token 值
  SharedPreferencesUtil.getToken().then((value) {
    if (value != null) {
      //参数
      parameters = {
        //商品 Id
        "goodsId": _goodsDetail.info.id,
        //规格 Id
        "productId": _goodsDetail.productList[0].id,
        //数量
        "number": _number
      };
      //调用购物车数据服务的 addCart 方法
      _cartService.addCart(parameters, (value) {
        ToastUtil.showToast(KString.ADD_CART_SUCCESS);
        //隐藏弹出框
        Navigator.of(context).pop();
        //通知刷新
```

```dart
      eventBus.fire(RefreshEvent());
    }, );
  } else {
    //如果没有token值则跳转至登录框
    NavigatorUtil.goLogin(context);
  }
});
}

//立即购买
_buy() {
  if (SharedPreferencesUtil.token != null) {
    //提交参数
    parameters = {
      //商品Id
      "goodsId": _goodsDetail.info.id,
      //商品规格Id
      "productId": _goodsDetail.productList[0].id,
      //商品数量
      "number": _number,
    };
    //调用商品购买方法
    _cartService.fastBuy(parameters, (success) {
      //进入商品结算页面
      NavigatorUtil.goFillInOrder(context, success);
    }, (error) {});
  } else {
    //未登录则跳转至登录页面
    NavigatorUtil.goLogin(context);
  }
}

//收藏商品
_collection() {
  SharedPreferencesUtil.getToken().then((value) {
    if (value == null) {
      //未登录则跳转至登录页面
      NavigatorUtil.goLogin(context);
    } else {
      token = value;
      //收藏和取消收藏商品
      _addOrDeleteCollect();
    }
  });
}

//收藏和取消收藏商品
_addOrDeleteCollect() {
  Options options = Options();
```

```dart
      //添加token至headers
      options.headers["X-Shop-Token"] = token;
      //参数
      var parameters = {"type": 0, "valueId": _goodsDetail.info.id};
      _collectService.addOrDeleteCollect(parameters, (onSuccess) {
        setState(() {
          _isCollection = true;
        });
      }, (error) {
        ToastUtil.showToast(error);
      });
    }

    //详情组件
    Widget _detailWidget() {
      return Stack(
        alignment: AlignmentDirectional.bottomCenter,
        children: <Widget>[
          ListView(
            children: <Widget>[
              //商品轮播图
              GoodsDetailGalleryWidget(_goodsDetail.info.gallery, _goodsDetail.info.gallery.length, 240.0),
              Divider(
                height: 2.0,
                color: Colors.grey,
              ),
              Padding(
                padding: EdgeInsets.only(top: 10.0),
              ),
              Container(
                margin: EdgeInsets.only(left: 10.0),
                child: Column(
                  crossAxisAlignment: CrossAxisAlignment.start,
                  children: <Widget>[
                    //商品名称
                    Text(
                      _goodsDetail.info.name,
                      style: TextStyle(
                        fontSize: 16.0,
                        color: Colors.black54,
                        fontWeight: FontWeight.bold),
                    ),
                    Padding(
                      padding: EdgeInsets.only(top: 6.0),
                    ),
                    //商品简介
                    Text(
                      _goodsDetail.info.brief,
```

```dart
          style: TextStyle(fontSize: 14.0, color: Colors.grey),
        ),
        Padding(
          padding: EdgeInsets.only(top: 4.0),
        ),
        Row(
          children: <Widget>[
            //商品原价
            Text(
              "原价: ${_goodsDetail.info.counterPrice}",
              style: TextStyle(
                  color: Colors.grey,
                  fontSize: 12.0,
                  decoration: TextDecoration.lineThrough),
            ),
            Padding(
              padding: EdgeInsets.only(left: 10.0),
            ),
            //商品现价
            Text(
              "现价: ${_goodsDetail.info.retailPrice}",
              style: TextStyle(
                  color: Colors.deepOrangeAccent, fontSize: 12.0),
            ),
          ],
        ),
      ],
    ),
  ),
  Padding(
    padding: EdgeInsets.only(top: 4.0),
  ),
  //商品属性
  _goodsDetail.attribute == null || _goodsDetail.attribute.length == 0
      ? Divider()
      : Container(
          //垂直布局
          child: Column(
            children: <Widget>[
              //商品属性标题
              Text(
                KString.GOODS_ATTRIBUTES,
                style: TextStyle(
                    color: Colors.black54,
                    fontSize: 20.0,
                    fontWeight: FontWeight.bold),
              ),
              Padding(
                padding: EdgeInsets.only(top: 6.0),
```

```dart
            ),
            //商品属性组件
            _attributeWidget(_goodsDetail),
          ],
        ),
      ),
      //商品详细信息
      Html(data: _goodsDetail.info.detail),
      //常见问题
      _goodsDetail.issue == null || _goodsDetail.issue.length == 0
          ? Divider()
          : Container(
              //垂直布局
              child: Column(
                children: <Widget>[
                  //常见问题标题
                  Text(
                    KString.COMMON_PROBLEM,
                    style: TextStyle(
                        color: Colors.black54,
                        fontSize: 20.0,
                        fontWeight: FontWeight.bold),
                  ),
                  Padding(
                    padding: EdgeInsets.only(top: 6.0),
                  ),
                  //常见问题组件
                  _issueWidget(_goodsDetail),
                ],
              ),
            ),
    ],
  ),
  ],
 );
}

//商品属性 Widget
Widget _attributeWidget(GoodsDetailModel goodsDetail) {
  print("${goodsDetail.attribute.length}");
  //商品属性通过列表一行一行地展示
  return ListView.builder(
    physics: NeverScrollableScrollPhysics(),
    shrinkWrap: true,
    //商品属性个数
    itemCount: goodsDetail.attribute.length,
    //商品属性项构建器
    itemBuilder: (BuildContext context, int index) {
      return _attributeItemWidget(goodsDetail.attribute[index]);
```

```
      });
}

//商品属性项
Widget _attributeItemWidget(AttributeModel attribute) {
  return Container(
      margin: EdgeInsets.only(left: 10, right: 10, top: 6, bottom: 6),
      decoration: BoxDecoration(
          color: Colors.grey[100], borderRadius: BorderRadius.circular(10.0)),
      padding: EdgeInsets.all(6.0),
      //水平布局
      child: Row(
        children: <Widget>[
          Expanded(
            flex: 2,
            //商品属性名称
            child: Text(
              attribute.attribute,
              style: TextStyle(color: KColor.attributeTextColor, fontSize: 14.0),
            ),
          ),
          Expanded(
              flex: 4,
              child: Container(
                alignment: Alignment.centerLeft,
                //商品属性值
                child: Text(
                  attribute.value,
                  style: TextStyle(color: KColor.attributeTextColor, fontSize: 14.0),
                ),
              )),
        ],
      ));
}

//常见问题
Widget _issueWidget(GoodsDetailModel goodsDetail) {
  //常用问题使用列表一行一行地展示
  return ListView.builder(
      physics: NeverScrollableScrollPhysics(),
      shrinkWrap: true,
      //常见问题条数
      itemCount: goodsDetail.issue.length,
      //常见问题项构建器
      itemBuilder: (BuildContext context, int index) {
        return _issueItemWidget(goodsDetail.issue[index]);
      });
}
```

```
//常见问题列表项
Widget _issueItemWidget(IssueModel issue) {
  return Container(
      margin: EdgeInsets.only(left: 10, right: 10, top: 6, bottom: 6),
      padding: EdgeInsets.all(6.0),
      //垂直布局,一问一答为一组
      child: Column(
        crossAxisAlignment: CrossAxisAlignment.start,
        children: <Widget>[
          //常见问题:问
          Text(
            issue.question,
            style: TextStyle(color: KColor.issueQuestionColor, fontSize: 14.0),
          ),
          Padding(
            padding: EdgeInsets.only(top: 10.0),
          ),
          Container(
              alignment: Alignment.centerLeft,
              //常见问题:答
              child: Text(
                issue.answer,
                style: TextStyle(color: KColor.issueAnswerColor, fontSize: 14.0),
              )),
        ],
      ));
}
```

其中,商品详情图片及后台提供的网页数据通常使用 Html 里的 <img/> 标签进行包裹,所以前端展示也需要使用 Html 组件来展示。

值得注意的是代码中还用到了 RefreshEvent,当用户单击添加至购物车时,触发该事件,通知购物车模块刷新其数据。打开 event/refresh_event.dart 文件,添加如下代码:

```
//event/refresh_event.dart 文件
import 'package:event_bus/event_bus.dart';

EventBus eventBus = EventBus();
//刷新事件
class RefreshEvent {}
```

商品详情运行效果如图 31-4 所示。

> 提示　商品详情页的布局比较复杂,首先要看页面的效果图,然后根据效果图进行页面拆分,再实现各个小组件,最后把它们组装起来即可。页面组装完成后,还需加上数据处理及方法的调用。

图 31-4　商品详情页面

第 32 章 购 物 车

用户浏览商品并选择好商品后,单击添加至购物车,然后可以继续浏览商品。当商品挑选完毕就可以打开购物车页面进行结算并提交订单。购物车是购物流程的中间环节,从这里我们可以知道购物车是与订单详情页面及填写订单页面有关联的。

本章将详细阐述购物车数据模型、购物车数据服务,以及购物车页面的实现过程,同时会说明购物车与其他模块的关系。

32.1 购物车列表数据模型

购物车列表数据模型 CartListModel 主要是用来给购物车列表提供渲染数据的,展示购物车下的所有商品及购物信息,其访问接口如下所示:

```
//获取购物车数据列表
http://localhost:8080/client/cart/index
```

打开 model/cart_list_model.dart 文件,添加如下代码:

```
//model/cart_list_model.dart 文件
import 'package:json_annotation/json_annotation.dart';
//扩展文件
part 'cart_list_model.g.dart';

//购物车列表数据模型
@JsonSerializable()
class CartListModel extends Object {
  //统计数据
  @JsonKey(name: 'cartTotal')
  CartTotalModel cartTotal;
  //商品数据列表
  @JsonKey(name: 'cartList')
  List<CartModel> cartList;
  //构造方法
```

```dart
  CartListModel(this.cartTotal, this.cartList,);

    factory CartListModel.fromJson(Map<String, dynamic> srcJson) => _
$CartListModelFromJson(srcJson);

  Map<String, dynamic> toJson() => _$CartListModelToJson(this);
}

//购物车统计数据模型
@JsonSerializable()
class CartTotalModel extends Object {

  //商品总数
  @JsonKey(name: 'goodsAmount')
  double goodsAmount;

  //选择的商品总数
  @JsonKey(name: 'checkedGoodsAmount')
  double checkedGoodsAmount;
  //构造方法
  CartTotalModel(this.goodsAmount, this.checkedGoodsAmount,);

    factory CartTotalModel.fromJson(Map<String, dynamic> srcJson) => _
$CartTotalModelFromJson(srcJson);

  Map<String, dynamic> toJson() => _$CartTotalModelToJson(this);
}

//购物车商品数据模型
@JsonSerializable()
class CartModel extends Object {
  //id
  @JsonKey(name: 'id')
  int id;
  //用户Id
  @JsonKey(name: 'userId')
  int userId;
  //商品Id
  @JsonKey(name: 'goodsId')
  int goodsId;
  //商品编号
  @JsonKey(name: 'goodsSn')
  String goodsSn;
  //商品名称
  @JsonKey(name: 'goodsName')
  String goodsName;
  //产品规格Id
  @JsonKey(name: 'productId')
  int productId;
```

```dart
//价格
@JsonKey(name: 'price')
double price;
//数量
@JsonKey(name: 'number')
int number;
//规格值
@JsonKey(name: 'specifications')
List<String> specifications;
//是否选择
@JsonKey(name: 'checked')
bool checked;
//图片
@JsonKey(name: 'picUrl')
String picUrl;
//添加时间
@JsonKey(name: 'addTime')
String addTime;
//更新时间
@JsonKey(name: 'updateTime')
String updateTime;
//是否删除
@JsonKey(name: 'deleted')
bool deleted;
//构造方法
CartModel(
  this.id,
  this.userId,
  this.goodsId,
  this.goodsSn,
  this.goodsName,
  this.productId,
  this.price,
  this.number,
  this.specifications,
  this.checked,
  this.picUrl,
  this.addTime,
  this.updateTime,
  this.deleted,
);

factory CartModel.fromJson(Map<String, dynamic> srcJson) => _$CartModelFromJson(srcJson);

Map<String, dynamic> toJson() => _$CartModelToJson(this);
}
```

数据模型里主要提供购物车的商品信息及一些统计数据,如购物车商品数量、商品总价,以及已选择的商品总数等。

接下来打开命令行工具,进入 shop-flutter 项目根目录并使用 build_runner 工具生成扩展文件 cart_list_model.g.dart。

32.2 购物车数据服务

购物车数据服务是给购物车页、商品详情页,以及填写订单页提供数据接口的类,提供以下几个方法:

- 添加商品至购物车:从商品详情添加商品至购物车使用;
- 查询购物车列表数据:供购物车页面查询使用;
- 删除购物车商品:供购物车列表项删除商品使用;
- 更新购物车商品:供购物车列表项更新购物数量使用;
- 购物车商品勾选处理:供购物车列表项是否勾选商品使用;
- 获取购物车数据:供填写订单获取购物车数据使用;
- 立即购买:供商品详情立即购买使用。

数据服务的实现可以参考前面几章的描述,打开 service/cart_service.dart 文件,添加如下代码:

```dart
//service/cart_service.dart 文件
import 'package:dio/dio.dart';
import 'package:shop/utils/http_util.dart';
import 'package:shop/config/server_url.dart';
import 'package:shop/config/index.dart';
import 'package:shop/model/cart_list_model.dart';
import 'package:shop/model/fill_in_order_model.dart';
//定义成功返回列表数据
typedef OnSuccessList<T>(List<T> list);
//定义成功返回数据
typedef OnSuccess<T>(T t);
//定义返回失败消息
typedef OnFail(String message);
//购物车数据服务
class CartService {

  //添加商品至购物车
  Future addCart(Map<String, dynamic> parameters, OnSuccess onSuccess, {OnFail onFail, Options options}) async {
    try {
      var response;
      if (options == null) {
```

```
              response = await HttpUtil.instance.post(ServerUrl.CART_ADD, parameters:
parameters);
        } else {
              response = await HttpUtil.instance.post(ServerUrl.CART_ADD, parameters:
parameters, options: options);
        }
        if (response['errno'] == 0) {
            //添加成功返回成功消息
            onSuccess(KString.SUCCESS);
        } else {
            //添加失败返回失败消息
            onFail(response['errmsg']);
        }
    } catch (e) {
        print(e);
        //返回服务器异常消息
        onFail(KString.SERVER_EXCEPTION);
    }
}

//查询购物车列表数据,购物车页面使用此方法
Future queryCart(OnSuccess onSuccess, {OnFail onFail, Options options}) async {
    try {
        var response;
        response = await HttpUtil.instance.get(ServerUrl.CART_LIST, options: options);
        if (response['errno'] == 0) {
            //将返回的Json数据转换成购物车列表数据
            CartListModel cartList = CartListModel.fromJson(response['data']);
            onSuccess(cartList);
        } else {
            //返回错误消息
            onFail(response['errmsg']);
        }
    } catch (e) {
        print(e);
        //返回服务器异常消息
        onFail(KString.SERVER_EXCEPTION);
    }
}

//删除购物车商品
Future deleteCart(OnSuccess onSuccess, OnFail onFail, Map<String, dynamic> parameters)
async {
    try {
        var response;
        response = await HttpUtil.instance.post(ServerUrl.CART_DELETE, parameters:
parameters);
        if (response['errno'] == 0) {
            //删除成功返回成功消息
```

```
          onSuccess(KString.SUCCESS);
        } else {
          //返回错误消息
          onFail(response['errmsg']);
        }
    } catch (e) {
      print(e);
      //返回服务器异常消息
      onFail(KString.SERVER_EXCEPTION);
    }
  }

  //更新购物车商品,可用于更新商品购买数量
  Future updateCart(OnSuccess onSuccess, OnFail onFail, Options options, Map<String, dynamic> parameters) async {
    try {
      var response;
      response = await HttpUtil.instance.post(ServerUrl.CART_UPDATE, options: options, parameters: parameters);
      if (response['errno'] == 0) {
        //返回更新成功消息
        onSuccess(KString.SUCCESS);
      } else {
        //返回错误消息
        onFail(response['errmsg']);
      }
    } catch (e) {
      print(e);
      //返回服务器异常消息
      onFail(KString.SERVER_EXCEPTION);
    }
  }

  //购物车商品勾选处理
  Future cartCheck(OnSuccess onSuccess, OnFail onFail, Map<String, dynamic> parameters) async {
    try {
      var response = await HttpUtil.instance.post(ServerUrl.CART_CHECK, parameters: parameters);
      if (response['errno'] == 0) {
        //将返回的Json数据转换成购物车列表数据
        CartListModel cartList = CartListModel.fromJson(response['data']);
        onSuccess(cartList);
      } else {
        //返回错误消息
        onFail(response['errmsg']);
      }
    } catch (e) {
      print(e);
```

```dart
        //返回服务器异常消息
        onFail(KString.SERVER_EXCEPTION);
    }
}

//获取购物车数据,填写商品订单页面使用
Future getCartDataForFillInOrder(OnSuccess onSuccess, OnFail onFail, Map<String, dynamic> parameters) async {
    try {
        var response = await HttpUtil.instance.get(ServerUrl.CART_BUY, parameters: parameters);
        if (response['errno'] == 0) {
            //将返回的Json数据转换成填写订单数据模型
            FillInOrderModel fillInOrderModel = FillInOrderModel.fromJson(response['data']);
            onSuccess(fillInOrderModel);
        } else {
            //返回错误消息
            onFail(response['errmsg']);
        }
    } catch (e) {
        print(e);
        //返回服务器异常消息
        onFail(KString.SERVER_EXCEPTION);
    }
}

//立即购买
Future fastBuy(Map<String, dynamic> parameters, OnSuccess onSuccess, OnFail onFail,) async {
    try {
        var response = await HttpUtil.instance.post(ServerUrl.FAST_BUY, parameters: parameters,);
        if (response['errno'] == 0) {
            //购买成功返回数据
            onSuccess(response["data"]);
        } else {
            //返回错误消息
            onFail(response['errmsg']);
        }
    } catch (e) {
        print(e);
        //返回服务器异常消息
        onFail(KString.SERVER_EXCEPTION);
    }
}
```

> 提示　这里的FillInOrderModel还尚未实现,它是为填写订单提供数据模型的,用于获取购物车中已选择的商品数据,同时还提供商品详情中需要使用的_addCart及fastBuy方法。

32.3　购物车页面实现

当数据模型和数据服务都准备好后,就可以实现购物车页面了。具体实现步骤如下:

步骤1:打开page/cart/cart_page.dart文件,添加CartPage组件,初始化以下几个变量:

```
//购物车数据服务
CartService _cartService = CartService();
//购物车列表数据
List<CartModel> _cartList;
//购物车列表数据模型
CartListModel _cartListModel;
//是否登录
bool _isLogin = false;
//是否全选
bool _isAllCheck = false;
//总计
double _totalMoney = 0.0;
//token
var token;
```

步骤2:由于购物车需要登录后才能使用,所以程序中需要判断登录状态,处理方式如下:

```
//获取token判断是否登录
SharedPreferencesUtil.getToken().then((onValue) {
  if (onValue == null) {
    //未登录则提示登录
    //...
  } else {
    //已登录则获取购物车数据
    token = onValue;
    _getCartData(onValue);
  }
});
```

步骤3:调用购物车数据服务获取购物车列表数据方法,获取数据后需要把数据赋值给购物车列表,同时要判断全选状态。大致代码如下:

```
_getCartData(token) {
  //查询购物车数据
  _cartService.queryCart((cartList) {
    //...
  }, options: options);
}
```

步骤4：当用户从商品详情页面单击添加至购物车时会触发刷新事件，购物车页面的build方法需要添加监听刷新事件。另外，当用户登录成功后，购物车数据也需要刷新，所以需要添加如下两行代码：

```
eventBus.on<RefreshEvent>().listen(...);
loginEventBus.on<LoginEvent>().listen(...)
```

步骤5：编写购物车页面交互方法，方法名称及作用如下：

- _updateCart：更新购物车，传入索引及数量；
- _checkCart：是否勾选商品，传入索引及是否勾选；
- _deleteGoods：根据索引删除购物车商品；
- _isCheckedAll：判断是否全选；
- _setCheckedAll：设置是否全选/全不选。

步骤6：通过上面的数据处理及交互方法实现后，就可以做界面渲染处理了。完整代码如下：

```
//page/cart/cart_page.dart 文件
import 'package:flutter/material.dart';
import 'package:dio/dio.dart';
import 'package:flutter_screenutil/flutter_screenutil.dart';
import 'package:shop/model/cart_list_model.dart';
import 'package:shop/utils/shared_preferences_util.dart';
import 'package:shop/config/index.dart';
import 'package:shop/widgets/cart_number_widget.dart';
import 'package:shop/utils/navigator_util.dart';
import 'package:shop/utils/toast_util.dart';
import 'package:shop/event/refresh_event.dart';
import 'package:shop/event/cart_number_event.dart';
import 'package:shop/widgets/cached_image_widget.dart';
import 'package:shop/event/login_event.dart';
import 'package:shop/service/cart_service.dart';
//购物车页面
class CartPage extends StatefulWidget {
  @override
  _CartPageState createState() => _CartPageState();
}
```

```dart
class _CartPageState extends State<CartPage> {
  //购物车数据服务
  CartService _cartService = CartService();
  //购物车列表数据
  List<CartModel> _cartList;
  //购物车列表数据模型
  CartListModel _cartListModel;
  //是否登录
  bool _isLogin = false;
  //是否全选
  bool _isAllCheck = false;
  //总计
  double _totalMoney = 0.0;
  //token
  var token;

  @override
  void initState() {
    super.initState();
    //获取token,判断是否登录
    SharedPreferencesUtil.getToken().then((onValue) {
      if (onValue == null) {
        //未登录则提示登录
        setState(() {
          _isLogin = false;
        });
      } else {
        //已登录则获取购物车数据
        token = onValue;
        _getCartData(onValue);
      }
    });
  }

  //获取购物车数据
  _getCartData(token) {
    Options options = Options();
    //查询购物车数据
    _cartService.queryCart((cartList) {
      setState(() {
        //是否登录,变量置为true
        _isLogin = true;
        _cartListModel = cartList;
        _cartList = _cartListModel.cartList;
      });
      //是否全选
      _isAllCheck = _isCheckedAll();
    }, options: options);
  }
```

```dart
//监听刷新事件,当用户从商品详情页面单击添加至购物车时会触发刷新事件
_refreshEvent() {
    eventBus.on<RefreshEvent>().listen((RefreshEvent refreshEvent) => _getCartData(token));
    loginEventBus.on<LoginEvent>().listen((LoginEvent loginEvent) {
        if (loginEvent.isLogin) {
            //刷新购物车数据
            _getCartData(SharedPreferencesUtil.token);
        } else {
            setState(() {
                _isLogin = false;
            });
        }
    });
}

@override
Widget build(BuildContext context) {
    //监听刷新事件
    _refreshEvent();
    return Scaffold(
        appBar: AppBar(
            //标题
            title: Text(KString.CART),
            centerTitle: true,
        ),
        //判断当前已登录并且购物车数据不为空
        body: _isLogin && _cartList != null
            ? Container(
                child: _cartList.length != 0
                    ? Stack(
                        alignment: Alignment.bottomCenter,
                        children: <Widget>[
                            //渲染购物车列表数据
                            ListView.builder(
                                //购物车列表项个数
                                itemCount: _cartList.length,
                                //购物车列表项构建器
                                itemBuilder: (BuildContext context, int index) {
                                    //根据索引返回列表项
                                    return _getCartItemWidget(index);
                                }),
                            Container(
                                height: ScreenUtil.getInstance().setHeight(120.0),
                                decoration: ShapeDecoration(
                                    shape: Border(
                                        top: BorderSide(
                                            color: Colors.grey,
                                            width: 1.0,
```

```
              ),
            ),
          ),
          //水平布局
          child: Row(
            crossAxisAlignment: CrossAxisAlignment.center,
            children: <Widget>[
              //全选复选框
              Checkbox(
                  value: _isAllCheck,
                  activeColor: KColor.defaultCheckBoxColor,
                  //选择改变事件回调
                  onChanged: (bool) {
                    //设置是否全选
                    _setCheckedAll(bool);
                  }),
              Container(
                width:
                    ScreenUtil.getInstance().setWidth(200.0),
                //全选价格
                child: Text(_isAllCheck
                    ? KString.TOTAL_MONEY + " $ {_cartListModel.cartTotal.checkedGoodsAmount}"
                    : KString.TOTAL_MONEY + " $ {_totalMoney}"),
              ),
              Expanded(
                  child: Container(
                margin: EdgeInsets.only(
                  right: ScreenUtil.getInstance().setWidth(30.0),
                ),
                alignment: Alignment.centerRight,
                //结算按钮
                child: RaisedButton(
                  //结算操作
                  onPressed: () {
                    //跳转到填写订单页面
                    _fillInOrder();
                  },
                  color: KColor.defaultButtonColor,
                  child: Text(
                    //结算标签
                    KString.SETTLEMENT,
                    style: TextStyle(
                        color: Colors.white,
                        fontSize: ScreenUtil.getInstance().setSp(26.0)),
                  ),
                ),
              )),
```

```dart
            ),
          ],
        ),
      )
    ],
  )
: Center(
    //垂直布局
    child: Column(
      mainAxisAlignment: MainAxisAlignment.center,
      children: <Widget>[
        //没有数据图片提示
        Image.asset(
          "images/no_data.png",
          height: 80,
          width: 80,
        ),
        Padding(
          padding: EdgeInsets.only(top: 10.0),
        ),
        Text(
          //没有数据文本提示
          KString.NO_DATA_TEXT,
          style: TextStyle(
              fontSize: 16.0,
              color: KColor.defaultTextColor
          ),
        )
      ],
    ),
  ),
)
: Container(
    child: Center(
      //登录按钮
      child: RaisedButton(
        color: KColor.defaultButtonColor,
        onPressed: () {
          //跳转至登录页面
          NavigatorUtil.goLogin(context);
        },
        //登录文本
        child: Text(
          KString.LOGIN,
          style: TextStyle(
              color: Colors.white,
              fontSize: ScreenUtil.getInstance().setSp(30.0)),
        ),
      ),
```

```dart
      ),
    ));
}

//跳转至填写订单页面
_fillInOrder() {
  NavigatorUtil.goFillInOrder(context, 0);
}

//根据索引获取购物车项组件
Widget _getCartItemWidget(int index) {
  return Container(
    height: ScreenUtil.getInstance().setHeight(180.0),
    width: double.infinity,
    child: InkWell(
      //长按打开"删除商品"对话框
      onLongPress: () => _deleteDialog(index),
      child: Card(
        //水平布局
        child: Row(
          crossAxisAlignment: CrossAxisAlignment.center,
          children: <Widget>[
            //是否勾选此商品
            Checkbox(
              //读取购物车列表数据中当前项的checked值
              value: _cartList[index].checked ?? true,
              activeColor: KColor.defaultCheckBoxColor,
              //改变回调方法
              onChanged: (bool) {
                _checkCart(index, bool);
              }),
            //缓存商品图片
            CachedImageWidget(
              ScreenUtil.getInstance().setWidth(140.0),
              ScreenUtil.getInstance().setWidth(140.0),
              //商品图片路径
              _cartList[index].picUrl,
            ),
            //垂直布局
            Column(
              crossAxisAlignment: CrossAxisAlignment.start,
              mainAxisAlignment: MainAxisAlignment.center,
              children: <Widget>[
                //商品名称
                Text(
                  _cartList[index].goodsName,
                  style: TextStyle(
                      fontSize: ScreenUtil.getInstance().setSp(24.0),
                      color: Colors.black54),
                ),
                Padding(
```

```dart
              padding: EdgeInsets.only(
                top: ScreenUtil.getInstance().setHeight(10.0),
              ),
            ),
            //商品价格
            Text(
              "￥ ${_cartList[index].price}",
              style: TextStyle(
                  fontSize: ScreenUtil.getInstance().setSp(24.0),
                  color: Colors.grey),
            )
          ],
        ),
        Expanded(
            child: Column(
          mainAxisAlignment: MainAxisAlignment.center,
          children: <Widget>[
            //购买商品数量
            Text(
              "X ${_cartList[index].number}",
              style: TextStyle(
                  color: Colors.black54,
                  fontSize: ScreenUtil.getInstance().setSp(24.0)),
            ),
            Padding(
              padding: EdgeInsets.only(
                top: ScreenUtil.getInstance().setHeight(10.0),
              ),
            ),
            //使用购物数量组件
            CartNumberWidget(_cartList[index].number, (value) {
              //根据返回的索引及数量更新购物车
              _updateCart(index, value);
            }),
          ],
        ))
      ],
    ),
  ),
);
}

//更新购物车,传入索引及数量
_updateCart(int index, int number) {
  Options options = Options();
  var parameters = {
    //规格 Id
    "productId": _cartList[index].productId,
    //商品 Id
    "goodsId": _cartList[index].goodsId,
```

```dart
      //商品数量
      "number": number,
      //id
      "id": _cartList[index].id,
    };
    _cartService.updateCart((success) {
      setState(() {
        _cartList[index].number = number;
      });
    }, (error) {
      ToastUtil.showToast(error);
      cartNumberEventBus.fire(CartNumberEvent(number - 1));
    }, options, parameters);
}

//是否勾选商品,传入索引及是否勾选
_checkCart(int index, bool isCheck) {
  var parameters = {
    //产品Id
    "productIds": [_cartList[index].productId],
    //是否选择
    "isChecked": isCheck ? 1 : 0,
  };
  //调用购物车数据服务方法
  _cartService.cartCheck((success) {
    setState(() {
      _cartListModel = success;
      _cartList = _cartListModel.cartList;
      //重新设置全选状态
      _isAllCheck = _isCheckedAll();
    });
    //计算总价
    _totalMoney = _cartListModel.cartTotal.goodsAmount;
  }, (error) {
    ToastUtil.showToast(error);
  }, parameters);
}

//删除对话框
_deleteDialog(int index) {
  return showDialog<void>(
      context: context,
      barrierDismissible: true,
      builder: (BuildContext context) {
        return AlertDialog(
          //提示
          title: Text(KString.TIPS),
          //是否确认删除
          content: Text(KString.DELETE_CART_ITEM_TIPS),
          actions: <Widget>[
```

```dart
                    //取消按钮
                    FlatButton(
                      onPressed: () {
                        Navigator.pop(context);
                      },
                      child: Text(
                        KString.CANCEL,
                        style: TextStyle(color: Colors.black54),
                      ),
                    ),
                    //删除按钮
                    FlatButton(
                      onPressed: () {
                        //删除商品
                        _deleteGoods(index);
                      },
                      child: Text(
                        KString.CONFIRM,
                        style: TextStyle(color: KColor.defaultTextColor),
                      ),
                    )
                  ],
                );
              });
}

//根据索引删除购物车商品
_deleteGoods(int index) {
  //通过索引获取产品 Id
  var parameters = {
    "productIds": [_cartList[index].productId]
  };
  //调用删除商品方法
  _cartService.deleteCart((success) {
    //删除成功提示
    ToastUtil.showToast(KString.DELETE_SUCCESS);
    setState(() {
      //本地列表移除数据
      _cartList.removeAt(index);
    });
    Navigator.pop(context);
  }, (error) {
    ToastUtil.showToast(error);
  }, parameters);
}

//判断是否全选
bool _isCheckedAll() {
  //迭代循环购物车列表所有 checked 属性,当全部为 true 时为全选状态
  for (int i = 0; i < _cartList.length; i++) {
    if (_cartList[i].checked == null || !_cartList[i].checked) {
```

```
        return false;
      }
    }
    return true;
  }

  //设置是否全选/全不选
  _setCheckedAll(bool checked) {
    setState(() {
      _isAllCheck = checked;
      for (int i = 0; i < _cartList.length; i++) {
        _cartList[i].checked = checked;
      }
    });
  }
}
```

购物车的列表项使用了购物车计数组件,整体布局使用的是一个列表展示,另外还要加上下面操作按钮部分。当用户长按列表项时可以删除购物商品。当单击结算按钮时可以跳转至填写订单页面,页面展示效果如图 32-1 所示。

图 32-1　购物车页面

第 33 章 订　　单

订单表示用户已确认要购买商品或已经付款购买完商品。订单是有状态的，如待付款、已付款、待发货、已发货、待确认、已确认和交易完成等。订单里包含了商品信息、总价、收货地址信息、运费，以及备注等数据。

本章将从以下几个模块来阐述订单：
- 填写订单模块；
- 我的订单模块；
- 订单详情模块。

33.1　填写订单数据模型

当用户从购物车单击结算，或者从商品详情单击立即购买就会跳转至填写订单页面。填写订单需要以下信息：
- 商品名称；
- 商品 Id；
- 商品价格；
- 订单价格；
- 收货人；
- 收货人电话；
- 收货人详细地址；
- 备注。

获取填写订单数据是通过下面的接口获取的，地址如下：

```
http://localhost:8080/client/cart/checkout
```

这个地址被划分为购物车数据服务里了，因为订单的数据主要是从购物车中获取的。用户从商品详情里不管是单击添加至购物车还是单击立即购买，都会把数据添加至购物车表里。接下来打开 model/fill_in_order_model.dart 文件，添加如下代码：

```dart
//model/fill_in_order_model.dart 文件
import 'package:json_annotation/json_annotation.dart';
//扩展文件
part 'fill_in_order_model.g.dart';
//填写订单数据模型
@JsonSerializable()
class FillInOrderModel extends Object {
  //实付价格
  @JsonKey(name: 'actualPrice')
  double actualPrice;
  //订单总价
  @JsonKey(name: 'orderTotalPrice')
  double orderTotalPrice;
  //购物车 Id
  @JsonKey(name: 'cartId')
  int cartId;
  //商品总价
  @JsonKey(name: 'goodsTotalPrice')
  double goodsTotalPrice;
  //地址 Id
  @JsonKey(name: 'addressId')
  int addressId;
  @JsonKey(name: 'checkedAddress')
  CheckedAddressModel checkedAddress;
  //运费
  @JsonKey(name: 'freightPrice')
  double freightPrice;
  //购物车选择的商品列表
  @JsonKey(name: 'checkedGoodsList')
  List<CheckedGoodsModel> checkedGoodsList;
  //构造方法
  FillInOrderModel(this.actualPrice,this.orderTotalPrice,this.cartId,this.goodsTotalPrice,this.addressId,this.checkedAddress,this.freightPrice,this.checkedGoodsList,);

  factory FillInOrderModel.fromJson(Map<String, dynamic> srcJson) => _$FillInOrderModelFromJson(srcJson);

  Map<String, dynamic> toJson() => _$FillInOrderModelToJson(this);

}
//已选择地址数据模型
@JsonSerializable()
class CheckedAddressModel extends Object {
  //id
  @JsonKey(name: 'id')
  int id;
  //用户名称
  @JsonKey(name: 'name')
  String name;
```

```dart
//用户Id
@JsonKey(name: 'userId')
int userId;
//省
@JsonKey(name: 'province')
String province;
//市
@JsonKey(name: 'city')
String city;
//国家
@JsonKey(name: 'county')
String county;
//地址详情
@JsonKey(name: 'addressDetail')
String addressDetail;
//地区编码
@JsonKey(name: 'areaCode')
String areaCode;
//电话
@JsonKey(name: 'tel')
String tel;
//是否为默认地址
@JsonKey(name: 'isDefault')
bool isDefault;
//添加时间
@JsonKey(name: 'addTime')
String addTime;
//更新时间
@JsonKey(name: 'updateTime')
String updateTime;
//是否删除
@JsonKey(name: 'deleted')
bool deleted;
//构造方法
CheckedAddressModel(this.id,this.name,this.userId,this.province,this.city,this.county,
this.addressDetail,this.areaCode,this.tel,this.isDefault,this.addTime,this.updateTime,
this.deleted,);

  factory CheckedAddressModel.fromJson(Map<String, dynamic> srcJson) => _$CheckedAddressModelFromJson(srcJson);

  Map<String, dynamic> toJson() => _$CheckedAddressModelToJson(this);

}
//已选择商品数据模型
@JsonSerializable()
class CheckedGoodsModel extends Object {
  //id
  @JsonKey(name: 'id')
```

```dart
int id;
//用户 Id
@JsonKey(name: 'userId')
int userId;
//商品 Id
@JsonKey(name: 'goodsId')
int goodsId;
//商品编号
@JsonKey(name: 'goodsSn')
String goodsSn;
//商品名称
@JsonKey(name: 'goodsName')
String goodsName;
//规格 Id
@JsonKey(name: 'productId')
int productId;
//价格
@JsonKey(name: 'price')
double price;
//数量
@JsonKey(name: 'number')
int number;
//规格值
@JsonKey(name: 'specifications')
List<String> specifications;
//是否勾选
@JsonKey(name: 'checked')
bool checked;
//图片地址
@JsonKey(name: 'picUrl')
String picUrl;
//添加时间
@JsonKey(name: 'addTime')
String addTime;
//更新时间
@JsonKey(name: 'updateTime')
String updateTime;
//是否删除
@JsonKey(name: 'deleted')
bool deleted;
//构造方法
CheckedGoodsModel(this.id, this.userId, this.goodsId, this.goodsSn, this.goodsName, this.productId, this.price, this.number, this.specifications, this.checked, this.picUrl, this.addTime, this.updateTime, this.deleted,);

  factory CheckedGoodsModel.fromJson(Map<String, dynamic> srcJson) => _$CheckedGoodsModelFromJson(srcJson);

  Map<String, dynamic> toJson() => _$CheckedGoodsModelToJson(this);

}
```

从上面的代码可以看出，填写订单数据模型中包含了已勾选的商品及选择的地址信息。接下来打开命令行工具，进入 shop-flutter 项目根目录并使用 build_runner 工具生成扩展文件 fill_in_order_model.g.dart。

33.2 订单数据服务

订单数据服务里提供了与订单相关的数据操作方法，方法及用途如下所示：

- submitOrder：提交订单，传入购物车 Id、地址 Id，以及备注等参数；
- queryOrder：查询我的订单数据，传入分页相关参数；
- deleteOrder：删除订单，传入订单 Id 参数；
- cancelOrder：取消订单，传入订单 Id 参数；
- queryOrderDetail：查询订单详情，传入订单 Id 参数。

参照前面章节，编写标准的订单数据操作方法。打开 service/order_service.dart 文件，添加如下代码：

```dart
//service/order_service.dart 文件
import 'package:shop/utils/http_util.dart';
import 'package:shop/config/server_url.dart';
import 'package:shop/model/order_detail_model.dart';
import 'package:shop/model/order_model.dart';
import 'package:shop/config/index.dart';
import 'package:dio/dio.dart';
//定义成功返回列表数据
typedef OnSuccessList<T>(List<T> list);
//定义成功返回数据
typedef OnSuccess<T>(T t);
//定义返回失败消息
typedef OnFail(String message);
//订单数据服务
class OrderService {
  //提交订单,传入购物车 Id、地址 Id,以及备注等参数
  Future submitOrder(Options options, Map<String, dynamic> parameters, OnSuccess onSuccess, OnFail onFail,) async {
    try {
      var response = await HttpUtil.instance.post(ServerUrl.ORDER_SUBMIT, parameters: parameters, options: options);
      if (response['errno'] == 0) {
        //提交成功提示
        onSuccess(KString.SUCCESS);
      } else {
        //返回错误消息
        onFail(response['errmsg']);
      }
```

```dart
    } catch (e) {
      print(e);
      //返回服务器异常消息
      onFail(KString.SERVER_EXCEPTION);
    }
  }
  //查询我的订单数据,传入分页相关参数
  Future queryOrder(Map < String, dynamic > parameters, OnSuccess onSuccess, OnFail onFail) async {
    try {
      var response = await HttpUtil.instance.get(ServerUrl.ORDER_LIST, parameters: parameters, );
      if (response['errno'] == 0) {
        //将Json数据转换成订单列表数据模型
        OrderListModel orderModel = OrderListModel.fromJson(response["data"]);
        onSuccess(orderModel.list);
      } else {
        onFail(response['errmsg']);
      }
    } catch (e) {
      print(e);
      //返回服务器异常消息
      onFail(KString.SERVER_EXCEPTION);
    }
  }
  //删除订单,传入订单Id参数
  Future deleteOrder(Map < String, dynamic > parameters, OnSuccess onSuccess, OnFail onFail) async {
    try {
      var response = await HttpUtil.instance.post(ServerUrl.ORDER_DELETE, parameters: parameters, );
      if (response["errno"] == 0) {
        onSuccess(KString.SUCCESS);
      } else {
        onFail(response["errmsg"]);
      }
    } catch (e) {
      print(e);
      //返回服务器异常消息
      onFail(KString.SERVER_EXCEPTION);
    }
  }
  //取消订单,传入订单Id参数
  Future cancelOrder(Map < String, dynamic > parameters, OnSuccess onSuccess, OnFail onFail) async {
    try {
      var response = await HttpUtil.instance.post(ServerUrl.ORDER_CANCEL, parameters: parameters, );
      if (response["errno"] == 0) {
```

```
            onSuccess(KString.SUCCESS);
          } else {
            onFail(response["errmsg"]);
          }
      } catch (e) {
        print(e);
        //返回服务器异常消息
        onFail(KString.SERVER_EXCEPTION);
      }
   }
   //查询订单详情,传入订单 Id 参数
   Future queryOrderDetail(Map < String, dynamic > parameters, OnSuccess onSuccess, OnFail onFail) async {
      try {
         var response = await HttpUtil.instance.get(ServerUrl.ORDER_DETAIL, parameters: parameters,);

         if (response['errno'] == 0) {
            //将 Json 数据转换成订单详情数据模型
            OrderDetailModel orderDetailModel = OrderDetailModel.fromJson(response["data"]);
            onSuccess(orderDetailModel);
         } else {
            onFail(response['errmsg']);
         }
      } catch (e) {
         print(e);
         //返回服务器异常消息
         onFail(KString.SERVER_EXCEPTION);
      }
   }
}
```

订单数据服务可供我的订单页面、订单详情页面,以及填写订单页面数据交互使用。

33.3 填写订单页面实现

填写订单页面首先需要提取与订单相关的数据,如购物车中勾选的商品信息和用户的默认地址等,然后再填写备注,用户确认订单后单击提交即可。

填写订单页面的实现步骤如下:

步骤 1:打开 page/order/fill_in_order_page.dart 文件,添加 FillInOrderPage 页面组件,然后初始化以下几个主要变量:

```
//填写订单数据模型
FillInOrderModel _fillInOrderModel;
```

```
//订单数据服务
OrderService _orderService = OrderService();
//购物车数据服务
CartService _cartService = CartService();
//文本编辑控制器
TextEditingController _controller = TextEditingController();
//token 值
var token;
//填写订单的 Future 对象
Future fillInOrderFuture;
```

由于填写订单需要从购物车中获取数据,所以这里需要初始化购物车数据服务。

步骤 2:首先判断用户是否登录,然后再从购物车数据服务中获取已勾选的商品数据。处理过程大致如下:

```
//获取 token 值
SharedPreferencesUtil.getToken().then((onValue) {
  //...
  //获取购物车勾选的数据
  _getFillInOrder();

}

//获取购物车勾选的数据
_getFillInOrder() {
  //参数 parameters...
  //获取购物车勾选的数据
  fillInOrderFuture = _cartService.getCartDataForFillInOrder((success) {
    //...
  }, (error) {}, parameters);
}
```

步骤 3:编写提交订单方法,当用户选择地址并填写好备注及确认好商品后,单击付款则会生成一个订单。方法结构如下:

```
//提交订单
_submitOrder() {
  //判断并提示收货地址……
  //提交参数……
  //提交订单
  _orderService.submitOrder(options, parameters, (success) {
    //提交成功后跳转至订单页面……
  }, (error) {
    //...
  });
}
```

当用户提交完订单后,就开始执行订单流程了。如填写订单→提交订单→付款→发货→确认收货等,当然这里的流程还可以细化。

步骤 4：接下来编写填写订单页面,整体采用垂直布局的方式,外面包裹了一个滚动组件,这样当所购商品较多时可以滚动查看。完整代码如下：

```dart
//page/order/fill_in_order_page.dart 文件
import 'package:flutter/material.dart';
import 'package:flutter_screenutil/flutter_screenutil.dart';
import 'package:flutter_spinkit/flutter_spinkit.dart';
import 'package:dio/dio.dart';

import 'package:shop/config/index.dart';
import 'package:shop/model/fill_in_order_model.dart';
import 'package:shop/widgets/item_text_widget.dart';
import 'package:shop/utils/shared_preferences_util.dart';
import 'package:shop/utils/navigator_util.dart';
import 'package:shop/utils/fluro_convert_util.dart';
import 'package:shop/utils/toast_util.dart';
import 'package:shop/widgets/cached_image_widget.dart';
import 'package:shop/service/cart_service.dart';
import 'package:shop/service/order_service.dart';
//填写订单页面
class FillInOrderPage extends StatefulWidget {
  //购物车 Id
  var cartId;
  //构造方法,传入购物车 Id
  FillInOrderPage(this.cartId);

  @override
  _FillInOrderPageState createState() => _FillInOrderPageState();
}

class _FillInOrderPageState extends State<FillInOrderPage> {
  //填写订单数据模型
  FillInOrderModel _fillInOrderModel;
  //订单数据服务
  OrderService _orderService = OrderService();
  //购物车数据服务
  CartService _cartService = CartService();
  //文本编辑控制器
  TextEditingController _controller = TextEditingController();
  //token 值
  var token;
  //填写订单的 Future 对象
  Future fillInOrderFuture;
  //Dio 的 Options 对象
  Options options = Options();

  @override
```

```
void initState() {
  super.initState();
  //获取 token 值
  SharedPreferencesUtil.getToken().then((onValue) {
    token = onValue;
    //获取购物车勾选的数据
    _getFillInOrder();
  });
}

//获取购物车勾选的数据
_getFillInOrder() {
  var parameters = {
    //购物车 Id
    "cartId": widget.cartId == 0 ? 0 : widget.cartId,
    //地址 Id
    "addressId": 0,
  };
  //获取购物车勾选的数据
  fillInOrderFuture = _cartService.getCartDataForFillInOrder((success) {
    setState(() {
      _fillInOrderModel = success;
    });
  }, (error) {}, parameters);
}

@override
Widget build(BuildContext context) {
  //异步构建组件
  return FutureBuilder(
      future: fillInOrderFuture,
      builder: (BuildContext context, AsyncSnapshot asyncSnapshot) {
        switch (asyncSnapshot.connectionState) {
          case ConnectionState.none:
          //等待状态
          case ConnectionState.waiting:
            return Container(
              child: Center(
                //等待状态组件
                child: SpinKitFoldingCube(
                  size: 40.0,
                  color: KColor.watingColor,
                ),
              ),
            );
          default:
            if (asyncSnapshot.hasError)
              return Container(
                child: Center(
```

```dart
                    //异常信息
                    child: Text(
                      KString.SERVER_EXCEPTION,
                      style: TextStyle(
                        fontSize: ScreenUtil.instance.setSp(26.0),
                      ),
                    ),
                  ),
                );
              else
                //返回订单内容
                return _contentWidget();
            }
          });
}

//订单内容
_contentWidget() {
  return Scaffold(
    appBar: AppBar(
      //填写订单标题
      title: Text(KString.FILL_IN_ORDER),
      centerTitle: true,
    ),
    body: SingleChildScrollView(
      child: Column(
        children: <Widget>[
          //收货地址
          _addressWidget(),
          Divider(
            height: ScreenUtil.instance.setHeight(1.0),
            color: Colors.grey[350],
          ),
          Divider(
            height: ScreenUtil.instance.setHeight(1.0),
            color: Colors.grey[350],
          ),
          //备注
          _remarkWidget(),
          Divider(
            height: ScreenUtil.instance.setHeight(1.0),
            color: Colors.grey[350],
          ),
          //总价
          ItemTextWidget(
              KString.GOODS_TOTAL, " ¥ $ {_fillInOrderModel.goodsTotalPrice}"),
          Divider(
            height: ScreenUtil.instance.setHeight(1.0),
            color: Colors.grey,
```

```dart
      ),
      //运费
      ItemTextWidget(
          KString.FREIGHT, "￥ ${_fillInOrderModel.freightPrice}"),
      Divider(
        height: ScreenUtil.instance.setHeight(1.0),
        color: Colors.grey[350],
      ),
      Divider(
        height: ScreenUtil.instance.setHeight(1.0),
        color: Colors.grey[350],
      ),
      //购物车选中的商品
      Column(
        children: _goodsItems(_fillInOrderModel.checkedGoodsList),
      )
    ],
  ),
),
//底部组件
bottomNavigationBar: BottomAppBar(
  child: Container(
    margin: EdgeInsets.only(left: ScreenUtil.instance.setWidth(20.0)),
    height: ScreenUtil.instance.setHeight(100.0),
    child: Row(
      children: <Widget>[
        //实付价格
        Expanded(
            child: Text("实付: ￥ ${_fillInOrderModel.orderTotalPrice}")),
        InkWell(
          //提交订单
          onTap: () => _submitOrder(),
          child: Container(
            alignment: Alignment.center,
            width: ScreenUtil.instance.setWidth(200.0),
            height: double.infinity,
            color: KColor.buyButtonColor,
            child: Text(
              //付款标题
              KString.PAY,
              style: TextStyle(
                  color: Colors.white,
                  fontSize: ScreenUtil.instance.setSp(28.0)),
            ),
          ),
        )
      ],
    ),
  ),
),
```

```dart
      ),
    );
}

//商品列表
List<Widget> _goodsItems(List<CheckedGoodsModel> goods) {
  List<Widget> list = List();
  //循环添加所有商品,用户可能挑选了多件商品
  for (int i = 0; i < goods.length; i++) {
    list.add(_goodsItem(goods[i]));
    list.add(Divider(
      height: ScreenUtil.instance.setHeight(1.0),
      color: Colors.grey[350],
    ));
  }
  return list;
}

//商品列表项
Widget _goodsItem(CheckedGoodsModel checkedGoods) {
  return Container(
    padding: EdgeInsets.only(
        left: ScreenUtil.instance.setWidth(20.0),
        right: ScreenUtil.instance.setWidth(20.0)),
    height: ScreenUtil.instance.setHeight(180.0),
    //水平布局
    child: Row(
      crossAxisAlignment: CrossAxisAlignment.center,
      children: <Widget>[
        //商品图片
        CachedImageWidget(ScreenUtil.instance.setWidth(140),
            ScreenUtil.instance.setWidth(140), checkedGoods.picUrl),
        Padding(
          padding: EdgeInsets.only(left: ScreenUtil.instance.setWidth(10.0)),
        ),
        //垂直布局
        Column(
          mainAxisAlignment: MainAxisAlignment.center,
          crossAxisAlignment: CrossAxisAlignment.start,
          children: <Widget>[
            //商品名称
            Text(
              checkedGoods.goodsName,
              style: TextStyle(
                  color: Colors.black54,
                  fontSize: ScreenUtil.instance.setSp(26.0)),
            ),
            Padding(
              padding:
```

```
                        EdgeInsets.only(top: ScreenUtil.instance.setHeight(6.0))),
                //商品规格
                Text(
                  checkedGoods.specifications[0],
                  style: TextStyle(
                      color: Colors.grey,
                      fontSize: ScreenUtil.instance.setSp(22.0)),
                ),
                Padding(
                    padding: EdgeInsets.only(
                        top: ScreenUtil.instance.setHeight(20.0))),
                //商品单价
                Text(
                  "¥ ${checkedGoods.price}",
                  style: TextStyle(
                      color: Colors.deepOrangeAccent,
                      fontSize: ScreenUtil.instance.setSp(26.0)),
                )
              ],
            ),
            //商品数量
            Expanded(
                child: Container(
                  alignment: Alignment.centerRight,
                  child: Text("X ${checkedGoods.number}"),
                ),
            ),
          ],
        ),
    );
}

//备注组件
Widget _remarkWidget() {
  return Container(
    height: ScreenUtil.instance.setHeight(80),
    width: double.infinity,
    alignment: Alignment.center,
    margin: EdgeInsets.only(top: ScreenUtil.instance.setHeight(10.0)),
    padding: EdgeInsets.only(
        left: ScreenUtil.instance.setWidth(20.0),
        right: ScreenUtil.instance.setWidth(20.0),
    ),
    //水平布局
    child: Row(
      crossAxisAlignment: CrossAxisAlignment.center,
      children: <Widget>[
        Text(
            //备注标签
```

```dart
                    KString.REMARK,
                style: TextStyle(
                    color: Colors.black54,
                    fontSize: ScreenUtil.instance.setSp(26.0)),
              ),
              //左侧间距
              Expanded(
                  child: Container(
                    margin: EdgeInsets.only(
                        left: ScreenUtil.instance.setWidth(10.0)
                    ),
                    height: ScreenUtil.instance.setHeight(80.0),
                    alignment: Alignment.centerLeft,
                    //备注
                    child: TextField(
                      maxLines: 1,
                      //输入框装饰器
                      decoration: InputDecoration(
                        //提示填写备注
                        hintText: KString.REMARK,
                        hintStyle: TextStyle(
                            color: Colors.grey[350],
                            fontSize: ScreenUtil.instance.setSp(26.0)),
                        hasFloatingPlaceholder: false,
                        enabledBorder: UnderlineInputBorder(
                            borderSide: BorderSide(
                                color: Colors.transparent,
                                width: ScreenUtil.instance.setHeight(1.0))),
                        focusedBorder: UnderlineInputBorder(
                            borderSide: BorderSide(
                                color: Colors.transparent,
                                width: ScreenUtil.instance.setHeight(1.0))),
                      ),
                      style: TextStyle(
                          color: Colors.black54,
                          fontSize: ScreenUtil.instance.setSp(26.0)),
                      controller: _controller,
                    ),
                  ))
            ],
          ),
        );
      }

      //地址组件
      Widget _addressWidget() {
        return Container(
          height: ScreenUtil.instance.setHeight(120.0),
          margin: EdgeInsets.all(ScreenUtil.instance.setWidth(10.0)),
```

```
            padding: EdgeInsets.only(
                left: ScreenUtil.instance.setWidth(20.0),
                right: ScreenUtil.instance.setWidth(20.0)),
            child: _fillInOrderModel.checkedAddress.id != 0
                ? InkWell(
                    onTap: () {
                      //跳转至地址页面,返回后获取地址信息
                      NavigatorUtil.goAddress(context).then((value) {
                        print(value.toString());
                        Map<String, dynamic> addressData = Map();
                        //将导航返回的地址数据转换成 Map
                        addressData = FluroConvertUtil.stringToMap(value);
                        setState(() {
                          //将返回地址数据转换成实体对象
                          _fillInOrderModel.checkedAddress = CheckedAddressModel.fromJson
(addressData);
                        });
                      });
                    },
                    //地址信息水平布局
                    child: Row(
                      crossAxisAlignment: CrossAxisAlignment.center,
                      mainAxisAlignment: MainAxisAlignment.start,
                      children: <Widget>[
                        Column(
                          mainAxisAlignment: MainAxisAlignment.center,
                          crossAxisAlignment: CrossAxisAlignment.start,
                          children: <Widget>[
                            Row(
                              mainAxisAlignment: MainAxisAlignment.start,
                              children: <Widget>[
                                //联系人姓名
                                Text(
                                  _fillInOrderModel.checkedAddress.name,
                                  style: TextStyle(
                                      color: Colors.black54,
                                      fontSize: ScreenUtil.instance.setSp(28.0)),
                                ),
                                Padding(
                                  padding: EdgeInsets.only(
                                      left: ScreenUtil.instance.setHeight(20.0)),
                                ),
                                //联系人电话
                                Text(
                                  _fillInOrderModel.checkedAddress.tel,
                                  style: TextStyle(
                                      color: Colors.black54,
                                      fontSize: ScreenUtil.instance.setSp(26.0)),
                                ),
```

```dart
                  ],
                ),
                Padding(
                  padding: EdgeInsets.only(
                      top: ScreenUtil.instance.setHeight(10.0)),
                ),
                //联系人地址信息
                Text(
                  _fillInOrderModel.checkedAddress.province +
                      _fillInOrderModel.checkedAddress.city +
                      _fillInOrderModel.checkedAddress.county +
                      _fillInOrderModel.checkedAddress.addressDetail,
                  style: TextStyle(
                      color: Colors.black54,
                      fontSize: ScreenUtil.instance.setSp(26.0)),
                ),
              ],
            ),
            Expanded(
              child: Container(
                alignment: Alignment.centerRight,
                //右侧箭头,提示可以选择地址
                child: Icon(
                  Icons.arrow_forward_ios,
                  color: Colors.grey,
                ),
              ),
            ),
          ],
        ),
      )
    : InkWell(
        //单击跳转至收货地址页面
        onTap: () {
          NavigatorUtil.goAddress(context);
        },
        child: Row(
          crossAxisAlignment: CrossAxisAlignment.center,
          children: <Widget>[
            //选择地址标题
            Text(
              KString.PLEASE_SELECT_ADDRESS,
              style: TextStyle(
                  color: Colors.grey,
                  fontSize: ScreenUtil.instance.setSp(30.0)),
            ),
            Expanded(
                child: Container(
              alignment: Alignment.centerRight,
```

```
          child: Icon(
            Icons.arrow_forward_ios,
            color: Colors.grey,
          ),
        ))
      ],
    ),
  ),
);
}

//提交订单
_submitOrder() {
  if (_fillInOrderModel.checkedAddress.id == 0) {
    //请选择地址提示
    ToastUtil.showToast(KString.PLEASE_SELECT_ADDRESS);
    return;
  }
  var parameters = {
    //购物车 Id
    "cartId": 0,
    //地址 Id
    "addressId": _fillInOrderModel.checkedAddress.id,
    //备注
    "message": _controller.text,
  };
  //提交订单
  _orderService.submitOrder(options, parameters, (success) {
    print(success);
    //提交成功后跳转至订单页面
    NavigatorUtil.goOrder(context);
  }, (error) {
    ToastUtil.showToast(error);
  });
}
}
```

填写订单页面布局较为复杂,一定要一部分一部分拆开来看。值得注意的是地址选择组件,当用户单击选择地址时会跳转到地址页面,选择好地址后再次返回填写订单页面,这里需要做路由的参数传递及数据转换,代码如下:

```
//跳转至地址页面,返回后获取地址信息
NavigatorUtil.goAddress(context).then((value) {
  print(value.toString());
  Map<String, dynamic> addressData = Map();
  //将导航返回的地址数据转换成 Map
```

```
      addressData = FluroConvertUtil.stringToMap(value);
      setState(() {
        //将返回地址数据转换成实体对象
        _fillInOrderModel.checkedAddress = CheckedAddressModel.fromJson(addressData);
      });
    });
```

> **提示** 这里的地址页面及地址数据模型均未实现,等这些部分实现完成后再结合填写订单页面详细分析数据是如何传递的。

填写订单运行效果如图 33-1 所示。

图 33-1 填写订单页面

33.4 我的订单数据模型

用户多次购买商品后就会生成一系列订单,可以通过用户中心选项下我的订单选项再次查看所有的订单。我的订单数据模型包含以下几个部分:

❑ 订单列表数据模型;
❑ 订单数据模型;

第33章 订单

❑ 订单商品数据模型。

我的订单数据查询地址如下：

```
http://localhost:8080/client/order/list
```

打开 model/order_model.dart 文件，添加如下完整代码：

```dart
//model/order_model.dart 文件
import 'package:json_annotation/json_annotation.dart';
//扩展文件
part 'order_model.g.dart';
//订单列表数据模型
@JsonSerializable()
class OrderListModel extends Object {
  //订单总数
  @JsonKey(name: 'total')
  int total;
  //页数
  @JsonKey(name: 'pages')
  int pages;
  //每页显示的数量
  @JsonKey(name: 'limit')
  int limit;
  //当前页
  @JsonKey(name: 'page')
  int page;
  //订单列表
  @JsonKey(name: 'list')
  List<OrderModel> list;
  //构造方法
  OrderListModel(this.total,this.pages,this.limit,this.page,this.list,);

  factory OrderListModel.fromJson(Map<String, dynamic> srcJson) => _$OrderListModelFromJson(srcJson);

  Map<String, dynamic> toJson() => _$OrderListModelToJson(this);

}
//订单数据模型
@JsonSerializable()
class OrderModel extends Object {
  //订单状态
  @JsonKey(name: 'orderStatusText')
  String orderStatusText;
  //订单编号
  @JsonKey(name: 'orderSn')
  String orderSn;
  //价格
```

```dart
  @JsonKey(name: 'actualPrice')
  double actualPrice;
  //商品列表
  @JsonKey(name: 'goodsList')
  List<OrderGoodsModel> goodsList;
  //订单 Id
  @JsonKey(name: 'id')
  int id;
  //构造方法
  OrderModel(this.orderStatusText,this.orderSn,this.actualPrice,this.goodsList,this.id);

  factory OrderModel.fromJson(Map<String, dynamic> srcJson) => _$OrderModelFromJson(srcJson);

  Map<String, dynamic> toJson() => _$OrderModelToJson(this);

}
//订单商品数据模型
@JsonSerializable()
class OrderGoodsModel extends Object {
  //商品数量
  @JsonKey(name: 'number')
  int number;
  //图片
  @JsonKey(name: 'picUrl')
  String picUrl;
  //价格
  @JsonKey(name: 'price')
  double price;
  //id
  @JsonKey(name: 'id')
  int id;
  //名称
  @JsonKey(name: 'goodsName')
  String goodsName;
  //规格
  @JsonKey(name: 'specifications')
  List<String> specifications;
  //构造方法

  OrderGoodsModel(this.number,this.picUrl,this.price,this.id,this.goodsName,this.specifications,);

  factory OrderGoodsModel.fromJson(Map<String, dynamic> srcJson) => _$OrderGoodsModelFromJson(srcJson);

  Map<String, dynamic> toJson() => _$OrderGoodsModelToJson(this);

}
```

接下来打开命令行工具，进入 shop-flutter 项目根目录并使用 build_runner 工具生成扩展文件 order_model.g.dart。

33.5 我的订单页面实现

我的订单页面用来展示用户下过的所有订单，这里通过列表的方式进行展示。展示的内容主要是商品图片、商品价格、商品购买数量，以及订单编号等简要信息。我的订单页面实现步骤如下：

步骤1：打开 page/order/order_page.dart 文件，添加 OrderPage 页面组件。定义订单服务及订单列表数据对象，代码如下：

```
//订单数据服务
OrderService _orderService = OrderService();
//订单列表数据
List<OrderModel> _orders = List();
```

步骤2：由于我的订单页面需要用户登录，所以需要获取本地 token 值，然后才能根据订单数据服务查询订单数据。代码结构如下：

```
SharedPreferencesUtil.getToken().then((value) {
  //...
  //查询订单
  _orderData();
});

//查询我的订单数据
_orderData() {
  //分页参数 parameters...
  //调用订单服务查询订单数据
  _orderService.queryOrder(parameters,(success) {
    //...
  }, (error) {
    //...
  });
}
```

步骤3：添加商品数量统计方法，每个订单简要信息里需要统计该订单下的商品数量。方法实现代码如下：

```
//计算商品数量
int goodNumber(OrderModel order) {
  int number = 0;
  //查询某个订单下的所有商品并累加处理
```

```
    order.goodsList.forEach((good) {
      number += good.number;
    });
    return number;
}
```

步骤4：根据查询的订单列表数据，使用ListView展示，完整代码如下：

```
//page/order/order_page.dart 文件
import 'package:flutter/material.dart';
import 'package:flutter_screenutil/flutter_screenutil.dart';
import 'package:shop/utils/navigator_util.dart';
import 'package:shop/widgets/no_data_widget.dart';
import 'package:shop/config/string.dart';
import 'package:shop/service/order_service.dart';
import 'package:shop/utils/shared_preferences_util.dart';
import 'package:shop/utils/toast_util.dart';
import 'package:shop/model/order_model.dart';
//我的订单页面
class OrderPage extends StatefulWidget {
  @override
  _OrderPageState createState() => _OrderPageState();
}

class _OrderPageState extends State<OrderPage> {
  //订单数据服务
  OrderService _orderService = OrderService();
  //订单列表数据
  List<OrderModel> _orders = List();
  //token 值
  var _token;
  //当前页码
  var _page = 1;
  //分页数量
  var _limit = 10;

  @override
  void initState() {
    super.initState();
    //获取本地 token 值
    SharedPreferencesUtil.getToken().then((value) {
      _token = value;
      //查询订单
      _orderData();
    });
  }

  //查询我的订单数据
```

```dart
_orderData() {
  //分页相关参数
  var parameters = {"page": _page, "limit": _limit};
  //调用订单服务,查询订单数据
  _orderService.queryOrder(parameters, (success) {
    setState(() {
      _orders = success;
    });
  }, (error) {
    ToastUtil.showToast(error);
  });
}

@override
Widget build(BuildContext context) {
  return Scaffold(
    appBar: AppBar(
      //我的订单标题
      title: Text(KString.MINE_ORDER),
      centerTitle: true,
    ),
    body: Container(
      height: double.infinity,
      margin: EdgeInsets.all(ScreenUtil.instance.setWidth(20.0)),
      child: _orders.length == 0
          //使用没有数据组件
          ? NoDataWidget()
          //使用 ListView 渲染订单列表
          : ListView.builder(
              //订单条数
              itemCount: _orders.length,
              //渲染订单
              itemBuilder: (BuildContext context, int index) {
                //根据索引返回订单项
                return _orderItemWidget(_orders[index]);
                //return Text("data");
              }),
    ),
  );
}

//订单列表项
Widget _orderItemWidget(OrderModel order) {
  return Card(
      child: InkWell(
        //单击中转至订单详情页
        onTap: () => _goOrderDetail(order.id),
        child: Container(
          margin: EdgeInsets.all(ScreenUtil.instance.setWidth(20.0)),
```

```dart
            //垂直布局
      child: Column(
        crossAxisAlignment: CrossAxisAlignment.start,
        children: <Widget>[
          Container(
            height: ScreenUtil.instance.setHeight(80.0),
            //水平布局
            child: Row(
              crossAxisAlignment: CrossAxisAlignment.center,
              children: <Widget>[
                //订单标题
                Text(
                  KString.ORDER_TITLE,
                  style: TextStyle(
                      color: Colors.black54,
                      fontSize: ScreenUtil.instance.setSp(26.0)),
                ),
                Padding(
                    padding: EdgeInsets.only(
                        left: ScreenUtil.instance.setWidth(10.0))),
                Expanded(
                    child: Container(
                      alignment: Alignment.centerRight,
                      child: Row(
                        mainAxisAlignment: MainAxisAlignment.end,
                        crossAxisAlignment: CrossAxisAlignment.start,
                        children: <Widget>[
                          //订单编号
                          Text(KString.MINE_ORDER_SN + "${order.orderSn}"),
                          Icon(
                            Icons.arrow_forward_ios,
                            size: ScreenUtil.instance.setWidth(40),
                            color: Colors.grey[350],
                          ),
                        ],
                      ))),
              ],
            ),
          ),
          //订单商品列表渲染
          ListView.builder(
              shrinkWrap: true,
              //订单商品个数
              itemCount: order.goodsList.length,
              physics: NeverScrollableScrollPhysics(),
              //订单商品项
              itemBuilder: (BuildContext context, int index) {
                return _goodItemWidget(order.goodsList[index]);
              }),
```

```dart
                    Container(
                      margin: EdgeInsets.only(
                          top: ScreenUtil.instance.setHeight(10.0)),
                      alignment: Alignment.centerRight,
                      //商品数量及总价展示
                      child: Text(KString.MINE_ORDER_TOTAL_GOODS +
                          " ${goodNumber(order)}" +
                          KString.MINE_ORDER_GOODS_TOTAL +
                          KString.MINE_ORDER_PRICE +
                          " ${order.actualPrice}"),
                    )
                  ],
                ),
              )));
}

//跳转至订单详情
_goOrderDetail(int orderId) {
  NavigatorUtil.goOrderDetail(context, orderId, _token).then((bool) {
    _orderData();
  });
}

//计算商品数量
int goodNumber(OrderModel order) {
  int number = 0;
  //查询某个订单下的所有商品并累加处理
  order.goodsList.forEach((good) {
    number += good.number;
  });
  return number;
}

//订单商品项组件
Widget _goodItemWidget(OrderGoodsModel good) {
  return Container(
    //水平布局
    child: Row(
      crossAxisAlignment: CrossAxisAlignment.center,
      children: <Widget>[
        //商品图片
        Image.network(
          good.picUrl ?? "",
          width: ScreenUtil.instance.setWidth(160.0),
          height: ScreenUtil.instance.setHeight(160.0),
        ),
        Container(
          margin: EdgeInsets.only(
              left: ScreenUtil.instance.setWidth(20.0),
```

```
                            top: ScreenUtil.instance.setHeight(20.0)),
                    child: Column(
                      crossAxisAlignment: CrossAxisAlignment.start,
                      children: <Widget>[
                        //商品名称
                        Text(
                            good.goodsName,
                            style: TextStyle(
                                color: Colors.black54,
                                fontSize: ScreenUtil.instance.setSp(26.0)),
                        ),
                        Padding(
                            padding: EdgeInsets.only(
                                top: ScreenUtil.instance.setHeight(10.0))),
                        //商品规格
                        Text(
                            good.specifications[0],
                            style: TextStyle(
                                color: Colors.grey,
                                fontSize: ScreenUtil.instance.setSp(26.0)),
                        ),
                        Padding(
                            padding: EdgeInsets.only(
                                top: ScreenUtil.instance.setHeight(10.0))),
                    ],
                  ),
                ),
                Expanded(
                    child: Container(
                      alignment: Alignment.centerRight,
                      margin: EdgeInsets.only(
                        left: ScreenUtil.instance.setWidth(20.0),
                        right: ScreenUtil.instance.setWidth(20.0),
                    ),
                    child: Column(
                      children: <Widget>[
                        //商品价格
                        Text(
                            "¥ ${good.price}",
                            style: TextStyle(
                                color: Colors.black54,
                                fontSize: ScreenUtil.instance.setSp(24.0)),
                        ),
                        Padding(
                            padding: EdgeInsets.only(
                                top: ScreenUtil.instance.setHeight(20.0))),
                        //商品数量
                        Text(
                            "X ${good.number}",
```

```
                    style: TextStyle(
                        color: Colors.black54,
                        fontSize: ScreenUtil.instance.setSp(24.0)),
                  ),
                ],
              ),
            ))
          ],
        ),
      );
    }
  }
```

这里布局的难点是列表套列表,外层的列表是订单列表,每个订单又套了一个商品列表,只要把这个嵌套关系理清再进行拆分即可实现其布局,我的订单页面展示如图 33-2 所示。

图 33-2　我的订单页面

33.6　订单详情数据模型

订单详情数据模型包含了更详细的订单数据。另外订单详情数据模型还包含了辅助操作,如下所示:

- 订单删除；
- 订单取消；
- 订单退款；
- 重新购买；
- 订单评论；
- 订单确认。

有了这些辅助数据，我们就可以在订单详情页面实现更多的功能。订单详情获取的接口如下：

```
http://localhost:8080/client/order/detail
```

打开 model/order_detail_model.dart 文件，添加如下代码：

```
//model/order_detail_model.dart 文件
import 'package:json_annotation/json_annotation.dart';
//扩展文件
part 'order_detail_model.g.dart';
//订单详情数据模型
@JsonSerializable()
class OrderDetailModel extends Object {
  //订单信息
  @JsonKey(name: 'orderInfo')
  OrderInfoModel orderInfo;
  //订单商品
  @JsonKey(name: 'orderGoods')
  List<OrderDetailGoodsModel> orderGoods;
  //构造方法
  OrderDetailModel(this.orderInfo,this.orderGoods,);

  factory OrderDetailModel.fromJson(Map<String, dynamic> srcJson) => _$OrderDetailModelFromJson(srcJson);

  Map<String, dynamic> toJson() => _$OrderDetailModelToJson(this);

}
//订单信息数据模型
@JsonSerializable()
class OrderInfoModel extends Object {
  //收货人
  @JsonKey(name: 'consignee')
  String consignee;
  //地址
  @JsonKey(name: 'address')
  String address;
  //添加时间
```

```dart
  @JsonKey(name: 'addTime')
  String addTime;
  //订单编号
  @JsonKey(name: 'orderSn')
  String orderSn;
  //实付价格
  @JsonKey(name: 'actualPrice')
  double actualPrice;
  //电话
  @JsonKey(name: 'mobile')
  String mobile;
  //订单状态
  @JsonKey(name: 'orderStatusText')
  String orderStatusText;
  //商品价格
  @JsonKey(name: 'goodsPrice')
  double goodsPrice;
  //订单 Id
  @JsonKey(name: 'id')
  int id;
  //运费
  @JsonKey(name: 'freightPrice')
  double freightPrice;
  //操作选项
  @JsonKey(name: 'handleOption')
  HandleOption handleOption;
  //构造方法
  OrderInfoModel(this.consignee,this.address,this.addTime,this.orderSn,this.actualPrice,
  this.mobile, this.orderStatusText, this.goodsPrice, this.id, this.freightPrice, this.
  handleOption);

  factory OrderInfoModel.fromJson ( Map < String, dynamic > srcJson ) = >  _
$OrderInfoModelFromJson(srcJson);

  Map<String, dynamic> toJson() => _$OrderInfoModelToJson(this);

}
//订单商品数据模型
@JsonSerializable()
class OrderDetailGoodsModel extends Object {
  //id
  @JsonKey(name: 'id')
  int id;
  //订单 Id
  @JsonKey(name: 'orderId')
  int orderId;
  //商品 Id
  @JsonKey(name: 'goodsId')
  int goodsId;
```

```dart
    //商品名称
    @JsonKey(name: 'goodsName')
    String goodsName;
    //商品编号
    @JsonKey(name: 'goodsSn')
    String goodsSn;
    //产品规格 Id
    @JsonKey(name: 'productId')
    int productId;
    //商品数量
    @JsonKey(name: 'number')
    int number;
    //商品价格
    @JsonKey(name: 'price')
    double price;
    //
    @JsonKey(name: 'specifications')
    List<String> specifications;
    //商品图片
    @JsonKey(name: 'picUrl')
    String picUrl;
    //评论
    @JsonKey(name: 'comment')
    int comment;
    //添加时间
    @JsonKey(name: 'addTime')
    String addTime;
    //更新时间
    @JsonKey(name: 'updateTime')
    String updateTime;
    //是否删除
    @JsonKey(name: 'deleted')
    bool deleted;
    //构造方法
    OrderDetailGoodsModel(this.id, this.orderId, this.goodsId, this.goodsName, this.goodsSn,
    this.productId, this.number, this.price, this.specifications, this.picUrl, this.comment, this.
    addTime, this.updateTime, this.deleted,);

    factory OrderDetailGoodsModel.fromJson(Map<String, dynamic> srcJson) => _
    $OrderDetailGoodsModelFromJson(srcJson);

    Map<String, dynamic> toJson() => _$OrderDetailGoodsModelToJson(this);

}
//操作选项
@JsonSerializable()
class HandleOption extends Object {
    //取消
```

```dart
@JsonKey(name: 'cancel')
bool cancel;
//删除
@JsonKey(name: 'delete')
bool delete;
//购买
@JsonKey(name: 'pay')
bool pay;
//评论
@JsonKey(name: 'comment')
bool comment;
//确认
@JsonKey(name: 'confirm')
bool confirm;
//退款
@JsonKey(name: 'refund')
bool refund;
//重新购买
@JsonKey(name: 'rebuy')
bool rebuy;
//构造方法

HandleOption(this.cancel, this.delete, this.pay, this.comment, this.confirm, this.refund, this.rebuy,);

factory HandleOption.fromJson(Map<String, dynamic> srcJson) => _$HandleOptionFromJson(srcJson);

Map<String, dynamic> toJson() => _$HandleOptionToJson(this);

}
```

商城项目里我们只实现了删除订单及取消订单功能,但保留其他操作字段对于扩展功能更有意义,建议保留。

接下来打开命令行工具,进入 shop-flutter 项目根目录并使用 build_runner 工具生成扩展文件 order_detail_model.g.dart。

33.7 订单详情页面实现

订单详情页面展示了订单的主要数据,它与填写订单的内容几乎一致,所以在数据字段及页面展示上非常接近。此页面在打开时需要传入订单 Id 及 token 值。具体实现步骤如下:

步骤 1:打开 page/order/order_detail_page.dart 文件,添加订单详情页面组件 OrderDetailPage。初始化如下变量及对象:

```
//订单数据服务
OrderService _orderService = OrderService();
//订单详情数据模型
OrderDetailModel _orderDetailModel;
//订单详情 Future 对象
Future _orderDetailFuture;
//订单动作,删除订单/取消订单
var orderAction;
//订单参数
var parameters;
```

其中,orderAction 有两个动作:1 表示取消订单,2 表示删除订单。parameters 表示订单参数,提供订单 Id 数据。

步骤 2:读取订单 Id 后,还需要读取该订单的详细数据,调用订单服务的 queryOrderDetail 方法,代码如下:

```
_queryOrderDetail() {
  _orderDetailFuture = _orderService.queryOrderDetail(parameters,(success) {
    //...
  }, (error) {});
}
```

步骤 3:取消订单及删除订单,方法如下:

❑ _deleteOrder:删除订单;

❑ _cancelOrder:取消订单。

步骤 4:参照填写订单页面布局实现方法,添加如下代码完成页面布局:

```
//page/order/order_detail_page.dart 文件
import 'package:flutter/material.dart';
import 'package:flutter_spinkit/flutter_spinkit.dart';
import 'package:flutter_screenutil/flutter_screenutil.dart';
import 'package:shop/service/order_service.dart';
import 'package:shop/config/index.dart';
import 'package:shop/widgets/item_text_widget.dart';
import 'package:shop/model/order_detail_model.dart';
import 'package:shop/widgets/divider_line_widget.dart';
import 'package:shop/widgets/no_data_widget.dart';
import 'package:shop/utils/toast_util.dart';
//订单详情页面
class OrderDetailPage extends StatefulWidget {
  //订单 Id
  var orderId;
  //token 值
  var token;
  //构造方法,传入订单 Id 及 token 值
```

```dart
    OrderDetailPage(this.orderId, this.token);

  @override
  _OrderDetailPageState createState() => _OrderDetailPageState();
}

class _OrderDetailPageState extends State<OrderDetailPage> {
  //订单数据服务
  OrderService _orderService = OrderService();
  //订单详情数据模型
  OrderDetailModel _orderDetailModel;
  //订单详情 Future 对象
  Future _orderDetailFuture;
  //订单动作,删除订单/取消订单
  var orderAction;
  //订单参数
  var parameters;

  @override
  void initState() {
    super.initState();
    //订单 Id 参数
    parameters = {"orderId": widget.orderId};
    _queryOrderDetail();
  }

  //查询订单详情数据
  _queryOrderDetail() {
    _orderDetailFuture =
        _orderService.queryOrderDetail(parameters,(success) {
      _orderDetailModel = success;
    }, (error) {});
  }

  @override
  Widget build(BuildContext context) {
    return Scaffold(
      appBar: AppBar(
        //订单详情标题
        title: Text(KString.MINE_ORDER_DETAIL),
        centerTitle: true,
      ),
      //异步构建组件
      body: FutureBuilder(
          future: _orderDetailFuture,
          builder: (BuildContext context, AsyncSnapshot asyncSnapshot) {
            //连接状态
            switch (asyncSnapshot.connectionState) {
              case ConnectionState.none:
```

```dart
            //等待状态
            case ConnectionState.waiting:
              return Container(
                child: Center(
                  //旋转组件
                  child: SpinKitFoldingCube(
                    size: 40.0,
                    color: KColor.watingColor,
                  ),
                ),
              );
            default:
              //错误提示
              if (asyncSnapshot.hasError) {
                return Container(
                  height: double.infinity,
                  //没有数据组件
                  child: NoDataWidget(),
                );
              } else {
                //内容展示
                return _contentWidget();
              }
          }
        }),
    );
  }
  //内容组件
  Widget _contentWidget() {
    return Container(
      margin: EdgeInsets.all(ScreenUtil.instance.setWidth(20.0)),
      //垂直布局
      child: Column(
        crossAxisAlignment: CrossAxisAlignment.start,
        children: <Widget>[
          //订单编号
          ItemTextWidget(KString.MINE_ORDER_SN, _orderDetailModel.orderInfo.orderSn),
          //分割线组件
          DividerLineWidget(),
          //订单创建时间
          ItemTextWidget(KString.MINE_ORDER_TIME, _orderDetailModel.orderInfo.addTime),
          DividerLineWidget(),
          Container(
            margin: EdgeInsets.only(left: ScreenUtil.instance.setWidth(20.0)),
            height: ScreenUtil.instance.setHeight(80.0),
            alignment: Alignment.centerLeft,
            //水平布局
            child: Row(
              crossAxisAlignment: CrossAxisAlignment.center,
```

```dart
            children: <Widget>[
                //商品信息
                Text(
                    KString.ORDER_INFORMATION,
                    style: TextStyle(
                        color: Colors.black54,
                        fontSize: ScreenUtil.instance.setSp(26)),
                ),
                //已取消订单提示
                Expanded(
                    child: Container(
                        alignment: Alignment.centerRight,
                        //是否显示已取消文本
                        child: Offstage(
                          offstage: _orderDetailModel.orderInfo.handleOption.cancel,
                          child: Text(
                              //已取消文本提示
                              KString.MINE_ORDER_ALREADY_CANCEL,
                              style: TextStyle(
                                  color: KColor.defaultTextColor,
                                  fontSize: ScreenUtil.instance.setSp(26.0)),
                          ),
                    )))
            ],
        )),
        DividerLineWidget(),
        //渲染商品列表数据
        ListView.builder(
            shrinkWrap: true,
            physics: NeverScrollableScrollPhysics(),
            //订单商品个数
            itemCount: _orderDetailModel.orderGoods.length,
            //构建商品列表项
            itemBuilder: (BuildContext context, int index) {
              return _goodItemWidget(_orderDetailModel.orderGoods[index]);
            }),
        DividerLineWidget(),
        Container(
          margin: EdgeInsets.only(
              left: ScreenUtil.instance.setWidth(20.0),
              top: ScreenUtil.instance.setHeight(20.0),
              bottom: ScreenUtil.instance.setHeight(20.0)),
          //垂直布局
          child: Column(
            mainAxisAlignment: MainAxisAlignment.center,
            crossAxisAlignment: CrossAxisAlignment.start,
            children: <Widget>[
              //水平布局
              Row(
```

```dart
                    children: <Widget>[
                      //收货人
                      Text(
                        _orderDetailModel.orderInfo.consignee,
                        style: TextStyle(
                            color: Colors.black54,
                            fontSize: ScreenUtil.instance.setSp(26.0)),
                      ),
                      Padding(
                        padding: EdgeInsets.only(
                            left: ScreenUtil.instance.setWidth(20.0)),
                      ),
                      //手机信息
                      Text(
                        _orderDetailModel.orderInfo.mobile,
                        style: TextStyle(
                            color: Colors.black54,
                            fontSize: ScreenUtil.instance.setSp(26.0)),
                      ),
                    ],
                  ),
                  Padding(
                      padding: EdgeInsets.only(
                          top: ScreenUtil.instance.setHeight(20.0))),
                  //收货地址信息
                  Text(
                    _orderDetailModel.orderInfo.address,
                    style: TextStyle(
                        color: Colors.black54,
                        fontSize: ScreenUtil.instance.setSp(26.0)),
                    softWrap: true,
                  ),
                ],
              ),
            ),
            DividerLineWidget(),
            //订单商品合计价格
            ItemTextWidget(KString.MINE_ORDER_DETAIL_TOTAL,
                KString.DOLLAR + "${_orderDetailModel.orderInfo.goodsPrice}"),
            DividerLineWidget(),
            //订单运费
            ItemTextWidget(KString.FREIGHT,
                KString.DOLLAR + "${_orderDetailModel.orderInfo.freightPrice}"),
            DividerLineWidget(),
            //订单实付价格
            ItemTextWidget(KString.MINE_ORDER_DETAIL_PAYMENTS,
                KString.DOLLAR + "${_orderDetailModel.orderInfo.actualPrice}"),
            DividerLineWidget(),
            Container(
```

```
                    height: ScreenUtil.instance.setHeight(100.0),
                    child: Row(
                      children: <Widget>[
                        //取消订单
                        Expanded(
                            child: MaterialButton(
                          color: KColor.defaultButtonColor,
                          //单击回调
                          onPressed: () {
                            //动作类型 1
                            orderAction = 1;
                            _showDialog();
                          },
                          child: Text(
                            //取消文本
                            KString.CANCEL,
                            style: TextStyle(
                                fontSize: ScreenUtil.instance.setSp(28.0),
                                color: Colors.white),
                          ),
                        )),
                        Padding(
                          padding:
                              EdgeInsets.only(left: ScreenUtil.instance.setWidth(60.0)),
                        ),
                        //删除订单
                        Expanded(
                            child: MaterialButton(
                          color: KColor.defaultButtonColor,
                          //单击处理
                          onPressed: () {
                            //动作类型 2
                            orderAction = 2;
                            _showDialog();
                          },
                          child: Text(
                            //删除按钮文本
                            KString.DELETE,
                            style: TextStyle(
                                fontSize: ScreenUtil.instance.setSp(28.0),
                                color: Colors.white),
                          ),
                        ))
                      ],
                    ),
                  )
              ],
            ),
          );
```

```dart
}
//商品组件
Widget _goodItemWidget(OrderDetailGoodsModel good) {
  return Container(
    //水平布局
    child: Row(
      crossAxisAlignment: CrossAxisAlignment.center,
      children: <Widget>[
        //商品图片
        Image.network(
          good.picUrl ?? "",
          width: ScreenUtil.instance.setWidth(160.0),
          height: ScreenUtil.instance.setHeight(160.0),
        ),
        Container(
          margin: EdgeInsets.only(
              left: ScreenUtil.instance.setWidth(20.0),
              top: ScreenUtil.instance.setHeight(20.0)),
          //垂直布局
          child: Column(
            crossAxisAlignment: CrossAxisAlignment.start,
            children: <Widget>[
              //商品名称
              Text(
                good.goodsName,
                style: TextStyle(
                    color: Colors.black54,
                    fontSize: ScreenUtil.instance.setSp(26.0)),
              ),
              Padding(
                padding: EdgeInsets.only(
                    top: ScreenUtil.instance.setHeight(10.0),
                ),
              ),
              //商品规格
              Text(
                good.specifications[0],
                style: TextStyle(
                    color: Colors.grey,
                    fontSize: ScreenUtil.instance.setSp(26.0)),
              ),
              Padding(
                padding: EdgeInsets.only(
                    top: ScreenUtil.instance.setHeight(10.0))),
            ],
          ),
        ),
        Expanded(
```

```
                    child: Container(
                  alignment: Alignment.centerRight,
                  margin: EdgeInsets.only(
                    left: ScreenUtil.instance.setWidth(20.0),
                    right: ScreenUtil.instance.setWidth(20.0),
                  ),
                  //垂直布局
                  child: Column(
                    children: <Widget>[
                      //商品价格
                      Text(
                        "￥${good.price}",
                        style: TextStyle(
                            color: Colors.black54,
                            fontSize: ScreenUtil.instance.setSp(24.0)),
                      ),
                      Padding(
                          padding: EdgeInsets.only(
                              top: ScreenUtil.instance.setHeight(20.0))),
                      //商品数量
                      Text(
                        "X${good.number}",
                        style: TextStyle(
                            color: Colors.black54,
                            fontSize: ScreenUtil.instance.setSp(24.0)),
                      ),
                    ],
                  ),
                ))
      ],
    ),
  );
}

//弹出操作对话框
_showDialog() {
  showDialog(
      context: context,
      barrierDismissible: false,
      builder: (BuildContext context) {
        return AlertDialog(
          //提示文本
          title: Text(
            KString.TIPS,
            style: TextStyle(
                color: Colors.black54,
                fontSize: ScreenUtil.instance.setSp(28.0)),
          ),
          //取消或删除提示文本
```

```dart
              content: Text(
                orderAction == 1
                    ? KString.MINE_ORDER_CANCEL_TIPS
                    : KString.MINE_ORDER_DELETE_TIPS,
                style: TextStyle(
                    color: Colors.black54,
                    fontSize: ScreenUtil.instance.setSp(28.0)),
              ),
              //动作按钮
              actions: <Widget>[
                //取消按钮
                FlatButton(
                    color: Colors.white,
                    //页面返回
                    onPressed: () {
                      Navigator.pop(context);
                    },
                    child: Text(
                      KString.CANCEL,
                      style: TextStyle(
                          color: Colors.black54,
                          fontSize: ScreenUtil.instance.setSp(24.0)),
                    )),
                //确认按钮
                FlatButton(
                    color: Colors.white,
                    onPressed: () {
                      Navigator.pop(context);
                      //取消订单
                      if (orderAction == 1) {
                        _cancelOrder();
                      } else {
                        //删除订单
                        _deleteOrder();
                      }
                    },
                    child: Text(
                      KString.CONFIRM,
                      style: TextStyle(
                          color: Colors.black54,
                          fontSize: ScreenUtil.instance.setSp(24.0)),
                    )),
              ],
            );
          });
    }
    //删除订单
    _deleteOrder() {
      //订单 Id 参数
```

```
    var parameters = {"orderId": widget.orderId};
    _orderService.deleteOrder(parameters, (success) {
      Navigator.of(context).pop(true);
    }, (error) {
      ToastUtil.showToast(error);
    });
  }
  //取消订单
  _cancelOrder() {
    //订单 Id 参数
    var parameters = {"orderId": widget.orderId};
    _orderService.cancelOrder(parameters, (success) {
      ToastUtil.showToast(KString.MINE_ORDER_CANCEL_SUCCESS);
      setState(() {
        _orderDetailModel.orderInfo.handleOption.cancel = false;
      });
    }, (error) {
      ToastUtil.showToast(error);
    });
  }
}
```

页面整体采用垂直布局的方式将订单编号、订单时间、订单商品、物流信息,以及商品价格等信息展示出来,订单详情页面打开后如图 33-3 所示。

图 33-3　订单详情页面

第 34 章 地 址

收货地址属于用户的个人数据,可以在商城 App 端进行增、删、改和查操作。从我的页面单击地址管理即可进入地址列表页面。

本章将详细阐述地址查询、新增地址、删除地址、更新地址,以及地址与填写订单的关系等内容。

34.1 地址数据模型

地址数据模型主要为地址的增、删、改和查提供数据支持,地址数据是通过下面的接口获取的,接口如下:

```
http://localhost:8080/client/address/list
```

地址主要包括地址列表数据模型及地址数据模型。接下来打开 model/address_model.dart 文件,添加如下代码:

```
//model/address_model.dart 文件
import 'package:json_annotation/json_annotation.dart';
//扩展文件
part 'address_model.g.dart';
//地址列表数据模型
@JsonSerializable()
class AddressListModel extends Object {
  //地址总数
  @JsonKey(name: 'total')
  int total;
  //总页数
  @JsonKey(name: 'pages')
  int pages;
  //分页数量
  @JsonKey(name: 'limit')
  int limit;
```

```dart
  //当前页
  @JsonKey(name: 'page')
  int page;
  //地址列表
  @JsonKey(name: 'list')
  List<AddressModel> list;
  //构造方法
  AddressListModel(this.total,this.pages,this.limit,this.page,this.list,);

  factory AddressListModel.fromJson(Map<String, dynamic> srcJson) => _$AddressListModelFromJson(srcJson);

  Map<String, dynamic> toJson() => _$AddressListModelToJson(this);

}
//地址数据模型
@JsonSerializable()
class AddressModel extends Object {
  //id
  @JsonKey(name: 'id')
  int id;
  //用户名
  @JsonKey(name: 'name')
  String name;
  //用户Id
  @JsonKey(name: 'userId')
  int userId;
  //省
  @JsonKey(name: 'province')
  String province;
  //市
  @JsonKey(name: 'city')
  String city;
  //国家
  @JsonKey(name: 'county')
  String county;
  //详情地址
  @JsonKey(name: 'addressDetail')
  String addressDetail;
  //地区编码
  @JsonKey(name: 'areaCode')
  String areaCode;
  //电话
  @JsonKey(name: 'tel')
  String tel;
  //是否为默认地址
  @JsonKey(name: 'isDefault')
  bool isDefault;
  //添加时间
```

```
@JsonKey(name: 'addTime')
String addTime;
//更新时间
@JsonKey(name: 'updateTime')
String updateTime;
//是否删除
@JsonKey(name: 'deleted')
bool deleted;
//构造方法
AddressModel(this.id, this.name, this.userId, this.province, this.city, this.county, this.
addressDetail, this.areaCode, this.tel, this.isDefault, this.addTime, this.updateTime, this.
deleted,);

factory AddressModel.fromJson(Map<String, dynamic> srcJson) => _$AddressModelFromJson
(srcJson);

Map<String, dynamic> toJson() => _$AddressModelToJson(this);
}
```

地址数据模型有一个字段需要注意，isDefault 为默认地址，在填写订单时首先使用的就是默认地址。接下来打开命令行工具，进入 shop-flutter 项目根目录并使用 build_runner 工具生成扩展文件 address_model.g.dart。

34.2 地址数据服务

地址数据服务里提供了与地址相关的增、删、改和查方法，方法及用途如下：
- getAddressList：获取地址列表；
- addAddress：新增地址；
- deleteAddress：删除地址；
- addressDetail：地址详细信息。

参照前面章节，编写标准的订单数据操作方法。打开 service/address_service.dart 文件，添加如下代码：

```
//service/address_service.dart 文件
import 'package:shop/utils/http_util.dart';
import 'package:shop/config/server_url.dart';
import 'package:shop/model/address_model.dart';
import 'package:shop/config/index.dart';
//定义成功返回列表数据
typedef OnSuccessList<T>(List<T> list);
//定义成功返回数据
typedef OnSuccess<T>(T t);
//定义返回失败消息
```

```dart
typedef OnFail(String message);
//地址数据服务
class AddressService {
  //获取地址列表
  Future getAddressList( OnSuccess onSuccessList, {OnFail onFail}) async {
    try {
      var response = await HttpUtil.instance.get(ServerUrl.ADDRESS_LIST,);
      if (response['errno'] == 0) {
        //将返回的Json转换成地址列表数据模型
        AddressListModel addressModel = AddressListModel.fromJson(response['data']);
        print(response['data']);
        //成功返回列表数据
        onSuccessList(addressModel.list);
      } else {
        //失败返回错误消息
        onFail(response['errmsg']);
      }
    } catch (e) {
      print(e);
      //返回服务器错误消息
      onFail(KString.SERVER_EXCEPTION);
    }
  }
  //新增地址
  Future addAddress( Map < String, dynamic > parameters, OnSuccess onSuccess, OnFail onFail) async {
    try {
      var response = await HttpUtil.instance.post(ServerUrl.ADDRESS_SAVE, parameters: parameters);
      if (response['errno'] == 0) {
        //返回添加成功消息
        onSuccess(KString.SUCCESS);
      } else {
        //失败返回错误消息
        onFail(response['errmsg']);
      }
    } catch (e) {
      print(e);
      //返回服务器错误消息
      onFail(KString.SERVER_EXCEPTION);
    }
  }
  //删除地址
  Future deleteAddress(Map< String, dynamic > parameters, OnSuccess onSuccess, OnFail onFail) async {
    try {
      var response = await HttpUtil.instance.post(ServerUrl.ADDRESS_DELETE, parameters: parameters);
      if (response['errno'] == 0) {
```

```
        //返回删除成功消息
        onSuccess(KString.SUCCESS);
      } else {
        //失败返回错误消息
        onFail(response['errmsg']);
      }
    } catch (e) {
      print(e);
      //返回服务器错误消息
      onFail(KString.SERVER_EXCEPTION);
    }
  }
  //地址详细信息
  Future addressDetail(
      Map<String, dynamic> parameters, OnSuccess onSuccess, {OnFail onFail}) async {
    try {
      var response = await HttpUtil.instance.get(ServerUrl.ADDRESS_DETAIL, parameters: parameters);
      if (response['errno'] == 0) {
        //将返回的Json数据转换成地址详情
        AddressModel addressDetail = AddressModel.fromJson(response["data"]);
        //返回数据
        onSuccess(addressDetail);
      } else {
        //失败返回错误消息
        onFail(response['errmsg']);
      }
    } catch (e) {
      print(e);
      //返回服务器错误消息
      onFail(KString.SERVER_EXCEPTION);
    }
  }
}
```

订单数据服务可供我的收货地址页面和编辑地址页面数据交互使用。

34.3 我的收货地址页面实现

我的收货地址页面用来展示当前用户的所有地址数据。可以从这个页面返回至填写订单页面,同时可以跳转到新增地址及编辑地址页面。

页面使用地址数据服务 AddressService 的 getAddressList 方法查询列表数据,展示使用列表组件 ListView 展示即可。打开 page/mine/address_page.dart 文件,添加如下完整代码:

```dart
//page/mine/address_page.dart 文件
import 'package:flutter/material.dart';
import 'package:flutter_screenutil/flutter_screenutil.dart';
import 'package:shop/config/index.dart';
import 'package:shop/model/address_model.dart';
import 'package:shop/service/address_service.dart';
import 'package:shop/utils/navigator_util.dart';
import 'package:shop/utils/fluro_convert_util.dart';
import 'package:shop/widgets/no_data_widget.dart';
//地址修改页面
class AddressPage extends StatefulWidget {
  @override
  _AddressPageState createState() => _AddressPageState();
}

class _AddressPageState extends State<AddressPage> {
  //地址列表数据
  List<AddressModel> _addressData;
  //地址数据服务
  AddressService addressService = AddressService();
  //地址 Future 对象
  var addressFuture;
  //token 值
  var token;

  @override
  void initState() {
    super.initState();
    //获取收货地址数据
    _getAddressData();
  }

  //获取收货地址数据
  _getAddressData() {
    addressService.getAddressList((addressList) {
      setState(() {
        _addressData = addressList;
      });
    });
  }

  @override
  Widget build(BuildContext context) {
    return Scaffold(
      appBar: AppBar(
        //标题
        title: Text(KString.MY_ADDRESS),
        centerTitle: true,
        //页面右上角操作按钮
```

```dart
          actions: <Widget>[
            InkWell(
              child: Container(
                margin: EdgeInsets.only(right: ScreenUtil.instance.setWidth(10.0)),
                alignment: Alignment.center,
                child: InkWell(
                  //跳转至添加地址页面,0 表示添加地址
                  onTap: () => _goAddressEdit(0),
                  child: Text(KString.ADD_ADDRESS),
                ),
              ))
          ],
        ),
        body: _addressData != null && _addressData.length > 0
            ? Container(
                //地址列表
                child: ListView.builder(
                  //地址列表长度
                  itemCount: _addressData.length,
                  itemBuilder: (BuildContext context, int index) {
                    return _addressItemView(_addressData[index]);
                  }))
                //没有数据提示组件
            : NoDataWidget(),
  );
}

//地址列表项组件,根据地址数据渲染地址信息
Widget _addressItemView(AddressModel addressData) {
  return Container(
    padding: EdgeInsets.only(left: ScreenUtil.instance.setWidth(10.0)),
    height: ScreenUtil.instance.setHeight(160.0),
    alignment: Alignment.center,
    //卡片布局
    child: Card(
      //单击返回至填写订单界面并使用地址数据
      child: InkWell(
      onTap: () => _goFillInOrder(addressData),
      //水平布局
      child: Row(
        crossAxisAlignment: CrossAxisAlignment.center,
        children: <Widget>[
          Expanded(
            //垂直布局
            child: Column(
              mainAxisAlignment: MainAxisAlignment.center,
              crossAxisAlignment: CrossAxisAlignment.start,
              children: <Widget>[
                //水平布局
```

```
            Row(
                children: <Widget>[
                  //姓名
                  Text(
                    addressData.name,
                    style: TextStyle(
                        color: Colors.black54,
                        fontSize: ScreenUtil.instance.setSp(26.0)),
                  ),
                  Padding(
                      padding: EdgeInsets.only(
                          left: ScreenUtil.instance.setWidth(10.0))),
                  //电话
                  Text(
                    addressData.tel,
                    style: TextStyle(
                        color: Colors.grey,
                        fontSize: ScreenUtil.instance.setSp(26.0)),
                  ),
                ],
              ),
              Padding(
                  padding: EdgeInsets.only(
                      top: ScreenUtil.instance.setWidth(20.0))),
              //省
              Text(
                addressData.province +
                    addressData.city +
                    addressData.county +
                    addressData.addressDetail,
                //最多两行展示
                maxLines: 2,
                overflow: TextOverflow.ellipsis, //显示不完,就在后面显示点点
                style: TextStyle(
                  color: Colors.black54,
                  fontSize: ScreenUtil.instance.setSp(26.0),
                ),
              )
            ],
          )),
          Container(
              width: ScreenUtil.instance.setWidth(120.0),
              margin:
                  EdgeInsets.only(right: ScreenUtil.instance.setWidth(10.0)),
              alignment: Alignment.center,
              decoration: ShapeDecoration(
```

```
                    shape: Border(
                        left: BorderSide(
                            color: Colors.grey[350],
                            width: ScreenUtil.instance.setWidth(1.0)))),
                    padding:
                        EdgeInsets.only(left: ScreenUtil.instance.setWidth(10.0)),
                //单击跳转至编辑页面
                    child: InkWell(
                      onTap: () => _goAddressEdit(addressData.id),
                      //编辑标签
                      child: Text(
                        KString.ADDRESS_EDIT,
                        style: TextStyle(
                            color: Colors.grey,
                            fontSize: ScreenUtil.instance.setSp(26.0)),
                      ),
                    ))
            ],
          ),
        )),
    );
  }
  //跳转至填写订单页面
  _goFillInOrder(AddressModel addressData) {
    Navigator.of(context).pop(FluroConvertUtil.objectToString(addressData));
  }
  //跳转至编辑页面,传入地址 Id 参数
  _goAddressEdit(var addressId) {
    NavigatorUtil.goAddressEdit(context, addressId).then((bool) {
      _getAddressData();
    });
  }
}
```

这里有段代码是和填写订单有关系的,若用户在填写订单时想选择一个新的地址则会进入此页面,用户选择好一个地址后会再次返回,这时需要携带 addressData 返回,代码如下:

```
onTap: () => _goFillInOrder(addressData),
```

由于页面采用的是列表展示布局并且不复杂,这里就不过多描述,页面显示效果如图 34-1 所示。

第34章　地址　507

图 34-1　我的收货地址页面

34.4　编辑地址页面实现

当用户单击地址列表里的编辑按钮即可进行编辑地址页面。编辑地址页面包含如下两种功能：

❑ 编辑地址；
❑ 新增地址。

这两项功能均在这一个页面实现，使用_addressId 变量区分，当其为 0 时则表示新增地址，当其为其他值时则表示编辑地址。页面实现步骤如下：

步骤 1：打开 page/mine/address_edit_page.dart 文件，添加 AddressEditPage 组件。初始化地址数据服务及地址数据模型，代码如下：

```
//地址数据服务
AddressService _addressService = AddressService();
//地址数据
AddressModel _addressData;
```

步骤 2：获取本地 Token 值，然后查询详细地址数据，当地址 Id 为 0 时新增地址，其他值时直接初始化组件即可，代码如下：

```
//获取本地 token
SharedPreferencesUtil.getToken().then((onValue) {
  //_addressId 为 0 表示新增,否则为编辑
  if (_addressId != 0) {
    //查询详细地址数据
    _queryAddressDetail(onValue);
  } else {
    //组件赋值...
  }
});
```

步骤 3：根据获取到的地址数据，初始化以下几个组件的值：
- 姓名输入框；
- 电话输入框；
- 详细地址输入框。

步骤 4：使用 CityPickers 组件弹出城市选择框，temp 即为选择的城市结果，可以提取出省、市和区等信息，调用代码如下：

```
Result temp = await CityPickers.showCityPicker(...);
```

步骤 5：编写界面操作的交互方法，方法名及作用如下：
- _deleteAddress：删除地址；
- _addAddress：新增地址提交；
- _checkAddressBody：判断地址信息是否完整。

步骤 6：编写编辑地址界面，整体采用垂直布局，添加地址的每一项操作即可。完整代码如下：

```
//page/mine/address_edit_page.dart 文件
import 'package:flutter/material.dart';
import 'package:flutter_screenutil/flutter_screenutil.dart';
import 'package:city_pickers/city_pickers.dart';
import 'package:dio/dio.dart';
import 'package:shop/config/index.dart';
import 'package:shop/model/address_model.dart';
import 'package:shop/service/address_service.dart';
import 'package:shop/utils/shared_preferences_util.dart';
import 'package:shop/utils/toast_util.dart';
//编辑地址页面
class AddressEditPage extends StatefulWidget {
  //地址 Id
```

```dart
  var addressId;
  //传入地址 Id 参数
  AddressEditPage(this.addressId);

  @override
  _AddressEditPageState createState() => _AddressEditPageState();
}

class _AddressEditPageState extends State<AddressEditPage> {
  //名称文本编辑控制器
  TextEditingController _nameController;
  //电话文本编辑控制器
  TextEditingController _phoneController = TextEditingController();
  //详细地址文本编辑控制器
  TextEditingController _addressDetailController = TextEditingController();
  //地址数据服务
  AddressService _addressService = AddressService();
  //地址列表数据
  AddressModel _addressData;
  var _cityText;
  //是否为默认地址
  var _isDefault = false;
  //地址 Id
  var _addressId;
  var token;
  //地址 Id
  var _areaId;
  //省
  var _provinceName;
  //城市
  var _cityName;
  //国家
  var _countryName;
  Options options = Options();

  @override
  void initState() {
    super.initState();
    //读取传入的地址 Id
    _addressId = widget.addressId;
    //获取本地 token
    SharedPreferencesUtil.getToken().then((onValue) {
      if (onValue != null) {
        token = onValue;
      }
      //_addressId 为 0 表示新增,否则为编辑
      if (_addressId != 0) {
        //查询详细地址数据
        _queryAddressDetail(onValue);
```

```dart
      } else {
        _initController();
      }
    });
}

//文本框初始化值
_initController() {
  //用户名初始化
  _nameController = TextEditingController(
    text: _addressData == null ? "" : _addressData.name,
  );
  //电话号码初始化
  _phoneController = TextEditingController(
    text: _addressData == null ? "" : _addressData.tel,
  );
  //详细地址初始化
  _addressDetailController = TextEditingController(
    text: _addressData == null ? "" : _addressData.addressDetail,
  );
}
//查询地址数据
_queryAddressDetail(var token) {
  //地址 Id 参数
  var parameters = {"id": _addressId};
  //查询地址详情
  _addressService.addressDetail(parameters, (addressDetail) {
    setState(() {
      //根据返回的数据赋值给本地变量
      _addressData = addressDetail;
      _areaId = _addressData.areaCode;
      _cityText = _addressData.province + _addressData.city + _addressData.county;
      _isDefault = _addressData.isDefault;
      _provinceName = _addressData.province;
      _cityName = _addressData.city;
      _countryName = _addressData.county;
    });
    _initController();
  });
}

@override
Widget build(BuildContext context) {
  return Scaffold(
    appBar: AppBar(
      //标题
      title: Text(KString.ADDRESS_EDIT_TITLE),
      centerTitle: true,
    ),
```

```
body: Container(
    padding: EdgeInsets.only(
        left: ScreenUtil.instance.setWidth(20.0),
        right: ScreenUtil.instance.setWidth(20.0)),
    child: Column(
      crossAxisAlignment: CrossAxisAlignment.start,
      children: <Widget>[
        //名称
        TextField(
            maxLines: 1,
            controller: _nameController,
            style: TextStyle(
                color: Colors.black54,
                fontSize: ScreenUtil.instance.setSp(26.0)),
            decoration: InputDecoration(
              //请输入名称提示
              hintText: KString.ADDRESS_PLEASE_INPUT_NAME,
              hintStyle: TextStyle(
                  color: Colors.grey,
                  fontSize: ScreenUtil.instance.setSp(26.0)),
              focusedBorder: UnderlineInputBorder(
                  borderSide: BorderSide(color: Colors.transparent)),
              enabledBorder: UnderlineInputBorder(
                  borderSide: BorderSide(color: Colors.transparent)))),
        Divider(
          color: Colors.grey[350],
          height: ScreenUtil.instance.setHeight(1.0),
        ),
        //电话
        TextField(
            maxLines: 1,
            controller: _phoneController,
            style: TextStyle(
                color: Colors.black54,
                fontSize: ScreenUtil.instance.setSp(26.0)),
            decoration: InputDecoration(
              //请输入地址提示
              hintText: KString.ADDRESS_PLEASE_INPUT_PHONE,
              hintStyle: TextStyle(
                  color: Colors.grey,
                  fontSize: ScreenUtil.instance.setSp(26.0)),
              focusedBorder: UnderlineInputBorder(
                  borderSide: BorderSide(color: Colors.transparent)),
              enabledBorder: UnderlineInputBorder(
                  borderSide: BorderSide(color: Colors.transparent)))),
        Divider(
          color: Colors.grey[350],
          height: ScreenUtil.instance.setHeight(1.0),
        ),
```

```dart
InkWell(
    //弹出城市选择组件
    onTap: () => this.show(context),
    child: Container(
      alignment: Alignment.centerLeft,
      height: ScreenUtil.instance.setHeight(100),
      child: Text(
        _cityText == null
        //请选择城市提示
            ? KString.ADDRESS_PLEASE_SELECT_CITY
            : _cityText,
        style: _cityText == null
            ? TextStyle(
                color: Colors.grey,
                fontSize: ScreenUtil.instance.setSp(26.0))
            : TextStyle(
                color: Colors.black54,
                fontSize: ScreenUtil.instance.setSp(26.0)),
      ),
    )),
Divider(
  color: Colors.grey[350],
  height: ScreenUtil.instance.setHeight(1.0),
),
//详细地址
TextField(
    maxLines: 1,
    controller: _addressDetailController,
    style: TextStyle(
        color: Colors.black54,
        fontSize: ScreenUtil.instance.setSp(26.0)),
    decoration: InputDecoration(
      //请输入详细地址提示
      hintText: KString.ADDRESS_PLEASE_INPUT_DETAIL,
      hintStyle: TextStyle(
          color: Colors.grey,
          fontSize: ScreenUtil.instance.setSp(26.0)),
      focusedBorder: UnderlineInputBorder(
          borderSide: BorderSide(color: Colors.transparent)),
      enabledBorder: UnderlineInputBorder(
          borderSide: BorderSide(color: Colors.transparent)))),
Divider(
  color: Colors.grey[350],
  height: ScreenUtil.instance.setHeight(1.0),
),
Container(
  height: ScreenUtil.instance.setHeight(100.0),
  //水平布局
  child: Row(
```

```dart
            crossAxisAlignment: CrossAxisAlignment.center,
            children: <Widget>[
                //默认地址提示
                Text(
                    KString.ADDRESS_SET_DEFAULT,
                    style: TextStyle(
                        color: Colors.black54,
                        fontSize: ScreenUtil.instance.setSp(26.0)),
                ),
                Expanded(
                    child: Container(
                    alignment: Alignment.centerRight,
                    //设置是否为默认地址
                    child: Switch(
                        value: _isDefault,
                        activeColor: KColor.defaultSwitchColor,
                        //选择改变回调方法
                        onChanged: (bool) {
                          setState(() {
                              //设置默认地址状态值
                              this._isDefault = bool;
                          });
                        }),
                )),
            ],
        ),
    ),
    Divider(
      color: Colors.grey[350],
      height: ScreenUtil.instance.setHeight(1.0),
    ),
    //当地址 Id 为 0 时不显示删除地址按钮
    Offstage(
        offstage: _addressId == 0,
        child: InkWell(
            //删除地址
            onTap: () => _deleteAddressDialog(context),
            child: Container(
              alignment: Alignment.centerLeft,
              height: ScreenUtil.instance.setHeight(100),
              child: Text(
                //删除地址文本
                KString.ADDRESS_DELETE,
                style: TextStyle(
                    color: KColor.defaultButtonColor,
                    fontSize: ScreenUtil.instance.setSp(26.0)),
              ),
            )),
    ),
```

```dart
                    //当地址 Id 为 0 时不显示删除地址按钮
                    Offstage(
                        offstage: _addressId == 0,
                        child: Divider(
                          color: Colors.grey[350],
                          height: ScreenUtil.instance.setHeight(1.0),
                        )),
                  ],
                ),
              ),
      bottomNavigationBar: BottomAppBar(
          child: Container(
            alignment: Alignment.center,
            width: double.infinity,
            color: KColor.defaultButtonColor,
            height: ScreenUtil.instance.setHeight(100.0),
            child: InkWell(
              //提交地址操作
              onTap: () => _addAddress(),
              child: Text(
                //提交文本
                KString.SUBMIT,
                style: TextStyle(
                    color: Colors.white,
                    fontSize: ScreenUtil.instance.setSp(28.0)),
              ),
            )),
        ),
    );
}

//弹出地址选择组件
show(context) async {
  //使用 CityPickers 组件弹出城市选择框
  Result temp = await CityPickers.showCityPicker(
    context: context,
    itemExtent: ScreenUtil.instance.setHeight(80.0),
    itemBuilder: (item, list, index) {
      return Center(
        child: Text(
            item,
            maxLines: 1,
            style: TextStyle(fontSize: ScreenUtil.instance.setSp(26.0)
            )
        ),
      );
    },
    height: ScreenUtil.instance.setHeight(400),
  );
```

```
      print(temp);
      setState(() {
        //设置选择好的地址信息
        _cityText = temp.provinceName + temp.cityName + temp.areaName;
        _areaId = temp.areaId;
        _provinceName = temp.provinceName;
        _cityName = temp.cityName;
        _countryName = temp.areaName;
      });
    }

//删除地址
_deleteAddressDialog(BuildContext context) {
  //弹出对话框
  showDialog<void>(
      context: context,
      barrierDismissible: true,
      builder: (BuildContext context) {
        return AlertDialog(
          //提示文本
          title: Text(
            KString.TIPS,
            style: TextStyle(
                fontSize: ScreenUtil.instance.setSp(30.0),
                color: Colors.black54),
          ),
          //删除收货地址提示文本
          content: Text(
            KString.ADDRESS_DELETE,
            style: TextStyle(
                fontSize: ScreenUtil.instance.setSp(30.0),
                color: Colors.black54),
          ),
          actions: <Widget>[
            //取消删除按钮
            FlatButton(
              color: Colors.white,
              //取消删除
              onPressed: () {
                Navigator.pop(context);
              },
              child: Text(
                KString.CANCEL,
                style: TextStyle(
                    color: KColor.defaultButtonColor,
                    fontSize: ScreenUtil.instance.setSp(26.0)),
              ),
            ),
            //删除按钮
```

```dart
              FlatButton(
                color: Colors.white,
                //删除收货地址
                onPressed: () {
                  Navigator.pop(context);
                  _deleteAddress();
                },
                child: Text(
                  KString.CONFIRM,
                  style: TextStyle(
                      color: KColor.defaultButtonColor,
                      fontSize: ScreenUtil.instance.setSp(26.0)),
                ),
              ),
            ],
          );
        });
  }

  //删除地址
  _deleteAddress() {
    //地址 Id 参数
    var parameters = {"id": _addressData.id};
    //调用地址服务执行删除
    _addressService.deleteAddress( parameters, (onSuccess) {
      //删除成功提示
      ToastUtil.showToast(KString.ADDRESS_DELETE_SUCCESS);
      Navigator.pop(context);
    }, (onFail) {
      ToastUtil.showToast(onFail);
    });
  }

  //新增地址提交
  _addAddress() {
    //首先判断地址数据
    if (_checkAddressBody()) {
      var parameters = {
        //详细地址
        "addressDetail": _addressDetailController.text.toString(),
        //地区编码
        "areaCode": _areaId,
        //城市
        "city": _cityName,
        //国家
        "county": _countryName,
```

```
      //地址 Id
      "id": _addressData == null ? 0 : _addressData.id,
      //是否为默认地址
      "isDefault": _isDefault,
      //姓名
      "name": _nameController.text.toString(),
      //省
      "province": _provinceName,
      //电话
      "tel": _phoneController.text.toString(),
    };
    //调用地址数据服务,添加地址
    _addressService.addAddress( parameters, (success) {
      //添加成功提示
      ToastUtil.showToast(KString.SUBMIT_SUCCESS);
      Navigator.of(context).pop(true);
    }, (error) {
      ToastUtil.showToast(error);
    });
  }
}

//判断地址信息是否完整
bool _checkAddressBody() {
  //判断详细地址是否为空
  if (_addressDetailController.text.toString().isEmpty) {
    ToastUtil.showToast(KString.ADDRESS_PLEASE_INPUT_DETAIL);
    return false;
  }
  //判断姓名是否为空
  if (_nameController.text.toString().isEmpty) {
    ToastUtil.showToast(KString.ADDRESS_PLEASE_INPUT_NAME);
    return false;
  }
  //判断电话是否为空
  if (_phoneController.text.toString().isEmpty) {
    ToastUtil.showToast(KString.ADDRESS_PLEASE_INPUT_PHONE);
    return false;
  }
  return true;
}
```

这里可以看到在提交数据之前,需要对文本的数据做一个空判断处理。另外在打开页面时,需要对文本框做一个初始数据的处理,编辑地址页面显示效果如图 34-2 所示。

图 34-2 编辑地址页面

第 35 章 收 藏

如果用户对商城中某件商品感兴趣就可以单击"收藏"按钮将商品收藏起来,再次购物时可以通过我的页面的收藏项查看收藏过的商品列表。

本章将详细说明收藏的数据模型、数据服务,以及展示页面是如何实现的。

35.1 收藏数据模型

收藏数据模型主要为我的收藏页面提供数据支持,获取收藏列表数据是通过下面的接口获取的,接口如下:

```
http://localhost:8080/client/collect/list
```

收藏主要包括收藏列表数据模型及收藏数据模型。接下来打开 model/collect_list_model.dart 文件,添加如下代码:

```
//model/collect_list_model.dart 文件
import 'package:json_annotation/json_annotation.dart';
//扩展文件
part 'collect_list_model.g.dart';
//收藏列表数据文件
@JsonSerializable()
class CollectListModel extends Object {
  //总数
  @JsonKey(name: 'total')
  int total;
  //总页数
  @JsonKey(name: 'pages')
  int pages;
  //每页条数
  @JsonKey(name: 'limit')
  int limit;
  //页码
```

```dart
  @JsonKey(name: 'page')
  int page;
  //收藏列表
  @JsonKey(name: 'list')
  List<CollectModel> list;
  //构造方法
  CollectListModel(this.total,this.pages,this.limit,this.page,this.list,);

  factory CollectListModel.fromJson(Map<String, dynamic> srcJson) => _$CollectListModelFromJson(srcJson);

  Map<String, dynamic> toJson() => _$CollectListModelToJson(this);

}
//收藏数据模型
@JsonSerializable()
class CollectModel extends Object {
  //描述
  @JsonKey(name: 'brief')
  String brief;
  //图片
  @JsonKey(name: 'picUrl')
  String picUrl;
  //商品 Id
  @JsonKey(name: 'valueId')
  int valueId;
  //名称
  @JsonKey(name: 'name')
  String name;
  //id
  @JsonKey(name: 'id')
  int id;
  //类型
  @JsonKey(name: 'type')
  int type;
  //价格
  @JsonKey(name: 'retailPrice')
  double retailPrice;
  //构造方法
  CollectModel(this.brief,this.picUrl,this.valueId,this.name,this.id,this.type,this.retailPrice,);

  factory CollectModel.fromJson(Map<String, dynamic> srcJson) => _$CollectModelFromJson(srcJson);

  Map<String, dynamic> toJson() => _$CollectModelToJson(this);

}
```

接下来打开命令行工具，进入 shop-flutter 项目根目录并使用 build_runner 工具生成扩展文件 address_model.g.dart。

35.2　收藏数据服务

收藏数据服务首先给收藏页面提供查询方法，其次给商品详情页面提供收藏或取消收藏方法。需要实现以下两个方法：

- addOrDeleteCollect：添加或删除收藏；
- queryCollect：查询我的收藏。

参照前面章节，编写标准的订单数据操作方法。打开 service/address_service.dart 文件，添加如下代码：

```dart
//service/collect_service.dart 文件
import 'package:shop/utils/http_util.dart';
import 'package:shop/config/server_url.dart';
import 'package:shop/config/index.dart';
import 'package:shop/model/collect_list_model.dart';
//定义成功返回列表数据
typedef OnSuccessList<T>(List<T> list);
//定义成功返回数据
typedef OnSuccess<T>(T t);
//定义返回失败消息
typedef OnFail(String message);

//收藏数据服务
class CollectService {
  //添加或删除收藏
  Future addOrDeleteCollect(Map<String, dynamic> parameters, OnSuccess onSuccess, OnFail onFail) async {
    try {
      var response = await HttpUtil.instance.post(ServerUrl.COLLECT_ADD_DELETE, parameters: parameters,);
      if (response['errno'] == 0) {
        //成功返回
        onSuccess(response["errmsg"]);
      } else {
        //失败返回
        onFail(response['errmsg']);
      }
    } catch (e) {
      print(e);
      //返回服务端异常消息
      onFail(KString.SERVER_EXCEPTION);
    }
  }
```

```
//查询我的收藏
Future queryCollect(Map<String, dynamic> parameters, OnSuccessList onSuccessList, OnFail onFail) async {
    try {
        var response = await HttpUtil.instance.get(ServerUrl.COLLECT_LIST, parameters: parameters, );
        if (response['errno'] == 0) {
            //将返回的Json数据转换成收藏列表数据模型
            CollectListModel collectModel = CollectListModel.fromJson(response["data"]);
            //成功返回列表数据
            onSuccessList(collectModel.list);
        } else {
            //失败返回
            onFail(response['errmsg']);
        }
    } catch (e) {
        print(e);
        //返回服务端异常消息
        onFail(KString.SERVER_EXCEPTION);
    }
}
```

35.3 我的收藏页面实现

我的收藏页面是从我的页面单击进去的，主要展示了收藏的商品列表数据。具体实现步骤如下：

步骤1：打开page/mine/collect_page.dart文件，添加CollectPage组件。初始化收藏数据服务及收藏列表数据，代码如下：

```
//收藏数据服务
CollectService _collectService = CollectService();
//收藏列表
List<CollectModel> _collects = List();
```

步骤2：调用收藏服务的queryCollect方法查询收藏商品数据，代码如下：

```
_queryCollect() {
    //参数...
    _collectService.queryCollect(parameters,(successList) {
        //...
    }, (error) {
        //...
    });
}
```

步骤 3：长按商品会弹出一个对话框，单击确认可以取消收藏该商品。添加取消收藏的方法如下：

```
_cancelCollect(int valueId, int index) {
    //type 为 0 表示取消收藏,商品 Id
    //参数 parameters...
    _collectService.addOrDeleteCollect(parameters, (onSuccess) {
      //...
    }, (error) {
      //...
    });
}
```

步骤 4：编写我的收藏页面布局代码，采用 GridView 网格布局，使用两列多行展示。完整的代码如下：

```
//page/mine/collect_page.dart 文件
import 'package:flutter/material.dart';
import 'package:flutter_screenutil/flutter_screenutil.dart';
import 'package:shop/service/collect_service.dart';
import 'package:shop/utils/shared_preferences_util.dart';
import 'package:shop/model/collect_list_model.dart';
import 'package:shop/utils/toast_util.dart';
import 'package:shop/config/index.dart';
import 'package:shop/widgets/no_data_widget.dart';
import 'package:shop/utils/navigator_util.dart';
//收藏页面
class CollectPage extends StatefulWidget {
  @override
  _CollectPageState createState() => _CollectPageState();
}

class _CollectPageState extends State<CollectPage> {
  //收藏数据服务
  CollectService _collectService = CollectService();
  //收藏列表
  List<CollectModel> _collects = List();
  //token 值
  var token;
  //当前页
  var _page = 1;
  //每页个数
  var _limit = 10;
  //类型
  var _type = 0;

  @override
  void initState() {
```

```dart
    super.initState();
    //获取token值
    SharedPreferencesUtil.getToken().then((value) {
      token = value;
      //查询
      _queryCollect();
    });
}

//查询收藏数据
_queryCollect() {
  //参数
  var parameters = {"type": _type, "page": _page, "limit": _limit};
  _collectService.queryCollect(parameters,(successList) {
    setState(() {
      _collects = successList;
    });
  }, (error) {
    ToastUtil.showToast(error);
  });
}

@override
Widget build(BuildContext context) {
  return Scaffold(
    appBar: AppBar(
      //我的收藏标题
      title: Text(KString.MINE_COLLECT),
      centerTitle: true,
    ),
    body: Container(
      margin: EdgeInsets.all(ScreenUtil.instance.setWidth(20.0)),
      child: _collects.length == 0
          //没有数据组件
          ? NoDataWidget()
          //GridView 展示商品
          : GridView.builder(
              //商品个数
              itemCount: _collects.length,
              gridDelegate: SliverGridDelegateWithFixedCrossAxisCount(
                //一行两列
                crossAxisCount: 2,
                mainAxisSpacing: ScreenUtil.instance.setWidth(10.0),
                crossAxisSpacing: ScreenUtil.instance.setHeight(10.0)),
                itemBuilder: (BuildContext context, int index) {
                  //根据索引返回商品项
                  return getGoodsItemWidget(_collects[index], index);
                },
            ),
    ),
  );
```

```dart
}
//返回商品项
Widget getGoodsItemWidget(CollectModel collect, int index) {
  return GestureDetector(
    child: Container(
      alignment: Alignment.center,
      child: SizedBox(
          width: 320,
          height: 460,
          child: Card(
            //垂直布局
            child: Column(
              children: <Widget>[
                //商品图片
                Image.network(
                  collect.picUrl ?? "",
                  fit: BoxFit.fill,
                  height: 100,
                ),
                Padding(
                  padding: EdgeInsets.only(top: 5.0),
                ),
                //商品名称
                Text(
                  collect.name,
                  maxLines: 1,
                  overflow: TextOverflow.ellipsis,
                  style: TextStyle(fontSize: 14.0, color: Colors.black54),
                ),
                Padding(
                  padding: EdgeInsets.only(top: 5.0),
                ),
                //商品价格
                Text(
                  "￥ ${collect.retailPrice}",
                  maxLines: 1,
                  overflow: TextOverflow.ellipsis,
                  style: TextStyle(
                      fontSize: 14.0, color: Colors.deepOrangeAccent),
                ),
              ],
            ),
          )),
    ),
    //单击商品处理
    onTap: () => _itemClick(collect.valueId),
    //长按商品处理
    onLongPress: () => _showDeleteDialog(collect, index),
  );
}
```

```dart
//弹出取消收藏商品对话框
_showDeleteDialog(CollectModel collect, int index) {
  showDialog(
      context: context,
      barrierDismissible: true,
      builder: (BuildContext context) {
        return AlertDialog(
          //提示文本
          title: Text(
            KString.TIPS,
            style: TextStyle(
                color: Colors.black54,
                fontSize: ScreenUtil.instance.setSp(28.0)),
          ),
          //取消提示文本
          content: Text(
            KString.MINE_CANCEL_COLLECT,
            style: TextStyle(
                color: Colors.black54,
                fontSize: ScreenUtil.instance.setSp(26.0)),
          ),
          //操作按钮
          actions: <Widget>[
            //取消按钮
            FlatButton(
              onPressed: () {
                Navigator.pop(context);
              },
              child: Text(
                KString.CANCEL,
                style: TextStyle(
                    color: Colors.grey,
                    fontSize: ScreenUtil.instance.setSp(26.0)),
              )),
            //确定按钮
            FlatButton(
              onPressed: () {
                Navigator.pop(context);
                //删除处理
                _cancelCollect(collect.valueId, index);
              },
              child: Text(
                KString.CONFIRM,
                style: TextStyle(
                    color: KColor.defaultTextColor,
                    fontSize: ScreenUtil.instance.setSp(26.0)),
              )),
          ],
        );
      });
}
```

```
//取消收藏处理
_cancelCollect(int valueId, int index) {
  //type 为 0 表示取消收藏,商品 Id
  var parameters = {"type": 0, "valueId": valueId};
  _collectService.addOrDeleteCollect(parameters, (onSuccess) {
    setState(() {
      //本地删除此商品
      _collects.removeAt(index);
    });
  }, (error) {
    ToastUtil.showToast(error);
  });
}
//单击跳转至商品详情页
_itemClick(int id) {
  NavigatorUtil.goGoodsDetails(context, id);
}
}
```

单击商品项可以跳转至商品详情页,长按商品项可以弹出对话框,单击确认可以取消收藏该商品,我的收藏页面展示效果如图 35-1 所示。

图 35-1 我的收藏页面

第 36 章 个人中心

个人中心在电商应用里通常包含的功能有账户管理、我的订单、我的足迹、收货地址、我的优惠券、我的收藏,以及我的红包等,即"我的"页面。

前面的章节已经实现了我的订单、收藏、地址管理,以及登录注册等内容,本章将会在前面实现的基础之上,再实现关于我们页面,然后将所有内容串联起来。

36.1 关于我们页面实现

关于我们页面包含了公司的名称、电话,以及邮箱等信息,是一个纯静态展示的页面,不需要数据模型及数据服务。打开 page/mine/mine_page.dart 文件,添加如下完整代码:

```dart
//page/mine/mine_page.dart 文件
import 'package:flutter/material.dart';
import 'package:flutter_screenutil/flutter_screenutil.dart';
import 'package:shop/config/index.dart';
import 'package:shop/widgets/item_text_widget.dart';
import 'package:shop/widgets/divider_line_widget.dart';
//关于我们页面
class AboutUsPage extends StatelessWidget {
  @override
  Widget build(BuildContext context) {
    return Scaffold(
      appBar: AppBar(
        //标题
        title: Text(KString.MINE_ABOUT_US),
        centerTitle: true,
      ),
      body: Container(
        margin: EdgeInsets.all(ScreenUtil.instance.setWidth(20.0)),
        //垂直布局
        child: Column(
          crossAxisAlignment: CrossAxisAlignment.start,
          children: <Widget>[
            //公司名称
            Text(
```

```
                KString.MINE_ABOUT_US_CONTENT,
                style: TextStyle(
                    color: Colors.black54,
                    fontSize: ScreenUtil.instance.setSp(26.0)),
              ),
              Padding(padding: EdgeInsets.all(ScreenUtil.instance.setHeight(10.0))),
              DividerLineWidget(),
              //名字
              ItemTextWidget(KString.MINE_ABOUT_NAME_TITLE, KString.MINE_ABOUT_NAME),
              DividerLineWidget(),
              //邮箱
              ItemTextWidget(KString.MINE_ABOUT_EMAIL_TITLE, KString.MINE_ABOUT_EMAIL),
              DividerLineWidget(),
              //联系方式
              ItemTextWidget(KString.MINE_ABOUT_TEL_TITLE, KString.MINE_ABOUT_TEL),
              DividerLineWidget(),
            ],
          ),
        ),
      );
    }
  }
```

页面展示的效果如图 36-1 所示。

图 36-1　关于我们页面

36.2 我的页面实现

我的页面是商城应用的一个核心模块,用户的个人数据都是通过这个页面导航查看或管理的。商城应用中我的页面包含以下几个功能:
- 头像昵称展示;
- 用户登录和登出;
- 我的订单导航;
- 收藏导航;
- 地址管理导航;
- 关于我们导航。

我的页面不需要数据模型,数据服务使用的是用户数据服务,因为它们实现用户登录及登出操作。实现步骤如下:

步骤1:打开 page/mine/mine_page.dart 文件,添加 MinePage 我的页面组件。初始化用户数据服对象,用于登录和登出操作,代码如下:

```
//用户数据服务
UserService _userService = UserService();
```

步骤2:获取 token 值用于判断用户是否登录,同时获取本地用户信息,代码如下:

```
_getUserInfo() {
  //获取 token
  SharedPreferencesUtil.getToken().then((token) {
    //设置登录状态
    //...
    //获取本地头像数据
    //...
    //获取本地用户名数据
    //...
  });
}
```

步骤3:监听登录事件,登录成功后使用登录后的用户数据,登录失败则设置登录状态为 false,处理代码大致如下:

```
loginEventBus.on<LoginEvent>().listen((LoginEvent loginEvent) {
  //登录成功处理
  //...
  //登录失败处理
```

```
    //...
});
```

步骤4：编写我的页面所使用的交互方法，包括登出方法及各个页面跳转的方法。方法如下：

- _loginOut：登出方法；
- _collect：跳转至我的收藏页面；
- _address：跳转至收货地址页面；
- _aboutUs：跳转至关于我们页面；
- _order：跳转至我的订单页面；
- _toLogin：跳转至登录页面。

步骤5：编写我的页面，布局整体采用垂直布局，上面为头像，下面为功能页面导航项，完整的代码如下：

```dart
//page/mine/mine_page.dart 文件
import 'package:flutter/material.dart';
import 'package:flutter_screenutil/flutter_screenutil.dart';
import 'package:shop/config/index.dart';
import 'package:shop/utils/shared_preferences_util.dart';
import 'package:shop/widgets/icon_text_arrow_widget.dart';
import 'package:shop/config/icon.dart';
import 'package:shop/utils/navigator_util.dart';
import 'package:shop/event/login_event.dart';
import 'package:shop/service/user_service.dart';
import 'package:shop/utils/toast_util.dart';
//我的页面
class MinePage extends StatefulWidget {
  @override
  _MinePageState createState() => _MinePageState();
}

class _MinePageState extends State<MinePage> {
  //是否登录变量
  bool isLogin = false;
  //头像地址
  var imageHeadUrl;
  //昵称
  var nickName;
  //用户数据服务
  UserService _userService = UserService();

  @override
  void initState() {
    super.initState();
```

```dart
    //获取用户信息
    _getUserInfo();
}
//刷新事件
_refreshEvent() {
    //登录事件监听
    loginEventBus.on<LoginEvent>().listen((LoginEvent loginEvent) {
      if (loginEvent.isLogin) {
        setState(() {
          //登录成功
          isLogin = true;
          //设置头像 url
          imageHeadUrl = loginEvent.url;
          //设置昵称
          nickName = loginEvent.nickName;
        });
      } else {
        setState(() {
          isLogin = false;
        });
      }
    });
}

//获取本地用户信息
_getUserInfo() {
    //获取 token
    SharedPreferencesUtil.getToken().then((token) {
      if (token != null) {
        setState(() {
          isLogin = true;
        });
        //获取本地头像数据
        SharedPreferencesUtil.getImageHead().then((imageHeadAddress) {
          setState(() {
            imageHeadUrl = imageHeadAddress;
          });
        });
        //获取本地用户名数据
        SharedPreferencesUtil.getUserName().then((name) {
          setState(() {
            nickName = name;
          });
        });
      }
    });
}

@override
```

```dart
Widget build(BuildContext context) {
    //刷新事件
    _refreshEvent();
    return Scaffold(
      appBar: AppBar(
        //标题
        title: Text(KString.MINE),
        centerTitle: true,
      ),
      //垂直布局
      body: Column(
        children: <Widget>[
          Container(
            height: ScreenUtil.getInstance().setHeight(160.0),
            width: double.infinity,
            alignment: Alignment.center,
            //判断是否登录
            child: isLogin
                ? Row(
                    mainAxisAlignment: MainAxisAlignment.start,
                    crossAxisAlignment: CrossAxisAlignment.center,
                    children: <Widget>[
                      Container(
                        width: 60,
                        height: 60,
                        margin: EdgeInsets.only(
                            left: ScreenUtil.getInstance().setWidth(20.0),
                        ),
                        //显示头像
                        child: CircleAvatar(
                          radius: 30,
                          foregroundColor: Colors.redAccent,
                          //头像图片
                          backgroundImage: NetworkImage(
                            imageHeadUrl,
                          ),
                        ),
                      ),
                      Padding(
                        padding: EdgeInsets.only(
                            left: ScreenUtil.getInstance().setWidth(10.0)),
                      ),
                      //昵称
                      Text(
                        nickName,
                        style: TextStyle(
                          fontSize: ScreenUtil.getInstance().setSp(26.0),
                          color: Colors.black),
                      ),
```

```dart
                        Expanded(
                            //退出按钮
                            child: InkWell(
                              //打开退出对话框
                              onTap: () => _loginOutDialog(),
                              child: Offstage(
                                //登录后显示此组件
                                offstage: !isLogin,
                                child: Container(
                                  padding: EdgeInsets.only(right: ScreenUtil.getInstance().setWidth(30)),
                                  alignment: Alignment.centerRight,
                                  //登出文本
                                  child: Text(
                                    KString.LOGIN_OUT,
                                    style: TextStyle(
                                        fontSize: ScreenUtil.getInstance().setSp(26),
                                        color: Colors.black54),
                                  ),
                                ),
                              )),
                        ),
                      ],
                    )
                  : InkWell(
                      //单击登录
                      onTap: () => _toLogin(),
                      //登录文本
                      child: Text(
                        KString.CLICK_LOGIN,
                        style: TextStyle(
                            color: Colors.black54,
                            fontSize: ScreenUtil.getInstance().setSp(30.0)),
                      ),
                    ),
            ),
            Padding(
              padding:
                  EdgeInsets.only(top: ScreenUtil.getInstance().setHeight(20.0)),
            ),
            Divider(
              height: ScreenUtil.getInstance().setHeight(1.0),
              color: Color(0xffd3d3d3),
            ),
            //我的订单
            IconTextArrowWidget(
                KIcon.ORDER, KString.ORDER, Colors.deepPurpleAccent, _order),
            Divider(
              height: ScreenUtil.getInstance().setHeight(1.0),
```

```dart
          color: Color(0xffd3d3d3),
        ),
        //收藏
        IconTextArrowWidget(
            KIcon.COLLECTION, KString.COLLECTION, Colors.red, _collect),
        Divider(
          height: ScreenUtil.getInstance().setHeight(1.0),
          color: Color(0xffd3d3d3),
        ),
        //地址管理
        IconTextArrowWidget(
            KIcon.ADDRESS, KString.ADDRESS, Colors.amber, _address),
        Divider(
          height: ScreenUtil.getInstance().setHeight(1.0),
          color: Color(0xffd3d3d3),
        ),
        //关于我们
        IconTextArrowWidget(
            KIcon.ABOUT_US, KString.ABOUT_US, Colors.teal, _aboutUs),
        Divider(
          height: ScreenUtil.getInstance().setHeight(1.0),
          color: Color(0xffd3d3d3),
        ),
      ],
    ),
  );
}

//登出对话框
_loginOutDialog() {
  showDialog(
      context: context,
      builder: (BuildContext context) {
        //对话框
        return AlertDialog(
          //提示
          title: Text(
            KString.TIPS,
            style: TextStyle(
                fontSize: ScreenUtil.getInstance().setSp(30),
                color: Colors.black54),
          ),
          //登出提示
          content: Text(
            KString.LOGIN_OUT_TIPS,
            style: TextStyle(
                fontSize: ScreenUtil.getInstance().setSp(30),
                color: Colors.black54),
          ),
          //操作按钮
          actions: <Widget>[
```

```dart
              //取消按钮
              FlatButton(
                onPressed: () {
                  Navigator.pop(context);
                },
                child: Text(
                  KString.CANCEL,
                  style: TextStyle(color: Colors.black54),
                ),
              ),
              //确定按钮
              FlatButton(
                //跳转到登录页面
                onPressed: () => _loginOut(),
                child: Text(
                  KString.CONFIRM,
                  style: TextStyle(color: KColor.defaultTextColor),
                ),
              )
            ],
          );
        });
}

//登出方法
_loginOut() {
  //调用用户数据服务的登出方法
  _userService.loginOut((success) {
    //触发 LoginEvent 事件
    loginEventBus.fire(LoginEvent(false));
  }, (error) {
    //触发 LoginEvent 事件
    loginEventBus.fire(LoginEvent(false));
    ToastUtil.showToast(error);
  });
  Navigator.pop(context);
}
//跳转至我的收藏页面
void _collect() {
  if (isLogin) {
    NavigatorUtil.goCollect(context);
  } else {
    _toLogin();
  }
}
//跳转至收货地址页面
void _address() {
  if (isLogin) {
    NavigatorUtil.goAddress(context);
  } else {
    _toLogin();
```

```
    }
  }
  //跳转至关于我们页面
  void _aboutUs() {
    if (isLogin) {
      NavigatorUtil.goAboutUs(context);
    } else {
      _toLogin();
    }
  }
  //跳转至我的订单页面
  void _order() {
    if (isLogin) {
      NavigatorUtil.goOrder(context);
    } else {
      _toLogin();
    }
  }
  //跳转至登录页面
  _toLogin() {
    NavigatorUtil.goLogin(context);
  }
}
```

上面的代码可以重点看一下登录与登出页面的变化,使用_isLogin 变量来判断是否登录,我的页面运行效果如图 36-2 所示。

图 36-2　我的页面

图书资源支持

感谢您一直以来对清华大学出版社图书的支持和爱护。为了配合本书的使用，本书提供配套的资源，有需求的读者请扫描下方的"书圈"微信公众号二维码，在图书专区下载，也可以拨打电话或发送电子邮件咨询。

如果您在使用本书的过程中遇到了什么问题，或者有相关图书出版计划，也请您发邮件告诉我们，以便我们更好地为您服务。

我们的联系方式：

地　　址：北京市海淀区双清路学研大厦 A 座 701

邮　　编：100084

电　　话：010-83470236　010-83470237

资源下载：http://www.tup.com.cn

客服邮箱：2301891038@qq.com

QQ：2301891038（请写明您的单位和姓名）

用微信扫一扫右边的二维码，即可关注清华大学出版社公众号。

科技传播·新书资讯

电子电气科技荟

资料下载·样书申请

书圈